T0344895

An Introduction To Ceramic Engineering Design

Related Titles Published by The American Ceramic Society:

Ceramic Armor Materials by Design (Ceramic Transactions Volume 134)
Edited by James McCauley, A. Rejendran, W. Gooch, S. Bless, S. Wax, and A. Crowson
©2002, ISBN 1-57498-148-X

Glass-Ceramic Technology
By Wolfram Höland and George Beall
©2002, ISBN 1-57498-107-2

Dielectric Materials and Devices
Edited by K.M. Nair, Amar S. Bhalla, Tapan K. Gupta, S.I. Hirano, Basavaraj V. Hiremath, Jau-Ho Jean, and Robert Pohanka
©2002, ISBN 1-57498-118-8

Microwaves: Theory and Application in Materials Processing V (Ceramic Transactions Volume 111)
Edited by David E. Clark, John G.P. Binner, and David A. Lewis
©2001, ISBN 1-57498-103-X

Boing-Boing the Bionic Cat and the Jewel Thief
By Larry L. Hench
©2001, ISBN 1-57498-129-3

Boing-Boing the Bionic Cat
By Larry L. Hench
©2000, ISBN 1-57498-109-9

The Magic of Ceramics
By David W. Richerson
©2000, ISBN 1-57498-050-5

Glazes and Glass Coatings
By Richard A. Eppler and Douglas R. Eppler
©2000, ISBN 1-57498054-8

Tape Casting: Theory and Practice
By Richard E. Mistler and Eric R. Twiname
©2000, ISBN 1-57498-029-7

Bioceramics: Materials and Applications III (Ceramic Transactions Volume 110)
Edited by Laurie George, Richard P. Rusin, Gary S. Fischman, and Vic Janas
©2000, ISBN 1-57498-102-1

Surface-Active Processes in Materials (Ceramic Transactions Volume 101)
Edited by David E. Clark, Diane C. Folz, and Joseph H. Simmons
©2000, ISBN 1-57498-079-3

Ceramic Innovations in the 20th Century
Edited by John B. Wachtman, Jr.
©1999, ISBN 1-57498-093-9

For information on ordering titles published by The American Ceramic Society, or to request a publications catalog, please contact our Customer Service Department at 614-794-5890 (phone), 614-794-5892 (fax), <customersrvc@acers.org> (e-mail), or write to Customer Service Department, 735 Ceramic Place, Westerville, OH 43081, USA.

Visit our on-line book catalog at <www.ceramics.org>.

An Introduction To Ceramic Engineering Design

Edited by
David E. Clark and Diane C. Folz
Virginia Polytechnic Institute and State University

and

Thomas D. McGee
Iowa State University

Published by
The American Ceramic Society
735 Ceramic Place
Westerville, Ohio 43081 USA

In Cooperation with
The National Institute of Ceramic Engineers.

The American Ceramic Society
735 Ceramic Place
Westerville, Ohio 43081
www.ceramics.org

ISBN: 1-57498-131-5

Cover Photos:

Front Cover: Top to bottom; from *Boing–Boing the Bionic Cat and the Jewel Thief*, by Larry L. Hench, illustrated by Ruth Denise Lear, and published by The American Ceramic Society; The space shuttle Atlantis, thermally protected by thousands of ceramic tiles, speeds toward space on mission STS-106. Photo courtesy of NASA.

Back Cover: Top to bottom; Saint-Gobain Advanced Ceramics silicon nitride components including Genuine CERBEC® balls and CERAMA™ roller blanks. Used with permission; Microwave combustion synthesis of ceramic composites. Research performed and photo courtesy of the Department of Materials Science and Engineering at Virginia Polytechnic Institute and State University (Blacksburg, Virginia); Cross-section of a hip joint design using sapphire bearing rings and ball. Photo courtesy of Thomas D. McGee; Using radioactive glass microspheres to deliver extremely large doses of localized beta radiation to inoperable tumors in the liver is a promising new treatment for liver cancer. Source: "Using Glass in the Body" by Delbert Day, *The American Ceramic Society Bulletin*, Vol. 74, No. 12, p. 67.

Library of Congress Cataloging-in-Publication Data

An introduction to ceramic engineering design / edited by David E. Clark and Diane C. Folz, and Thomas D. McGee.
p. cm.
Includes bibliographical references and index.
ISBN 1-57498-131-5
1. Ceramic engineering. 2. Ceramics. 3. Engineering design. I. Clark, David E. II. Folz, Diane C. III. McGee, Thomas D. (Thomas Donald), 1925-

TP807 .I655 2002
666--dc21

2002034170

For information on ordering titles published by The American Ceramic Society, or to request a publications catalog, please call 614-794-5890, or visit www.ceramics.org.

Contents

Preface

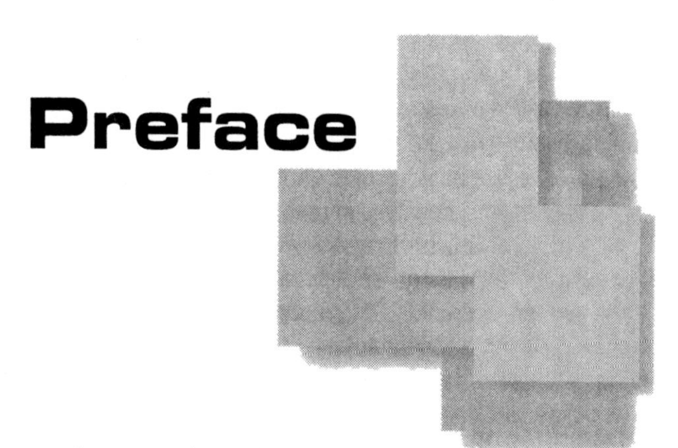

The engineer uses science, mathematics, engineering design principles, and experience to achieve useful objectives that benefit society. The major focus of this book is *ceramic design*, and its purpose is to serve as a text or reference book for ceramic science and engineering students and faculty involved in undergraduate teaching and learning.

This book is based primarily on a workshop entitled "Designing with Engineering Ceramics" sponsored by the National Institute of Ceramic Engineers (NICE) and the Engineering Ceramics Division of the American Ceramic Society. The workshop was held January 2001, in Cocoa Beach, Florida, as part of the 25th Annual Conference on Advanced Ceramics and Composites. Additional papers were solicited from other experts in the field to provide a comprehensive overview of this subject. All of the papers were peer-reviewed in addition to being reviewed by the editors. Not all papers submitted were accepted for publication.

The Accreditation Board for Engineering and Technology (ABET) requires a major design experience in both materials science and engineering (MSE) and ceramic science and engineering (CSE) programs. With respect to these programs, design can refer to any one of the following tasks: selecting an existing material based on its properties for a specific application; creating a new material with specific properties; or designing a process to produce an existing material or a completely new material.

Many MSE and CSE departments attempt to fulfill the design requirement by integrating design issues and principles throughout the curriculum and by including a capstone design experience (one or two courses in the senior year). Unfortunately, many of the general textbooks used in these programs do not contain open-ended design problems that require students to pull information from a wide range of sources. Furthermore, to our knowledge, there is not a textbook suitable for the capstone design experience and, consequently, many of the faculty who teach these courses rely on their personal experiences—gained either through having worked in industry or from consulting. Although valuable, these personal experiences may provide a biased and limited perspective. The MSE and CSE communities would be better served by a series of design textbooks available for teaching as well as for reference. In addition to being useful to faculty, students, and practicing engineers, these books also would be a valuable resource for designers working in other disciplines (e.g., automotive, aerospace, communications).

A unique feature of this book is the set of design questions and problems. Each chapter contains several questions of varying complexity levels, provided by the author. Some of the questions can be answered based on the material presented in that chapter, while others require the student to use additional reference sources. Since many "real world" engineering problems require experts from different disciplines, some of the problems in this book are designed to be solved through a team approach.

This book is intended to serve only as a first step in strengthening the design experience for ceramic engineers. It reviews the essential elements of the design process with specific emphasis on ceramic materials. The connection between what a ceramic engineer needs to know and what is required from ABET is discussed in Chapter 1 and also is integrated into the other chapters.

The issues of ethics and steps required to legally protect designs also are addressed. Although several case histories and illustrative examples are presented, more of these need to be included in future texts on design.

Acknowledgments

The editors appreciate the contributions of the presenters and the authors. The NICE has developed an excellent working relationship with several of the divisions within The American Ceramic Society and, especially with the Engineering Ceramics Division (ECD). The NICE thanks the ECD for allowing it to hold this important symposium during its internationally acclaimed Cocoa Beach conference and hopes to be able to continue sponsoring additional symposia at this event.

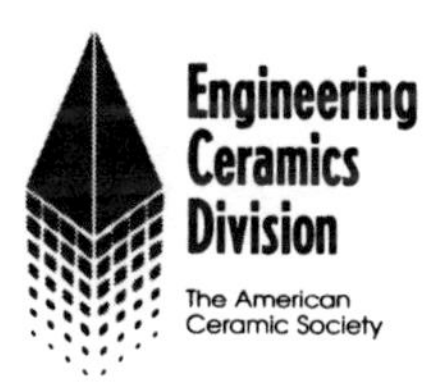

Concepts and Principles of Engineering Design

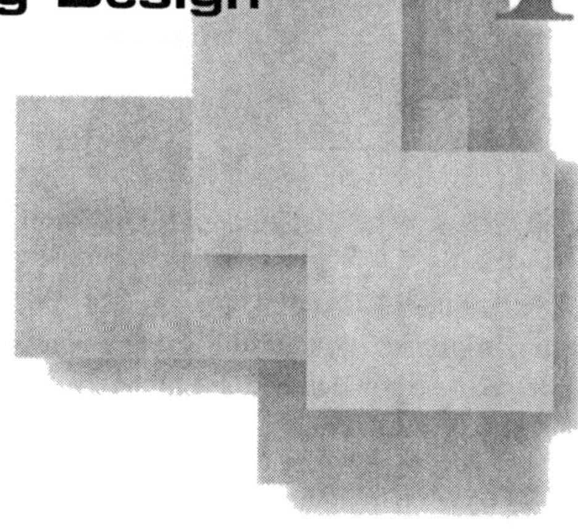

D.E. Clark

D.C. Folz

T.D. McGee

CONCEPTS AND PRINCIPLES OF ENGINEERING DESIGN

D.E. Clark and D.C. Folz
Virginia Polytechnic Institute & State University
Dept. of Materials Science & Engineering
Blacksburg, VA 24061

T.D. McGee
Iowa State University
Dept. of Materials Science & Engineering
Ames, IA 50011

ABSTRACT

This chapter serves as an introduction to the topic of engineering design, especially how it relates to materials and ceramic engineering. Essential components for general engineering design as well as for materials and ceramic engineering design are discussed. Illustrative examples and problems are provided to clarify design issues and to help the engineering student develop design skills.

INTRODUCTION

Engineering is the application of science and mathematics to achieve useful objectives for the benefit of society. Engineering education strives to educate engineers effectively in the application of science and mathematics to solve engineering problems. Engineering societies, such as the National Institute of Ceramic Engineers (NICE), the American Institute for Mining and Metallurgical Engineers (AIME) and the American Society of Civil Engineers (ASCE), representing different engineering disciplines have cooperated through the American Association of Engineering Societies (AAES) to support elements of practice and public policy common to these disciplines. They also have cooperated in setting standards for engineering education through the Accreditation Board for Engineering and Technology (ABET). The ABET is recognized by the U.S. Department of Education as the sole agency responsible for accreditation of educational programs leading to degrees in engineering. One of the primary concerns for engineering education has been to teach the student the theory, philosophy and methods of design. The purpose of this chapter is to broadly define engineering design with respect to the ABET engineering criteria and to describe specifically the role of design in materials and ceramic engineering.

ENGINEERING DESIGN

According to ABET, design is "the process of devising a system, component or process to meet desired needs. It is a decision-making process (often iterative), in which the basic sciences and mathematics and engineering sciences are applied to convert resources optimally to meet a stated objective. Among the fundamental elements of the design process are the establishment of objectives and criteria, synthesis, analysis, construction, testing and evaluation. The engineering design components of a curriculum must include most of the following features: development of student creativity, use of open-ended problems, development and use of modern design theory and methodology, formulation of design problem statements and specifications, consideration of alternative solutions, feasibility considerations, production processes, concurrent engineering design and detailed system descriptions. Further, it is essential to include a variety of realistic constraints, such as economic factors, safety, reliability, aesthetics, ethics and social impact." (From Criteria for Accrediting Programs in Engineering in the United States (1996-97), Engineering Accreditation Commission, ABET, Inc., Baltimore, MD.)

As defined by the Engineering Accreditation Commission of ABET, the engineering design component of the program criteria states, "Engineering design, with some treatment of engineering economics, must be an integral part of the curriculum. An important aspect of this requirement in all programs must be the design function as applied to processing. The creative and original effort required for an effective design component can be met in several ways, such as through portions of courses, projects or research programs, or special problems that go beyond the limited activity of observation and analysis. However, a capstone engineering design experience in the final year of the program is required in order to integrate the various curricular components."

In the "real world," an engineer must manipulate engineering knowledge to produce a desired outcome, considering economic, human and, in some cases, aesthetic values. Experience leading to solid judgement is required to organize the tremendous body of knowledge and the many possible methods that could be employed to achieve the desired result.

Figure 1 illustrates the central concept of the design process. Based on the drivers for the specific case under consideration, the basic problem is defined.

Design concepts, incorporating constraints such as social impact, economics, safety and others, then are developed as potential solutions to the problem. The design concept stage itself can cost from a few hundred dollars to several hundred thousand, depending on the size of the project. At this point, decisions are made as to which design concepts merit the financial resources to advance to the next stage, namely the proof of concept, otherwise known as feasibility studies. This is the basic procedure for proceeding with all engineering design development, whether driven by private industry, the needs of society or the government, a "quantum leap" in science or technology or other types of drivers.

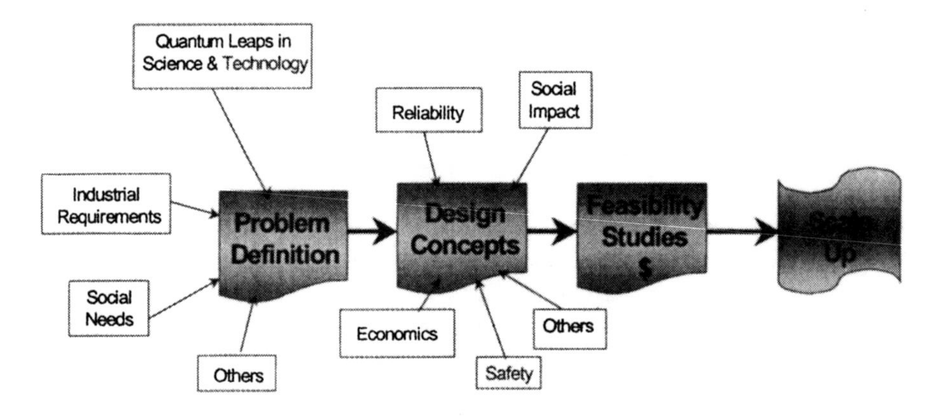

Figure 1. Components of the general design process.

A classic example is from civil engineering in which the engineer designs a bridge. The project would be driven by the needs of society (i.e., local community, county, state or federal). First, the problem is defined in detail, for example, to design a bridge in a certain location with a specific strength and capacity within a specific budget. Then, the engineer(s) must obtain data about the size required; the probable funding; the loads to be imposed, including conditions such as traffic, wind, snow; the load-bearing capacity of the local geology; the clearance of the traffic or obstacles underneath; and other impacting factors. He/she must select basic structure possibilities such as a suspension bridge or an arch. Preliminary designs must be conceived and calculations for the size of members, foundations and components must be made to evaluate the preliminary design concepts. Cost and availability of materials, labor and services also must be estimated. After completing each preliminary design, their attributes and deficiencies must be evaluated. Modifications can be made to satisfy deficiencies which, in turn, can lead to development of a more practical preliminary design. Additionally, while engineering

designs often have natural symmetry and beauty, the need for an aesthetically pleasing result may be inherent in evaluating alternatives to some engineering design problems, as would be the case in designing a new bridge. The elements of design are apparent:

- Problem definition or purpose
- Development of design concepts
- Proof of concept(s) or feasibility studies
- Evaluation/analysis and re-design of concepts, if required
- Client approval
- Application of solution (scale-up)

These elements are common to all engineering design. Judgement is required in all stages and conception of viable solutions requires imagination and knowledge. Basic understanding of the discipline(s) involved is essential. Design is, inherently, an iterative process, requiring one or more initial design concepts followed by evaluation and improvements. While evaluation/analysis is a required component of engineering design, it must not be confused with design itself. Analysis is only one tool in the creative design process.

Because design requires solution of practical problems, it requires more than basic science and mathematics; it also requires an understanding of the application. This is an important distinction between science and engineering, one not often recognized by scientists who have excellence in science but who have only cursory knowledge of how to apply this knowledge to real problems.

MATERIALS ENGINEERING DESIGN

In materials engineering, the feasibility studies shown in Figure 1 would include one or more of the following:

- materials selection (existing and new materials),
- process development,
- performance evaluation, and
- process/product scale-up.

The relationship between these components is illustrated in Figure 2. These components have very different requirements, but each incorporates its own design process, from defining the problems to developing the proof of concepts. The role of the materials engineer in the design process will determine which of these components will be used. In many cases, his/her role will be limited to materials

selection (Chapter 3). For example, if he/she is part of an engineering team commissioned to design a high-speed monorail transit system, he/she most likely would limit input to materials selection since a wide array of acceptable materials already are available. If the overriding requirements for the monorail system were hardness and wear resistance, specific metal alloys, high-tech ceramics or metal-ceramic composites would be selected.

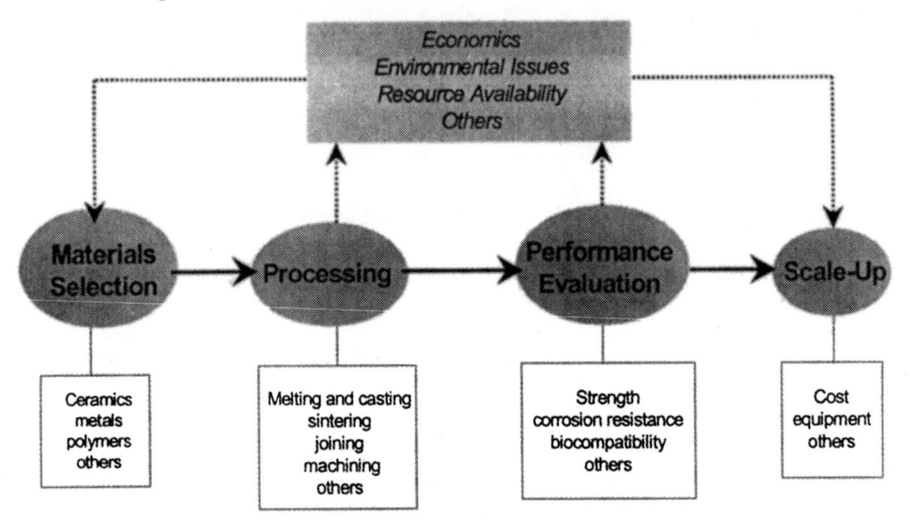

Figure 2. Proof of concept specific to materials and ceramic engineers.

Many general textbooks in materials science and engineering provide examples of design problems based entirely on materials selection. For example, when designing a product subjected to a tensile stress, fracture mechanics might be used to select a suitable material according to the fracture toughness equation for plane strain,

$$a_c = \frac{N^2}{Y^2 \pi}\left(\frac{K_{IC}}{\sigma}\right)^2 \qquad (1)$$

where,

a_c = critical crack length required for fracture, (m),

Y = dimensionless parameter that depends on specimen size and loading conditions (typically about 1),

K_{IC} = plane strain fracture toughness, $MPa{\cdot}m^{1/2}$, where the subscript "I" refers to tension;

σ = applied stress, MPa, and

N = safety factor, dimensionless.

By replacing K_{IC} with K_{IIC} or K_{IIIC}, equation (1) applies to failure in one of the two possible modes of deformation in shear.

For a given safety factor, a known applied stress, values of fracture toughness from reference data books, and consideration of other design properties, several materials can be ranked based on critical crack length. If cost were not a consideration, the materials engineer would select the material with the longest critical crack length. However, if there was a large cost penalty with the use of this material, the engineer could opt for a material with a smaller critical crack length and recommend regularly scheduled non-destructive evaluations (NDE), such as radiography, dye penetration, microscopy and ultrasonics. Of course, the maximum acceptable crack length must be large enough to be detected with the NDE technique.

Metals generally will be favored over ceramics because of their higher values for fracture toughness. However, there are many applications where other properties, such as stiffness, hardness, chemical durability or high-temperature stability are required in addition to fracture toughness. In these applications, ceramics with acceptable fracture toughness values may have to be used.

The engineer also should be fully aware of the boundary conditions for the equations used in design. Furthermore, the value of fracture toughness itself needs to be understood fully. Fracture toughness is related to the area under the stress/strain curve. In fiber-reinforced ceramics, this area can be increased significantly due to crack bridging that produces resistance to crack growth and fiber pullout which results in higher strain prior to catastrophic failure. In these materials, matrix cracking can occur at much smaller strain levels than that required for complete failure and, in certain applications, it might be appropriate to use lower values of fracture toughness. Even so, these composites provide prior warning to pending failure and their failure occurs in a more graceful fashion than do pure ceramics.

Values of materials properties will change with use environments, such as high temperature, the presence of radiation, high humidity and/or aggressive/reactive gases. In some materials, slow crack growth can lead to catastrophic failure (referred to as delayed failure, static fatigue), even at low stresses. In such cases, the flaw size would need to be monitored on a regular basis to ensure that it does not exceed the value dictated by equation (1).

An appropriate safety factor should be selected based on economics, previous experience and the consequences of failure in terms of loss of life, personal injury and/or property damage. Typical values can range from 1.2 to greater than 5. The more confidence the engineer has in the materials parameters and use conditions, the smaller the safety factor will need to be. Due to their brittle nature and the absence of plastic deformation (i.e., yield point), ceramics provide little, if any, warning prior to failure and most likely will require the use of higher safety factors than do metals.

In equation (1), the safety factor is selected with respect to the stress. For example, if the maximum applied stress is expected to be 100 MPa and a safety factor of 2 is selected, the actual stress used to calculate the critical flaw size would be 200 MPa. In this case, the safety factor would reduce the maximum acceptable flaw size by a factor of 4.

It should be understood that selection of existing materials is only one part of design. For certain applications, it may be determined that the material(s) required to solve the problem are not yet available and an entirely new concept based on materials development is warranted. In these cases, the role of the materials engineer is expanded significantly. Excellent examples of this include the thermal protection system for the space shuttle (Chapter 9), specialized borosilicate glasses for long-term nuclear waste disposal (Chapter 11) and biomaterials (Chapter 13) and. In any case, the engineer must understand their limitations and merits in order to properly select existing materials. A thorough understanding of processing/structure/property relationships is needed if development of new materials is required.

Process development is critical to the materials industry since processes control the resulting composition, structure and properties of the material. Whether the route to the appropriate material performance is melting and casting, drying and sintering, joining, machining or one of hundreds of other processing methods, the selection of a process will determine the successful application of a specific material for the problem. Excellence in processing caused steel production to be shifted from the United States to Asia. Now, even textbooks on refractories used in the production of steel products are being written in Asia. It is critical for the materials engineer to understand the role of processing in engineering design concepts if they are to be useful to industry.

In order to evaluate the processing and performance of the materials product, characterization and testing must be conducted. It is important to understand the actual application for a material in order to evaluate its properties and performance under specific conditions. While a two inch square block of the material under consideration may exceed performance criteria, the same material may not provide the required strength if the dimensions are altered. Basic design calculations should be made on the realistically estimated exposure conditions which are likely to be imposed upon the whole part in service, rather than on the concept of an average exposure on an average section. The concepts of materials reliability, service life, variability and effects of more than two parameters operating simultaneously on a material, inevitably lead to the need for statistical calculations. Tools such as computer modeling, statistical analysis and probabilistic design (Chapters 4, 5 and 14) are available to assist the materials engineer in predicting performance when use conditions are varied. For example, Weibull plots in combination with finite element analysis and/or proof testing are used widely for predicting the probability of failure for high-performance metals and ceramics under various levels of stress, size and configuration (see Figure 3).

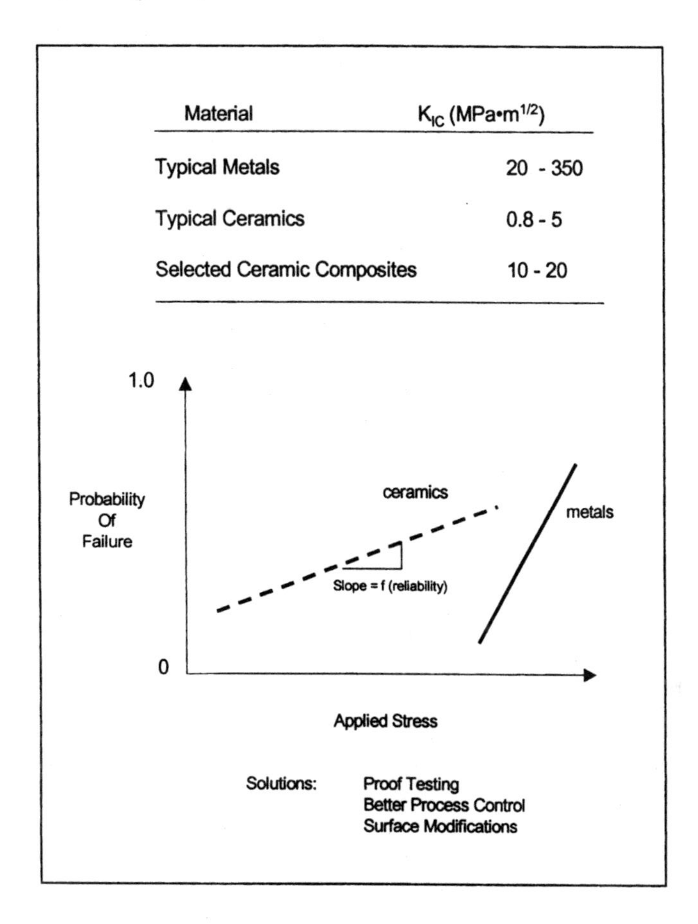

Material	K_{IC} (MPa•$m^{1/2}$)
Typical Metals	20 - 350
Typical Ceramics	0.8 - 5
Selected Ceramic Composites	10 - 20

Figure 3. Example of the use of performance evaluation tools, in this case, for predicting probability of failure. Greater line slopes represent higher reliability.

CERAMIC ENGINEERING DESIGN

Ceramics are manufactured and used in a variety of ways, as shown in Table 1. Prior to the 20th century, ceramic products were manufactured by individual ceramic specialists or small industrial operations and sold directly to the consumer. Today, it is more likely that these products are manufactured by large corporations that sell only a small fraction directly to the consumer. The majority of ceramic

products are sold to the mainstream industries who then incorporate them into more sophisticated and complex products that are sold then to the consumer. Thus, as shown in Figure 4, it is likely that ceramic engineers are employed in many of the mainstream industries as well as the major ceramic industries.

Table 1. Some Applications of Ceramic Materials.

Common Uses
Floor and Wall Tiles
Bulbs (lights, TV tubes)
Containers (food and beverage)
Construction Materials (concrete blocks, brick, plaster, gypsum)
Windows (home and automotive)
Stove Tops (glass-ceramics)
Cookware (glass-ceramics, Pyrex®)
Sanitary Ware (kitchen, bathroom)
Porcelain Enamels (appliances, chemical processing vessels)
Abrasives (sandpaper, grinding and cut-off wheels)
Architectural Windows
High-Tech Uses
Nuclear Fuels (commercial electric power plants)
Glass Fibers (communications)
Packages for Integrated Circuits (electronics)
Capacitors (electronics)
Shuttle Tiles (thermal protection)
Cutting Tools
Varistors (lightning and voltage surge protectors)
Catalytic Converters (pollution control for autos)
Senors (analytical and pollution control)
Rotors (turbochargers)
Waste Glass (immobilization of nuclear waste)
Composites (high-temperature structural applications)
Batteries and Fuel Cells (energy storage/generation)
Biomaterials
High-Speed Electronic Circuitry
Magnetic Components

Glass is a good example of a ceramic material that is sold both directly to the public as well as being integrated into mainstream industrial products. The earliest uses of glass were as art, jewelry, window panes and beverage containers. Today,

glasses are used extensively as vital components in products such as light bulbs, fiber optics, fibers for insulation (Chapter 10) and nuclear waste containment (Chapter 11).

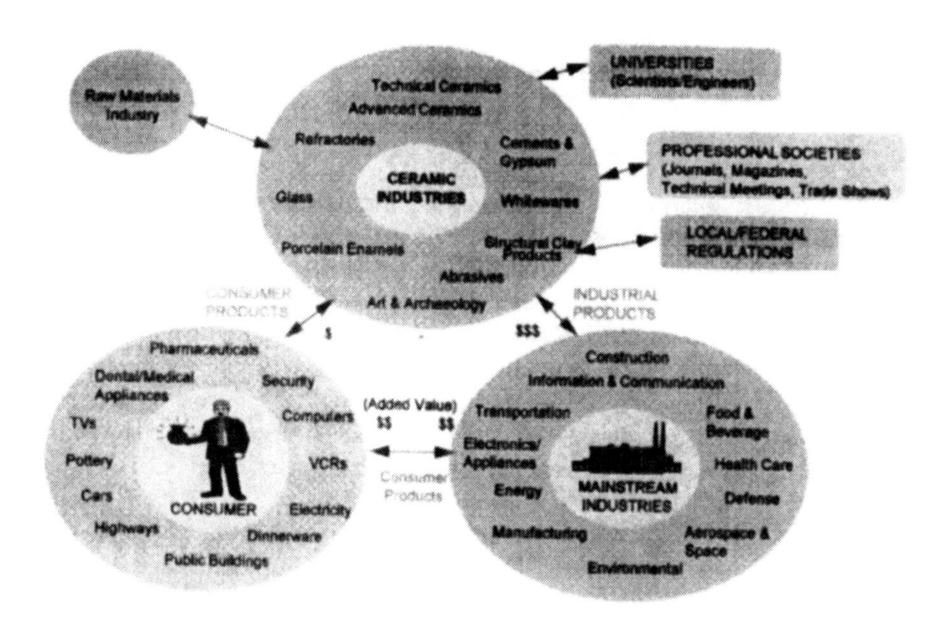

Figure 4. Integration of ceramics into society. Some ceramic products (dinnerware, pottery) are provided directly to the consumer by the ceramic industries. The majority of ceramic products reach the consumer indirectly through the mainstream industries. Mainstream industries provide the consumer with two types of products: (1) hardware such as televisions and computers that contain numerous ceramic components or require the use of ceramics during manufacturing, and (2) services such as electricity, health care and national security where ceramics provide essential functions.

A ceramic engineer designs materials systems and processes in order to obtain economically competitive products with durable properties that will benefit society. He or she must communicate between customers and the various personnel in the company to determine the most efficient and economical methods of manufacturing, product testing and quality control.

Essential skills for the ceramic engineer include:

- good foundation in basic science and engineering
- ability to apply basic science concepts to engineering and manufacturing problems
- oral and written communication skills
- knowledge of statistics and computers
- ability to work as part of a team
- understanding of other engineering disciplines
- well-grounded in the principles of design

Property requirements play a central role in the design of materials for any application. Properties for which ceramics are noted are listed in Table 2. Specific applications involving one or several of these properties are discussed in detail in subsequent chapters. In some cases, the other two broad categories of materials, metals and polymers, also exhibit these same properties. For example, polymers exhibiting excellent chemical durability are used in many applications where this property is important (e.g. food and beverage containers). Many metals are used in applications where wear resistance is required (e.g. bearings, gears). However, when combinations of these properties are required, which usually is the case, ceramics excel (Chapter 6). Water heaters and pharmaceutical reaction vessels require better chemical durability and high-temperature stability. In these applications, the metal vessels must be lined with special glass. Moreover, in applications requiring temperatures over 1200°C, only ceramics will survive (Figure 5).

Table 2. Important properties of ceramic materials.

- ❑ Chemical
- ❑ Wear
- ❑ Hardness
- ❑ Optical (color, transparency)
- ❑ Electrical/Magnetic
- ❑ Thermal
- ❑ High-temperature Stability
- ❑ Density

In contrast, as seen in Figure 3, ceramics generally exhibit poor fracture toughness and reliability in comparison to most metals (and polymers). Thus for applications requiring high-temperature stability (>1200°C) and high fracture toughness, new materials involving combinations of ceramics with either themselves or metals must be developed. Designing these composite materials will provide a challenge to the ceramic engineer for many years to come.

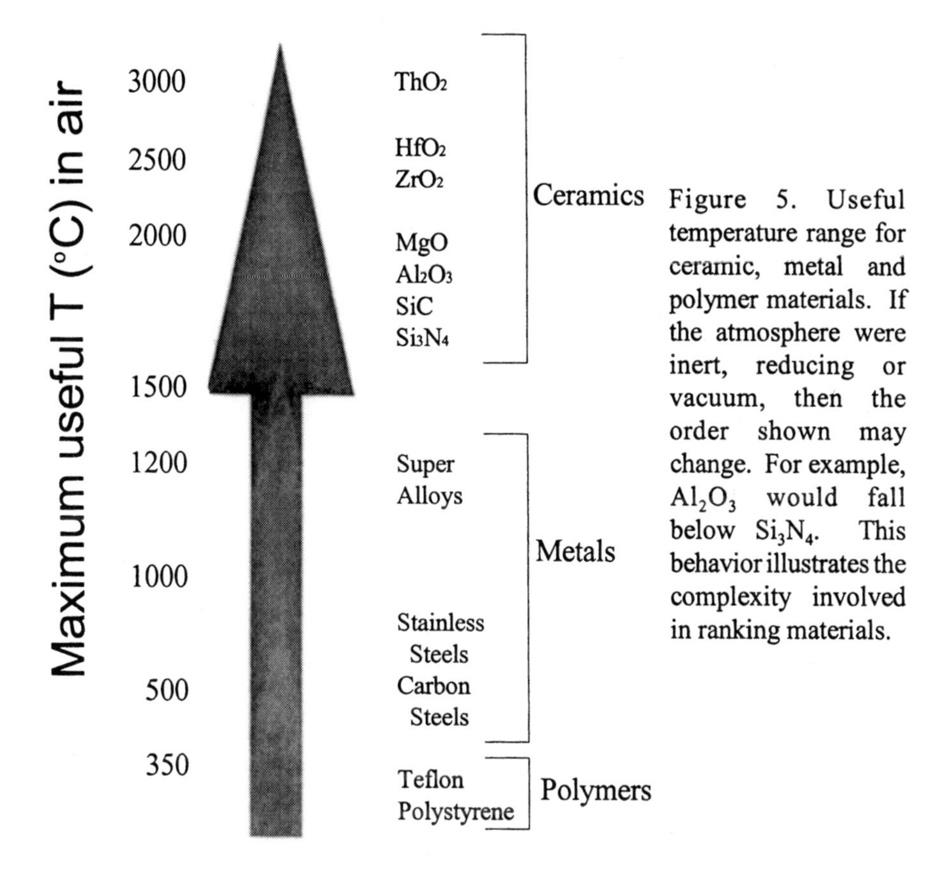

Figure 5. Useful temperature range for ceramic, metal and polymer materials. If the atmosphere were inert, reducing or vacuum, then the order shown may change. For example, Al_2O_3 would fall below Si_3N_4. This behavior illustrates the complexity involved in ranking materials.

Today, there are many characterization tools available to assist the ceramic engineer in processing and performance evaluation. Table 3 provides a partial list of some of the most commonly used tools as well as the type of information that is obtained from each. The engineer needs to be aware of costs associated with these tools. In some cases, two different tools may provide the same type of information (e.g., optical microscope and SEM). What determines which of the tools should be used is the level of detail required and the costs. Scanning electron microscopy will provide the greatest detail with respect to microstructure, but it also will cost about a factor of five more than optical microscopy. Similarly, TEM will yield even greater detail, but again at a significantly higher cost. When large numbers of parameters are required to be evaluated, the statistical design techniques discussed in Chapter 4 can provide a high level of confidence with a minimum number of samples.

Table 3. Commonly used characterization and processing tools of ceramic engineers.

Tool	Information Provided
x-ray diffraction	phase, composition
x-ray fluorescence	composition, trace element analysis
Weibull analysis/proof testing	expected product performance
x-ray sedigraph/laser scattering	particle size analysis
ultrasonics, radiography	non-destructive flaw size determination
Inductively Coupled Plasma Spectroscopy (ICP)	composition, trace element analysis
optical microscopy	fracture origins, quantitative microstructural analysis
scanning electron microscopy with energy dispersive microscopy (SEM-EDS)	microstructure, rough composition
transmission electron microscopy (TEM)	nanostructure
thermogravimetric analysis (TGA)	processing - weight loss
differential thermal analysis (DTA)	processing - phase changes
dilatometry	expansion, densification
Archimedes apparatus	bulk and apparent density
pycnometry	true density
microindentation	hardness, fracture toughness
mechanical tester	strength, elastic modulus, modulus of rupture
gas adsorption	surface area and porosity
Auger electron spectroscopy (AES)	surface composition
Fourier transform infrared spectroscopy (FTIR)	composition and structure of glasses
phase diagrams	equilibrium phases
Ellingham charts	thermodynamically favored phases under various redox conditions

In addition to costs, the engineer should be aware of the limitations of the tools. Significant errors in analyses resulting from an improper use of the data can lead to failure in the field or added expense in the process design.

Finally, scale-up must be accomplished to apply the design concept on an industrial scale. The ceramic engineer should realize that processes and performance do not always scale linearly and that costs may be driven up when larger parts are manufactured. For example, if a 1 cm^3 specimen of an alumina part can be fired to full density at 1500°C in 1 hour, 15 to 20 hours might be required to fire a 10 cm^3 part due to poor thermal conductivity of the ceramic as well as the need to cool the larger part more slowly to prevent thermal shock (Chapters 7 and 8). In addition to the increased cost related to longer firing time, the performance of the part may be influenced significantly due to differences in microstructures over the cross-section (i.e., interior will have more pores and smaller grain size as compared to the exterior).

Environmental issues, resource availability and safety also are important parameters in any manufacturing scheme. Besides affecting costs, these parameters can have a significant impact on plant location, air and water quality and community acceptance.

As an engineer in charge of process and product design, you also will most certainly be involved in legal issues related to intellectual property rights and product liability. These topics will be discussed in more detail in Chapter 19.

Chapter 18 presents an example of a university program developed specifically to integrate design into the materials engineering curriculum.

ILLUSTRATIVE EXAMPLES

Ceramic engineering design can involve any one or all of the following:

1) Selecting an existing material, based on properties, for a specific application. The application could require one or several properties that, in combination, dictate the performance of the material. The engineer first needs to assess the most important property and the range of values expected for this property in the use environment(s). If the primary property requirements cannot be met with existing materials, then a new material must be developed.

2) Designing a new material for a specific application. This could involve the development of a new composition, a tailored microstructure with controlled grain size and phase distribution, or a nano-structure with controlled pore size and species functionality.

3) Designing a process. Processing is the key to success. The engineer might be able to design a new material that will, theoretically, meet the performance requirements. However, to be successful he/she also must design a process for manufacturing the material. This process may involve selecting raw materials, determining a suitable forming technique, evaluating heating schedules, assessing post-firing requirements and implementing a quality control program.

   Examples are provided to illustrate each of these types of design. Additional and more in depth examples can be found in subsequent chapters.

Selecting an existing material, based on properties, for a specific application.

Example provided in the Case Studies section of Chapter 3, by Dr. Jack Lackey, Georgia Institute of Technology, George W. Woodruff School of Mechanical Engineering, Atlanta, Georgia

Designing a new material for a specific application.

Example provided by Dr. Delbert E. Day, Ceramic Engineering Department and Graduate Center for Materials Research, University of Missouri-Rolla, Rolla, MO.

Design of Radiation Delivery Materials for Treating Cancer

Introduction

Within medicine, radiation is an accepted method of treating patients suffering from various types of cancer. Unfortunately, radiation has not proven to be effective in treating patients suffering from liver cancer, since the amount of radiation that can be delivered to malignant tumors in the liver from an eternal source is limited by the damage done to nearby healthy tissue. This example describes how glass microspheres were designed to allow irradiation of malignant tumors in the liver in situ. The advantage of in situ radiation is that the radiation is more localized so that less damage is done to adjacent healthy tissue, thereby permitting much larger doses of radiation with less patient discomfort.

Concept

The basic idea was to use glass microspheres that would contain a radioactive isotope that emitted beta radiation, which is more localized and

short range than gamma emitters. The glass microspheres would be injected into the artery feeding blood to the liver so that the microspheres would become trapped in the capillary bed of the liver where each radioactive microsphere would irradiate a small volume of the liver over a period of several days until the radioisotope had decayed.

To implement this concept, it was necessary to develop or design a glass that possessed the needed properties. This example, therefore, focuses on how to design a material for a particular function or application. Glass was the initial material of choice since it can be made into microspheres by several methods, the chemical composition can vary over wide limits, it can possess good chemical durability and it has been used in the body for other applications.

Materials Requirements

The glass to be used as an in situ radiation delivery vehicle had to possess the following characteristics:

- *Biocompatible and non toxic.* The glass could not be harmful to the body, for obvious reasons.
- *Chemically insoluble in body fluids (blood).* This was one of the most important requirements, since it was critical to prevent the radioisotope from escaping from the liver and moving to other parts of the body. To prevent the leakage of radioactivity from the target site, it was decided that the radioisotope should be dissolved chemically into an insoluble glass. This requirement restricted the chemical composition of the glasses that could be considered, since components that reduced the chemical durability could not be present.
- *Free of unwanted, neutron activatable elements.* For ease of manufacture, it was desired to first prepare the glass microspheres from non-radioactive materials and then, as the last step in the manufacturing process, make them radioactive by the process of neutron activation, i.e. bombarding the microspheres with neutrons inside a nuclear reactor. Radioactive microspheres could have been made by using a radioactive material in the glass batch, but neutron activation avoided the potential risk and expense of handling radioactive materials during the melting and microsphere manufacturing process.

  However, the neutron activation requirement greatly limited the chemical compositions that could be considered, since a high percentage of the elements normally used in glass become radioactive during neutron activation. This behavior produces unwanted

radiation, especially the longer range gamma radiation that could damage nearby healthy tissue.

In addition to the three primary requirements listed above, the glass also had to be capable of being melted at reasonable temperatures (below 1600°C), formed into glass microspheres of the desired size (25 to 35 micrometers in diameter) and have a reasonable cost. These requirements, however, were less restrictive and were expected to be satisfied easily.

Designing the Glass

The first step was to survey existing glasses that might satisfy the above requirements along with the additional requirement that the glass contain as much yttrium oxide (Y_2O_3), also referred to as yttria, as possible (at least 20 wt%). For medical reasons, the radioactive element of choice was the beta-emitting radioisotope 90-Y, which can be formed by neutron bombardment (activation) of naturally occurring 89-Y. A high yttrium content was necessary to produce a glass that would become highly radioactive (high specific activity) during neutron activation so that a sufficiently large dose of radiation could be delivered to the patient by only a small quantity (about 100 mg) of glass microspheres.

As expected no existing glasses were found that met the above requirements. For maximum chemical durability, a glass high in silica was considered desirable and neither silicon nor oxygen formed objectionable radioisotopes during neutron bombardment. Alumina also was considered acceptable, since aluminosilicate glasses are known to have a high chemical durability and aluminum does not form objectionable radioisotopes. Thus the initial decision was to investigate the Al_2O_3-Y_2O_3-SiO_2 system, which contains only four elements, even though all of these oxides melt above 1700°C and the compositional limits for glass formation were unknown, as were the chemical durability and maximum Y_2O_3 content.

Based on laboratory scale melts, a fairly large region of glass formation was found for alumina-yttria-silica (YAS) compositions that could be melted at 1600°C and below. Glasses were obtained that contained up to 50 wt% yttria. Glass microspheres of the proper size also were prepared easily. Chemical durability tests on glass microspheres in distilled water and saline solutions at 37°C (body temperature) established that the leaching of yttrium from the glass was extremely small and acceptable for use in humans.

In short, the YAS glasses satisfied all of the design requirements for in situ radiation delivery vehicles in humans. Aluminosilicate glasses containing rare earths such as beta-emitting Sm, Ho, Nd or Dy also have been found to satisfy the above design requirements as radiation delivery vehicles. These glasses also offer a flexibility in use, since a rare earth can be chosen that offers the optimum radiation energy, range in tissue and duration of treatment (half life) for a particular application. Thus rare earth aluminosilicate glass microspheres can function as vehicles to deliver a wide range of specific radiation to selected targets.

Practical Use

After extensive animal and human testing, YAS glass microspheres now are being used under the trade name TheraSphere™ to treat patients with inoperable, primary liver cancer. Patients are treated on an outpatient basis and are given an injection of about 5 million glass microspheres through a catheter inserted into the hepatic artery. The beta-emitting, YAS microspheres become trapped within the liver in just a few seconds where they irradiate the surrounding tissue to an average distance of about 3 mm from each microsphere. After observing the patient for a few hours, the patient is released from the hospital while their liver continues to be irradiated for a period of between three and four weeks. The beta radiation is too weak to escape outside their bodies, so they are no danger to others. After four weeks, the radioactivity has decayed and the treatment is over, although the glass microspheres remain in the liver indefinitely (they do not cause any harm).

There are two major advantages of this method of in situ radiation treatment. First, the YAS glass microspheres safely deliver nearly 10 times more radiation – up to 15,000 rads – than any other known technique. Since the glass microspheres are inside the liver where the weaker beta radiation can be used, there is much less damage to healthy tissue than in general radiation because of the localized radiation. Therefore, larger amounts of radiation, which increase the probability of destroying the tumors, can be used safely.

Glass microspheres now are being investigated as radiation delivery vehicles for other applications. Potentially, they can be used to irradiate any organ which has a capillary bed capable of trapping the radioactive microspheres. Glass microspheres containing samarium(Sm) have been tested in animals for the purpose of treating kidney cancer while microspheres containing rhenium (Re) have been used to irradiate sites in the brain where tumors have been removed surgically. Another potential application is in the treatment of rheumatoid arthritis, where radioactive particles are injected into a joint to reduce pain and swelling, a treatment called radiation synovectomy. In several animal experiments, glass microspheres (10 to 15 microns in diameter) containing dysprosium (Dy) have been injected into the joint capsule with no adverse effects. The laboratory results are promising.

Designing a process.

Example provided by Dr. Carlos Suchicital, Department of Materials Science and Engineering, Virginia Polytechnic Institute and State University, Blacksburg, VA.

Design of a Rapid Prototyping Process

Introduction

As costs for developing new technologies from concept to production increase, rapid development of prototypes has become a priority for small and large-scale manufacturers. Over the past ten years, *rapid prototyping* technologies have been the subject of much research. An excellent historical perspective was compiled by Beaman et al in which it is noted that the first attempts at developing dimensional topographical relief maps were accomplished as early as 1890. Soon thereafter, engineers realized that, with some modifications, this technique offered industry the potential to produce parts whose shapes were difficult to make with traditional machining methods.

Most products under consideration to be introduced into the marketplace will go from a prototype stage to a commercially viable one. Such products range from the very technologically complex (i.e., airplanes, oil tankers, automobiles, cellular phones, cardiac pace makers) to the relatively simple ones (i.e., small medical implants, shop tools, cookware). It is clear that prototyping is an expensive process. Complex shapes require more prototype iterations, driving up the cost for prototyping and increasing the time required in the product development stage. To increase the efficiency of production of widely varying parts, machining technologies have advanced to provide multiple axis flexibility and software compatibility. However, even with the improvements in machining equipment and processes, the complexity of shapes that can be formed is limited.

The engineering problem is to find a method for producing prototypes for complex, three-dimensional structures with clearly defined interior and exterior surfaces in order to minimize post-production processing (machining).

Concept

The goal for this process design is to produce a three-dimensional structure, with all its complexity, in one single piece. To achieve this goal, the designer will have to take advantage of present day computing power capabilities to manipulate data from high-resolution scanning systems and convert these images to data that can be processed by prototyping systems. It also will be necessary to build on the work that specialists in the field have

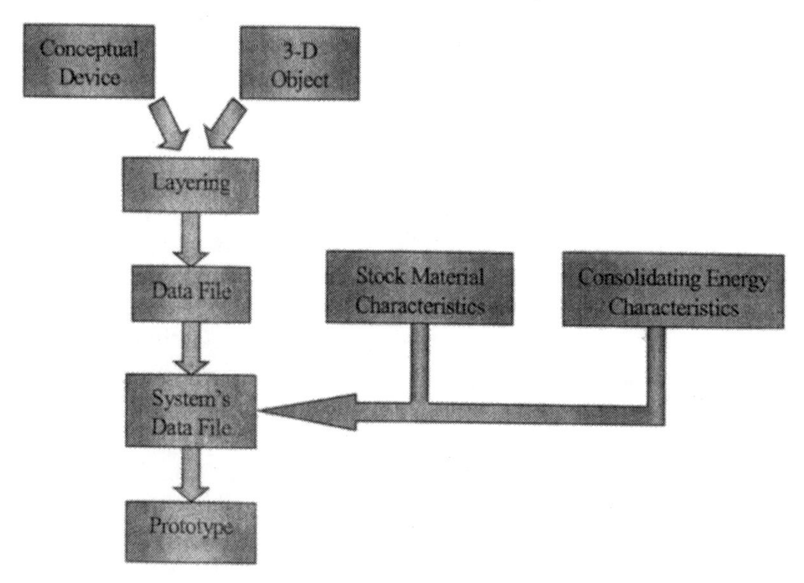

done in areas such as the concepts of layering, three-dimensional object scanning, stock material[1] addition to prototyping cavities and consolidation or sintering methods for stock materials. Current rapid prototyping systems follow the same generic concepts (Figure 6), but are tailored to a specific stock material and the type of energy used for its consolidation.

Figure 6. Concept process for designing a rapid prototyping system.

Process Requirements

Fundamental characteristics of a commercial rapid prototyping system include

---

[1]Stock material is the material out of which the prototype is made. It may be in the form of a fine powder of aggregates, paste, liquid or even gas.

- *Scanning the Object.* The system must be able to scan the surfaces of three-dimensional objects and manipulate the data in such a way as to represent said surfaces as a sequence of layers, creating a "layering data file."

  In order to create these layers for internal surfaces, it may be necessary to

  a) partition the object into smaller sections and then compile the scanned sections using a computer to obtain the final 3-D object; or
  b) scan the object using x-rays or acoustic methods to define the internal space.

  For conceptual objects under design, digitalized graphics data may be generated by a computer in a layered object file to include complex external and internal surfaces.

  This complex computing capability is expensive, both for hardware and software. When preparing the Layering Data File, it takes into account the type of stock material, including the energy required for consolidation. This Layering Data File then is converted into a System Data File that is used by the system to build the actual prototype. The design engineer must consider this expense when determining the level of complexity required by the system for the application under consideration.

- *System Computer Capability.* The system's on-board computer must be capable of handling files of the level of complexity required for the specific type of application. In addition to being able to hold large amounts of data, the system will be required to manipulate the data at a rapid rate.
- *Stock Material.* Most rapid prototyping processes are additive; stock material is added gradually, one layer at a time, to build the final 3-D shape. In order to build the prototype, a database of the characteristics of the stock material is essential for generation of the Data File. The stock material may be tailored to the process by using it as a liquid, powder or paste. In some instances, the stock material can be mixed with a sacrificial binder to provide mechanical stability of the prototype prior to final processing (i.e., heat treatment to gradually eliminate the binder and solidify the prototype).

Depending on the application, the physical and mechanical properties of the stock material will determine the preferred method for building the prototype layers. In other cases, the method available for building the prototype may dictate the stock materials that may be used.

- *Consolidating Energy*. Available data on the consolidation methods for stock materials indicate that light sources are very versatile means for conveying energy quickly to localized regions (i.e., CO_2 laser light).

  Other consolidation methods utilize resistance heating to fluidize a composite stock material delivered through a fine nozzle to specific locations on a plane and to subsequently solidify the deposited material. A similar, but more elaborate system, uses multiple nozzles to generate the x-y patterns in a dot-matrix print-head fashion before thermally treating the deposed material.

  Regardless of the preferred method for building the sequential planes of stock material, most of the methods require further thermal processing of the stock materials to achieve the mechanical stability required of the prototype.

- *Prototype*. In general, the desired characteristics of the prototype are determined by its intended application, e.g., transparency to show internal structures. The size of a manufactured prototype with presently available instrumentation can range from several millimeters to several meters. However, due to design considerations and equipment costs, most systems produce a prototype that is several decimeters in size that is then assembled into a larger, 3-D object.

Designing the Process

The desired process will take advantage of present-day prototyping technologies to implement the general concept described in Figure 6 while meeting the characteristics described above. A rapid prototyping system will allow for the "quick" manufacture of a 3-D object from a System Data File for further product development or final product manufacturing. Whatever the application, the process must incorporate safety features to protect the operator from potentially dangerous effects from high voltages and currents, toxic by-products, radiation, etc.

Table 4 enumerates rapid prototyping technologies with products presently on the market, based on the generic process presented in Figure 6.

Table 4. Commercial rapid prototyping technologies.*

Technology	Maximum Part Size (cm)	Strengths	Weaknesses
Stereolithography	51x51x61	Large part size; accuracy	Post processing required
Wide area inkjet	25x20x20	Speed, office friendly	Size and weight; fragile parts, limited materials
Single jet inkjet	30x15x23	Accuracy; finish; office friendly	Speed; materials; part size
Laminated object manufacturing	81x56x51	Large part size good for castings; material cost	Part stability; presence of smoke and potential fire hazard
Selective laser sintering	38x43x46	Accuracy; materials availability	System price; surface finish
Fused deposition modeling	61x61x61	Office friendly; price	Limited materials; low speed
3-D printing	20x25x20	Speed; office friendly	Fragile parts; materials availability; finish

* Published by permission of Castle Island Company. "The Worldwide Guide to Rapid Prototyping," E.P. Grenda.

Applicability and New Uses

The commercial systems based on this process can produce very complex prototypes in hours instead of days or weeks. Modifications to prototypes are easier to implement and the results can be seen immediately. However, it is not yet a high-volume manufacturing process. Rather, it is one that is very well-suited to development of new, complex shapes. In the near term, new stock materials are being developed to allow for production of "ready to use" objects. However, niche areas already exist for low-volume rapid prototype production of 3-D objects, e.g., replacement body parts for humans and animals in which the products are produced according to individual specifications.

H. Koukka has assembled a list of rapid prototyping methods that provides an overview of the types of systems and applications where this process is being used. Commercial systems presently make use of stock materials in the form of liquids, powders, solids, sheets, gases and atomic level structures. New materials amenable to rapid prototyping technologies are being developed at research centers throughout the world. Rapid prototyping is not seen as a replacement for current fabrication methods, but is viewed as a new process for very specific areas of manufacturing.

Example

Engineers at Stanford University's Rapid Prototyping Laboratory have been working on a process to produce functional prototypes from computer models. The process, named Mold Shape Deposition Manufacturing, relies on known principles of technologies such as Solid Freeform Fabrication, mold casting, and computer numeric control machining to create a complex fugitive mold. The stock material, in a gel form, is cast into the mold and cured or allowed to set, depending on the type of material used (Figure 7)

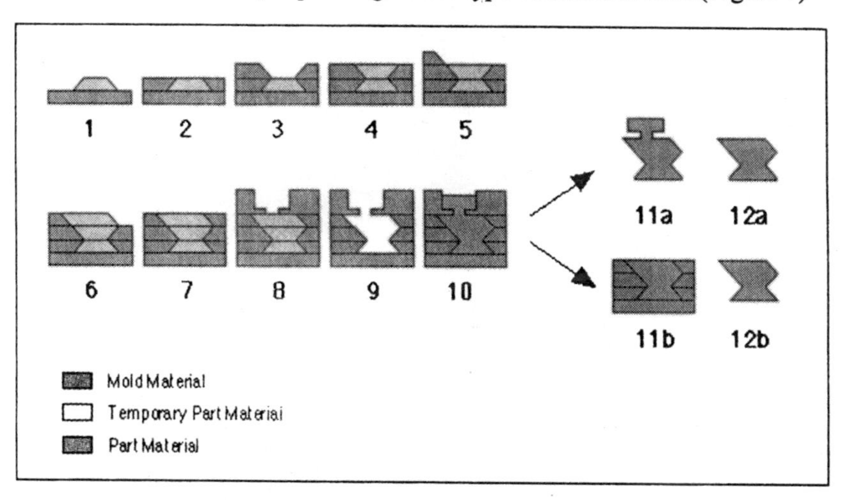

Figure 7. Pictorial flow diagram for Mold Shape Deposition Manufacturing (SDM). Taken from Cooper et al (34 in bibliography).

Examples of sintered ceramic devices manufactured via the mold SDM process are shown in Figure 8.

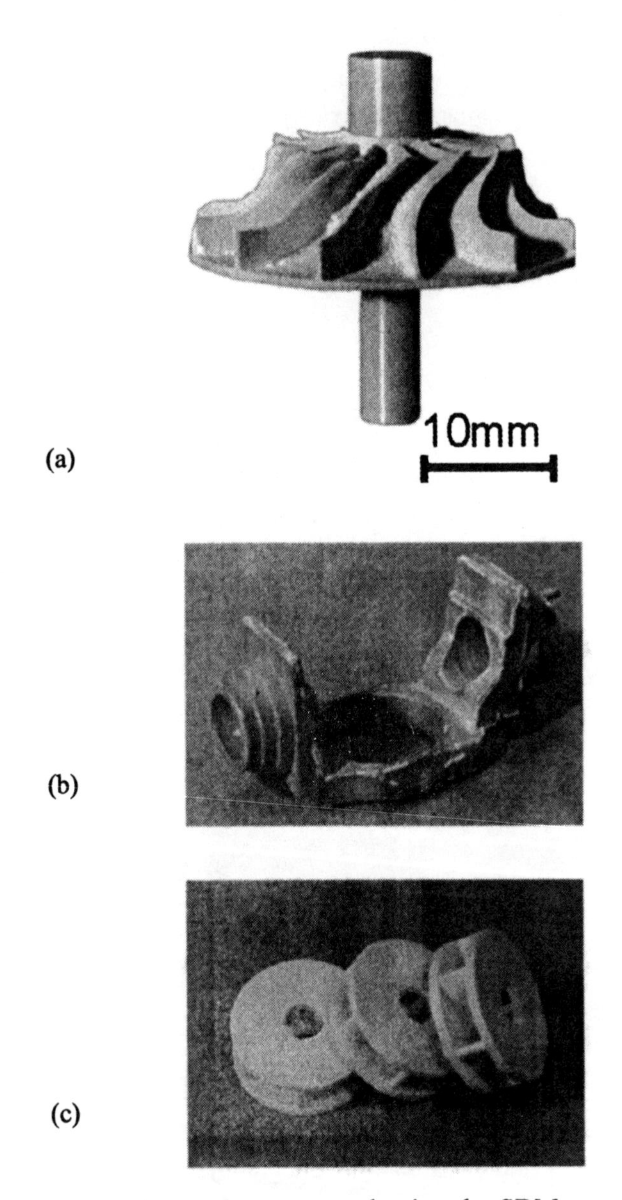

Figure 8. Fired ceramic parts cased using the SDM process: (a) silicon nitride turbine, (b) silicon nitride pitch shaft, and (c) alumina pump impellers [34, 35, 36].

SUMMARY

This chapter has introduced students, faculty and practicing engineers to the basic concepts of design as they relate to ceramic engineering products and processes. Illustrative examples have been provided to help the reader to understand ceramic materials design problems, ranging from selecting existing materials, developing new materials or developing new processes for producing new or existing materials.

The chapters that follow will provide insights into design by reviewing these concepts as they are used to address specific ceramic engineering problems. Each chapter provides a bibliography with suggestions for further study in this area as well as a set of questions/problems that will reinforce these concepts for the reader.

BIBLIOGRAPHY/SUGGESTED READING

1. Criteria for Accrediting Programs in Engineering in the United States (1996-97), Engineering Accreditation Commission, Accreditation Board for Engineering and Technology, Inc., Baltimore, MD.

2. William D. Callister, Jr., Materials Science and Engineering, An Introduction, 4th Edition, John Wiley & Sons, Inc., New York, 1996.

3. W.E.C. Creybe, I.E.J. Sainsbury and R. Morrell, Design with Non-Ductile Materials, Applied Science Publishers, London and New York, 1982.

4. James F. Shackelford, Introduction to Materials Science for Engineers, MacMillan Publishing Company, New York, 1992.

5. E.H. Cornish, Materials and the Designer, Cambridge University Press, New York, 1987.

6. D.L. Hartsock and A.E. McLean, "What the Designer with Ceramics Needs," Ceramic Bulletin, Vol. 63, No. 2, pp. 266-270 (1984).

7. N.E. Dowling, Mechanical Behavior of Materials: Engineering Methods for Deformation, Fracture, and Fatigue, 2nd Edition, Prentice Hall, Upper Saddle River, NJ, 1999.

8. R. Morrell, Handbook of Properties of Technical & Engineering Ceramics: Parts 1 and 2, Her Majesty's Stationery Office, London, England, 1985.

9. T. Richardson, Composites: A Design Guide, Industrial Press, Inc., New York, NY, 1987.

10. J.B. Wachtman, Mechanical Properties of Ceramics, John Wiley & Sons, Inc., New York, NY, 1996.

11. L.S. O'Bannon, Dictionary of Ceramic Science and Engineering, Plenum Press, New York, NY, 1984.

12. J.C. Phillips, Physics of High-Tc Superconductors, Academic Press, Inc., Murray Hill, NJ, 1989.

13. D.P. Woodruff and T.A. Delchar, Modern Techniques of Surface Science, Cambridge University Press, New York, NY, 1986.

14. D.C. Boyd and J.F. MacDowell, Commercial Glasses, Advances in Ceramics, Vol. 18, American Ceramic Society, Inc., Columbus, OH, 1986.

15. S. Musikant, What Every Engineer Should Know about Ceramics, Marcel, Dekker, Inc., New York, NY, 1991.

16. J.T. Jones and M.F. Berard, Ceramics: Industrial Processing and Testing, 2nd Edition, Iowa State University Press, Ames, IA, 1993.

17. V.D. Frechette, Failure Analysis of Brittle Materials, Advances in Ceramics, Vol. 28, American Ceramic Society, Inc., 1990.

18. Refractories Handbook, The Technical Association of Refractories, Tokyo, Japan, 1998.

19. Ceramic Innovations in the 20th Century, J.B. Wachtman, Jr., ed., American Ceramic Society, Inc., Westerville, OH, 1999.

20. A. Dodd and D.Murfin, Dictionary of Ceramics, 3rd Edition, The Institute of Materials, London, England, 1994.

21. M. F. Ashby and D.R.H. Jones, Engineering Materials 1: An Introduction to Their Properties and Applications, Pergamon Press, Oxford, England, 1980.

22. M. F. Ashby and D.R.H. Jones, Engineering Materials 2: An Introduction to microstructure, Processing and Design , Pergamon Press, Oxford, England, 1980.

23. Particle Size Analysis , J.D. Stockham and E.G. Fochtman, eds. Ann Arbor Science Publishers, Inc., Ann Arbor, MI, 1977.

24. Processing for Improved Productivity, K.M. Nair, ed., Advances in Ceramics, Vol. 11, American Ceramic Society, Columbus, OH, 1984.

25. Advanced Technical Ceramics, S. Somiya, ed., Academic Press, Inc., San Diego, CA, 1984.

26. R.F. Speyer, Thermal Analysis of Materials, Marcel Dekker, Inc., New York, NY, 1994.

27. J.P. Schaffer, A. Saxena, S.D. Antolovich, T.H. Sanders, Jr. and S.B. Warner, The Science and Design of Engineering Materials, 2nd Edition, WCW/McGraw-Hill, New York, NY, 1999.

28. High-Technology Ceramics: Past, Present and Future, Ceramics and Civilization, Vol. III, W.D. Kingery, ed., American Ceramic Society, Inc., Westerville, OH, 1986.

29. Materials Processing in Space, Advances in Ceramics, Vol. 5, B.J. Dunbar, ed., American Ceramic Society, Inc., Westerville, OH, 1983.

30. C.B. Fleddermann, Engineering Ethics, Prentice Hall, Upper Saddle River, NY, 1999.

31. "Solid Freeform Fabrication," J.J. Beaman, J.W. Barlow, D.L. Bourell, R.H. Crawford, H.L. Marcus and K.P. McAlea, Kluwar Academic Publishers, London, 1997.

32. "The Worldwide Guide to Rapid Prototyping," E.P. Greneda, [http://home.att.net/~castleisland/home.htm]. Copyright 1999, 2000, 2001, Castle Island Co., Arlington, MA.

33. "The Hole RP Family Tree," H. Koukka, [http://ltk.hut.fi/~koukka/RP/rptree.html] Helsinki University of Technology, June 2001.

34. "Fabrication of Ceramic Parts for a Miniature Jet Engine Application Using Mold SDM," A.G. Cooper, S. Kang, J. Stampfl and F.B. Prinz, Ceramic Transactions, Vol 108, 389-398, 2000.

35. "Mesoscopic Assemblies with SDM Processing," B.H. Park and F.B. Prinz, Proceedings fo the Solid Freeform Fabrication Symposium, University of Texas at Austin, August 2000.

36. "μ-Mold Shape Deposition Manufacturing," S.W. Nam, H.C. Liu, J. Stampfl, S. Kand and F.B. Prinz, MRS Spring Meeting Proceedings, April 2000, in print.

PROBLEMS

1. Using Figure 1, suggest three possible "others" as input into the various components of the general design process, problem definition and design concepts.

2. (a) Select at least three ceramics from existing materials that will meet the following performance criteria:

   Materials Criteria: $K_{IC} \geq 10\ \text{MPa}\cdot\text{m}^{1/2}$
   $\sigma \geq 100\ \text{MPa}$
   $T_{use} \leq 500°\text{C}$

   (b) From this list of candidate materials, select the best based on cost.

3. (a) In designing a ceramic cutting tool, what are the important materials properties to be considered?

b. How would you design an evaluation scheme to assess the performance of this material?

4. Outline an approach for developing a delivery system capable of treating carcinogenic tumors. The system should allow for controlled release of a chemotherapeutic drug and there should be no residue of the carrier material remaining after six months.

5. The average oxide composition of specimens brought back from the moon by the Apollo astronauts is

Oxide	Wt%
SiO2	43.0
TiO2	11.0
Al2O3	7.7
FeO	21.0
MnO	0.3
MgO	0.5
CaO	9.0
Na2O	0.4
K2O	0.2

a. What tools listed in Table 3 were most likely used to determine this composition?

b. What phases are most likely present in these specimens? What tool shown in Table 3 would you use to verify the phases?

c. Determine the properties of a structural material capable of protecting the occupants of a research station from the moon's harsh environment.

d. Design a process suitable for manufacturing building blocks from the material with the properties determined in part (c). The manufacturing facility will be build and operated on the moon.

2 Ethical Issues In Design

M.L. Cummings

ETHICAL ISSUES IN DESIGN

Mary L. Cummings
Virginia Polytechnic Institute & State University
Engineering Fundamentals Division
Blacksburg, VA 24061

INTRODUCTION

Often when engineers and students of engineering hear the word ethics, they wrinkle their faces in distasteful expressions. "I became an engineer to work with absolutes and objectivity, not touchy feely philosophy," is a common response. Those engineers, both in industrial and academic settings who dismiss ethics as an integral part of engineering, have not embraced all aspects of their professional responsibilities. All engineers will face many ethical dilemmas during their careers. Whether it is a patent issue, a design question, or an application concern, ethics are a fundamental and extremely important aspect of all engineering decisions.

This chapter will explain basic applied ethical theory, ethical decision-making and the design process, the role of professional societies and ethical codes, as well as many other areas where future ethical problems may be encountered in materials engineering. When reading this chapter, especially the case studies, keep in mind that even in traditional engineering problems, very rarely does a clear and absolute answer exist. Just as critical, non-linear thinking is essential to the materials design process, these same analytical skills are needed for difficult ethical decisions.

BASIC ETHICAL THEORIES AND PRINCIPLES

What is the difference between morals and ethics?

Webster's dictionary defines *moral* as pertaining to a standard of right or wrong and *morality* as a system of moral principles and conduct. *Ethical*, often used as a synonym for moral, relates more as a subtle question as to the rightness or wrongness of an act, a person, or some kind of decision. For engineers who prefer unambiguous clear definitions, ethics is essentially the application of morals. Most people believe in certain *prima facie* rules in life: don't hurt others;

don't take what is not yours, etc. Ethics and ethical behavior is essentially taking these basic beliefs and applying them in both our personal and professional lives.

Because the typical engineer has little or no philosophical training, we will begin with two major frameworks for ethical decision-making, *utilitarianism* and the *respect-for-persons* approach. While there are many complex and multifaceted issues in the field of ethics, these two provide a more straightforward foundation by which an engineer-in-training can begin to grasp the application of ethics in engineering.

1. Utilitarianism

Utilitarians believe that the most ethical viewpoint is one that promotes the overall goodness, well-being, or utility for those affected. As an outcomes-based approach, utilitarianism supports those decisions and acts that maximize human welfare for the whole, while minimizing any harm that may result. Engineers often feel very comfortable with the utilitarian concept because it is one that they use on a daily basis. Engineers, economists, and business professionals refer to utilitarianism in the workplace as *cost-benefit analysis*. The realities for most engineers faced with design decisions quite often revolve around cost. A civil engineer who is tasked to design a bridge may wish to use the latest advances in materials and technology for the design but will have to make many tradeoffs to keep the design within a prescribed budget.

As appealing as this ethical framework is, especially in the budget driven world of engineering, there are many problems with this approach. First, who defines goodness, well-being, and utility? Should the majority decide and if so, would not this be a conflict of interest? As any engineer knows, measuring outcomes is a critical part of any analysis. Keeping that in mind, how do we measure what the greatest good is and just as important, when do we stop measuring?

For example, the use of placebo drugs in clinical trials is a topic of recent intense debate. Many human rights activists maintain that in clinical trials using placebos, it is not ethical or fair for some individuals to receive an experimental drug that could cure a disease, while others are unknowingly given a sugar pill that has no therapeutic benefit. Utilitarians maintain that this is an ethical approach because even though a few people might suffer or perhaps die due to a lack of treatment, exponentially more humans will benefit from the medical knowledge gained in such clinical trials. This argument over deception, and especially the use of placebo clinical trials in third world nations, is not easily settled.

2. Respect for Persons

If utilitarianism is thought of as promotion of the greatest good for many, then the respect-for-persons approach could be thought of as promotion of what is best for the individual. The respect-for-persons ethical viewpoint insists that each and every person should be accorded equal respect and rights, in effect treating every person as an autonomous moral agent. In this sense, an autonomous person is a self-governing individual who should be allowed basic inalienable rights: not to be killed, not to be deceived, and freedom to pursue their own happiness.

One manifestation of the respect-for-persons approach in our every day lives is a saying most people have heard in some fashion, despite their cultural background (see inset) [1]. This saying is known as the Golden Rule, most commonly expressed in this country as "do unto others as you would do unto yourself."

The Golden Rule: A Universal Edict?

Christian: "And as ye would that men should do to you, do ye also to them likewise" (*Holy Bible*, King James Version, St. Luke 6:31).

Jewish: "What is hateful to yourself do not do to your fellow man" (*Babylonian Talmud*, Shabbath 31a).

Muslim: "No man is a true believer unless he desires for his brother that which he desires for himself" (*Hadith, Muslim, imam* 71-72).

Buddhist: "Hurt not others with that which pains yourself." (*Udanavarga*, v. 18)

Hindu: "Let not any man do unto another any act that he wisheth not done to himself by others" (*Mahabharata*, Shanti Parva, cclx.21).

Confucian: "Do not do to others what you would not want them to do to you" (*Analects*, Book xii, #2).

Often the utilitarian and respect-for-persons viewpoints will agree on similar outcomes in ethical dilemmas, but the individual nature of the respect-for-persons framework will often bring the two theories into discord. One of the drawbacks to the respect-for-persons approach is defining whose individual rights should supersede in cases of conflict.

For example, suppose the construction of a new interstate will cause the forcible relocation of several hundred people. From one respect-for-persons viewpoint, it could be said that the relocation is infringing on the rights of the individuals forced to move. However, the hundreds of personnel employed for the project could also assert their individual rights for employment, which would lead to better economic conditions for their families. Thousands more could protest and say that their lives will be directly enhanced by the improved road conditions. Whose individual rights are more important and who is qualified to make that judgment?

From a professional standpoint, the question should be asked, "Is the best and most ethical course of action the one that puts individual rights above all else? Since by its very nature, engineering advancements typically benefit large groups of people in all aspects of life, engineers tend to gravitate to towards the utilitarian perspective. However, the respect-for-persons approach should not be discounted and there are many aspects which should be considered whenever faced with an ethical decision

3. Ethical Relativism

Bribery in many international cultures is an accepted part of everyday business, and in many situations is thought of as a local custom. An engineer assigned to an international contract might find it easy to justify bribing local officials for operating permits by rationalizing that since bribery is both an accepted custom and an expectation of the culture, then bribing local officials *in this case* is ethical. This is an example of ethical relativism, which is the viewpoint that all morals and ethics are relative to a particular religion or cultural custom, and thus there is no universal truth. In essence, if a culture feels that a particular course of action is correct, then society's acceptance of this tradition or custom makes that act morally justifiable.

On the surface, the general concept of relativism might seem to be an attitude that promotes tolerance and a greater understanding of foreign cultures. While embracing cultural differences and the promotion of diverse viewpoints are especially important for those engineers working in a global setting, it is equally important to understand the impact of ethical relativism, especially in an engineering context. Misguided applications of relativism can provide individuals, cultures, nations, and companies with avenues to promote unethical and abusive acts under the guise of the statement, "Well, it is all relative!"

It is very tempting to rationalize a particular engineering decision because of a cultural norm. Many other countries do not have the stringent environmental and occupational health and safety standards that are required of companies in the United States. As a result of the cost savings, many U.S. companies have built industrial complexes overseas. Is it ethical then to build a plant in a foreign nation with reduced safety measures because this is a cultural norm? This relativistic attitude caused the deaths of thousands in Bhopal, India when lacking safety measures in a Union Carbide plant caused the accidental release of cyanide gas into the surrounding community.

Cultural forces are extremely important considerations, especially in international engineering projects. It will be paramount to understand local customs, and the challenge engineers face is incorporating culturally different business practices without compromising their ethical principles. It will be very tempting to apply the "When in Rome, do as the Romans do" principle, but the ethical and professional engineer will recognize the impending relativist dilemma, and then be able to find options to circumnavigate this potential minefield. The responsibility of engineers to promote the health, safety, and welfare of the public *is not* relative, and applies in every situation, regardless of the culture or custom.

Case Study #1

To Dam Or Not To Dam? [2, 3]

In 1994, China undertook its biggest construction project since the Great Wall's completion during the Ming dynasty, the Three Gorges Dam. In an attempt to harness the tremendous power of the Yangtze River, China's leadership decided to build the 1.4 mile long dam across the river, which was one of China's most awe-inspiring landscapes. The dam, which will be the world's largest hydroelectric dam, is scheduled to be partially online in 2003, with the intent to be fully operational by 2009. Once completed, the dam will provide China with almost 10% of its total energy requirements and will provide for flood control along the Yangtze River, which will possibly save thousands of lives. The new waterway will allow cargo ships and cruise liners the ability to navigate 1500 miles inland, deep into the heart of China, having a tremendous impact on the economies along the river.

Once the Yangtze is dammed, the river will create a reservoir that is 390 miles long and 575 feet deep. When the reservoir is full, it will submerge 395 square miles, including 30 million acres of cultivated land. Approximately 30 cities, 140 towns and 1352 villages will be lost, and innumerable archeological & cultural sites will be destroyed. An estimated 1.2 million people, mostly from rural areas, will be resettled.

Environmentalists have strongly criticized the Three Gorges Dam project for not only the tremendous impact it will have on animal and marine life, but human life as well. The Sierra Club maintains that the water supply of the city of Shanghai, with a population of eight million, will be seriously affected by salt-water intrusion. The river already used as a dumping site for raw sewage and toxic waste, once dammed will not be able to maintain an effective ability to self-cleanse. With plans for major urban areas to be built along the river, the level of pollution will only increase, creating not only a human health hazard but animal and marine life dangers as well. Many endangered animal species will be threatened by the construction, specifically the Siberian White Crane and Yangtze River dolphin.

The environmental impact of the Three Gorges Dam extends far beyond the immediate riverbanks of the Yangtze, and even beyond Chinese borders. There is scientific concern that once the Yangtze River flow is substantially reduced, it will not dump adequate levels of fresh water into the East China Sea, causing the temperature of the Sea of Japan to rise. During peak times at the mouth near Shanghai, the Yangtze delivers 30,000 cubic meters per second of fresh water to the East China Sea. The Yangtze fresh water, which currently dilutes the salinity of the Sea of Japan, would be drastically reduced once the dam is in operation. The surface layer salt content of the Sea of Japan would increase due to the reduction in fresh water, causing sea surface layer temperatures to rise, which would heat the atmosphere over Japan.

On the other hand, China accounts for one fourth of the world's population, and is constantly in need of more energy sources, which will lead to improved food production. The Three Gorges Dam will generate as much electricity as 18 nuclear power plants. However, the construction of the dam will negatively impact over a million people, and potentially severely impact not only the immediate Chinese environment, but that of neighboring countries as well.

1. Do you think the dam should have been constructed? Why?
2. How would the utilitarian viewpoint differ from the respect-for-persons approach?
3. What if the Sea of Japan is affected in the future? If the goal is to achieve the greatest good for the greatest number, then how do we measure this in complicated cases such as this?

ETHICAL DECISION MAKING AND THE DESIGN PROCESS

In Chapter One, you were introduced to the basic steps of the engineering design process: "Among the fundamental elements of the design process are the establishment of objectives and criteria, synthesis, analysis, construction, testing and evaluation" and "It is a decision-making process which is often iterative." These design elements apply not only to decisions concerning the physical design process, but they also apply to the ethical decisions that often occur throughout the design process as well. Let's look at the design steps and see how they might apply to a materials engineer.

Suppose you are a joint replacement designer for a major biomedical devices corporation. Through your job, you are solely responsible for the development of a new ceramic/plastic joint design that has revolutionized the hip joint replacement industry. Typically hip replacement joints have to be replaced every 7-10 years, but your new design has at least doubled the replacement interval and could eliminate it all together.

At a professional conference, a rival biomedical firm approaches you. They offer to double your salary if you will work for them, plus they offer a generous stock option package. You are tempted but you suspect the offer is based primarily on the rival company's desire to hire you because of your invention of the new replacement joint. You explain to them that your current employer owns the patent on the new replacement joint design, and you would not be able to divulge your current employer's proprietary information if you took their offer.

The rival company assures you they have no intention of violating U.S. patent laws but they would like you to work in their international division. They tell you it is not illegal to develop a slightly modified version of your design for markets outside of the United States. In addition, they point out that they will make the replacement joints available to third world nations at significantly reduced prices (compared to your current employer's prices.) If you work for this new company, your new design has the potential to help millions of people worldwide, especially those in underdeveloped and poverty-stricken nations. What should you do?

This is a complicated ethical decision and one that cannot be made without an in-depth analysis of all the information available. So where to begin? We can use the engineering design elements to guide us to a decision.

1) The establishment of objectives and criteria:
    - This step is often the hardest to both recognize and verbalize. In ethically murky waters, it is sometimes difficult to distinguish what the real debate concerns. Should the utilitarian view of the greater good be paramount or do individual rights take precedence? Should you remain loyal to your company? Just what are your professional responsibilities?
    - The main objective for this decision is whether or not it is ethical for you to take company proprietary information (that you were personally responsible for developing), and using it for another potential employer. In this particular example, you would also want to check on the legal consequences as well.

2) Synthesis:
    - This step is essentially the information gathering process. To make any decision whether technical or ethical, you want to make sure you have all the relevant information. Knowing where to look is often a major stumbling block, and experience is a valuable commodity in this area.
    - For this example, you would want to review the legal terms of your company's patent, any other company intellectual property policies, any non-compete contract clauses that may apply, and U.S. versus international patent law.
    - It is important when researching your predicament to take a close look at company policies regarding ethical and/or potentially unsafe actions by company personnel. Many companies such as Lockheed Martin have very proactive policies regarding potential ethical dilemmas, and many companies encourage open communication through such methods as hot-line phones or ombudsmen to help employees "do the right thing" without fear of reprisal.

3) Analysis:
    - Once you have gathered all of the pertinent information, you would want to analyze it to the greatest extent possible. Just as in typical engineering design, you need to formulate all of your available options. Then you want to look at the benefits and drawbacks to each, keeping in mind the basic ethical frameworks that were previously discussed.
    - The analysis of our example is complicated. The utilitarian view points to the greater good, i.e. if you work for the rival company, more people are likely to benefit from worldwide distribution of the hip replacement joint. The respect-for-persons viewpoint is not so clear. You could say that the individual rights of third world persons are violated who are not given equal access to the new replacement joints. However, the rights of the company are violated if you give their trade secret to another company. Just some of the decision options for our example are:
        - Do nothing and continue in your current job.
        - Quit and work for the rival company exactly as they asked.
        - Negotiate with your current company for a higher salary and better benefits.
        - Negotiate with the rival company for a more clear job description that would not require you to divulge any proprietary information.

4) Construction:
   - In this step, you construct a detailed solution to your problem. Previous analysis will hopefully show you that there are many avenues, some which will be more ethical than others. However, for complicated ethical decisions, the answer is not always easy to see, even after all the data has been analyzed.
   - In this particular example, there is no clear decision, which is often the case. The decision that you make will be dependent on your morals and your particular life situation at the time.

5) Testing and evaluation:
   - Just as this step is critical in the design of any new material, testing and evaluation is just as important in deciding a particular ethical course of action. For ethical decisions, the most useful form of testing and evaluation is to talk to other people and get their perspectives concerning not only the situation in general, but especially the chosen solution. While this may seem like a simple step, for many, it can be the most difficult. Often many people will make decisions in isolation for fear of appearing weak or not in control of a particular situation. Though sometimes difficult, it is crucial to discuss an ethical dilemma with someone who is not close to a situation to gain their insights, and possibly show you some options that you previously did not consider.
   - In our example, there are many people who could help in evaluating the best course of action. A patent or intellectual property lawyer would be an excellent source of information, especially for any potential legal issues. Your professional society is another resource for information (see the next section for a more in-depth discussion.) Objective peers not related to either company would be helpful, or you can always contact a former college professor or some other industry expert for additional guidance. Important professional decisions such as detailed in our example should never be made without input from qualified, objective sources.

PROFESSIONAL SOCIETIES AND CODES OF ETHICS

Almost all fields of engineering have a professional society that they can look to for guidance in a number of areas: technological advancements, career advice, employment trends, current issues in design, etc. The major professional engineering societies go a step further in the field of ethics and provide their members with ethical guidance in the form of formal codes, discussion forums, case studies (including real world examples), and general guidance for possible ethical dilemmas that an engineer may encounter.

For every engineer who graduates from an accredited engineering institution, the Accreditation Board for Engineering and Technology (ABET) provides the initial Codes of Ethics of Engineers based on the following principles:

Engineers uphold and advance the integrity, honor and dignity of the engineering profession by:

I. using their knowledge and skill for the enhancement of human welfare;
II. being honest and impartial, and servicing with fidelity the public, their employers and clients;
III. striving to increase the competence and prestige of the engineering profession; and
IV. supporting the professional and technical societies of their disciplines.

From these principles, engineering ethical fundamental canons were developed that should apply to all practicing engineers, regardless of field or licensing (see inset.) It is interesting to note that first fundamental canon, to hold

ABET Code of Ethics for Engineers

Fundamental Canons

Engineers, in the fulfillment of their professional duties, shall:

1. Hold paramount the safety, health and welfare of the public in the performance of their professional duties.
2. Perform services only in areas of their competence.
3. Issue public statements only in an objective and truthful manner.
4. Act for each employer or client as faithful agents or trustees, and shall avoid conflicts of interest.
5. Build their professional reputation on the merit of their services and shall not compete unfairly with others.
6. Act in such a manner as to uphold and enhance the honor, integrity and dignity of the profession.
7. Continue their professional development throughout their careers and shall provide opportunities for the professional development of those engineers under their supervision.

paramount the safety, health and welfare of the public sounds very much like the utilitarian ethical framework. Because of the very nature of engineering as a cost-benefit-analysis profession, the greatest good for the greatest number theory often is a guiding principle for engineers. However, after reading through all the canons, it should be clear that the ends do not always justify the means. Engineers are expected to uphold the highest standards of both personal and professional integrity. The Code of Ethics for the National Society for Professional Engineers (NSPE), an umbrella professional organization for licensed professional engineers, is almost identical to the ABET code, as are the specific ethical codes for all other major engineering societies. There are some slight differences however, which reflect each society's particular concerns and areas of specialized interest.

For instance, the Code of Ethics Fundamental Canons for the American Society of Civil Engineers (ASCE) is almost identical to the ABET code. The most noticeable difference is the first canon, which for ASCE reads "Engineers shall hold paramount the safety, health and welfare of the public and shall strive to comply with the principles of sustainable development in the performance of their professional duties." Because of the close link between civil engineering and the environment, which is not critical in some fields of engineering (like software), the addition of the "sustainable development" element to this canon demonstrates ASCE's commitment to environmentally ethical engineering problem solving.

All professional societies' codes of ethics are important in the sense that they provide an overall mission statement concerning the incorporation of ethics into professional engineering. However, it is important to remember that a code of ethics is simply a framework for ethical decision-making. The codes are not comprehensive and are not the final moral authorization for a particular act. Codes are a set of basic guidelines developed by professionals with years of experience, and as such should be reflected on and taken very seriously as an engineer's primary responsibility.

As previously discussed, the role of the professional society cannot be over-emphasized for the engineer who faces a difficult ethical decision. Codes of ethics provide basic guidelines but because of their limited nature, they cannot be expected to cover every possibility. This is why it is so important to remember that engineering professional societies have a large number of objective experienced engineers on hand who have either faced similar situations or know someone who has. Before making an ethical decision that could result in either professional censure or a lawsuit, it is critical to remember the testing and evaluation stage of an ethical decision and talk to someone.

SAFETY AND RISK

Four thousand years ago in Babylon, in some of the earliest known written professional codes, engineers were given very clear instructions on their ethical and professional responsibilities in the Code of Hammurabi,

"If a builder build a house for a man and do not make its construction firm, and the house which he has built collapse and cause death of the owner of the house, that builder shall be put to death.

If it cause the death of the son of the owner of the house, they shall put to death a son of that builder.

If it cause the death of a slave of the owner of the house, he shall give to the owner of the house a slave of equal value.

If it destroy property, he shall restore whatever it destroyed, and because he did not make the house which he built firm and it collapsed, he shall rebuild the house which collapsed from his own property."

Fortunately for engineers today the penalties for accidents and failures are not as deadly as in ancient times, but the message is the same: Engineers must make safety paramount in their designs. As you will recall, this is the first fundamental canon of engineering ethics. However, few designs are ever 100% safe and consequently, risk in engineering is a fact of life. As technology and the resulting applications grow more complex, the more risk and safety become crucial issues in engineering.

As previously discussed, the engineering design process typically follows a basic utilitarian cost benefit approach, and this is especially true when considering acceptable levels of risk in a design. When considering risk, safety, and design, an engineer may look to prescribed government codes concerned with safety or industry standards for acceptable levels of risk. Unfortunately though, when an engineer or a group of engineers embarks on previously untried technology, there are no accepted practices or even previous experience from which to draw. Sometimes with cutting edge technologies and designs, which are usually high-risk endeavors to begin with, mistakes are made and failure occurs.

When an engineering failure occurs, whether it is a bridge that collapses or a power plant that shuts down due to a software glitch, some kind of human error occurred in the design or manufacturing engineering processes that allowed the accident to happen. What is critical to remember in such cases is that humans will err and it is unrealistic to think that in any human endeavor, especially one as complex as engineering, that no mistakes will be made. Henry Petroski, an engineer who has written extensively on the culture of engineering, states, *"Because man is fallible, so are his constructions, however. The history of engineering in general, may be told in its failures as well as in its triumphs. Success may be grand, but disappointment can often teach us more.*[4] "

While no engineer strives to build something that will catastrophically fail, it is just such failures that elevate an entire field of engineering to a higher level of competence. The Tacoma Narrows Bridge example is just such a case. Often used as an engineering ethics case study, the Tacoma Narrows Bridge is actually a much better example of engineering culture and failure. In 1940, a narrow two-lane suspension bridge, the first of its kind, was built between Washington State and the Olympic Peninsula. Instead of the typical deep stiffening truss design used up to that point in history, shallow plate girders were used instead.

Just a few months after its opening, the bridge catastrophically failed in high wind conditions, twisting and undulating with such force that the bridge seemed to be made of rubber instead of concrete and steel. While the final failure seemed quite extreme, the bridge had communicated its instability long before its collapse, and even did so during the construction phase.

Even up to the moment of final collapse, the Tacoma Narrows Bridge engineers were hard at work trying to solve the motion problems. After the failure, they were able to determine that bridge essentially was subjected to tremendous aerodynamic forces that were previously not accounted for in earlier designs. Did the engineers make mistakes? Yes, but they did not act in an unethical manner. In 1940, suspension bridge design was still in its infancy, and the lessons learned from the Tacoma Narrows Bridge collapse will not be repeated in the future. Bridges today are now tested in wind tunnels to prevent such mistakes from reoccurring.

The collapse of the Tacoma Narrows Bridge on November 7th, 1940. Gig Harbor Peninsula Historical Society, Photo by James Bashford.

So if engineering design and failure are not necessarily unethical and human error is to some degree unavoidable, then how do questionable ethical situations arise in engineering?

Unethical decisions in design engineering often result when an engineer or manager is faced with a decision concerning a bad design.

A Questionable Design + A Human Decision = Potential Ethical Dilemma

In many cases, the decision will involve money, and the unavoidable cost-benefit aspect of engineering only complicates the already complex process in bringing a design from a concept into production. While many engineering decisions concerning risk and safety certainly straddle ethical gray areas, history shows us that some were made without appropriate ethical considerations.

The infamous Ford Pinto case of the 1970's is one such example. During production, engineers discovered that the rear-mounted gas tank of the Pinto, which was built to industry standards, could possibly explode in a rear-end collision. When management of Ford was notified of the gas tank problem, they applied a cost-benefit approach and determined the cost to fix the problem as opposed to the cost that could be paid out in lawsuits. The repair would have only cost $11 per Pinto, but with over 12.5 million Pintos sold worldwide, the recall costs would have been $137 million. For the expected 180 deaths and 180 serious injuries that would result, the cost for potential lawsuits was estimated to be only $49 million.

When faced with a questionable design and a decision concerning money, Ford chose to save money and not recall the Pintos. Consumer action groups maintain that almost 500 people died in Pinto crashes, and Ford endured more than 50 subsequent lawsuits. Ford was eventually forced to recall and fix the Pintos. Did Ford make an unethical decision by making a choice based upon a cost-benefit analysis? Was this decision process utilitarian? Can you put a price on a human life, and do you think this situation could happen today?

Case Study #2

A Deadly Tire Design [5-7]

On August 9th, 2000 the tire manufacturer Firestone recalled 6.5 million AT, ATX, and Wilderness AT tires which were primarily used on light trucks and sport utility vehicles. The tread belts of these tires had a tendency to separated from the walls of the tires at high speeds, often causing the vehicle to roll over as a result. The vehicle affected most by this tire problem was the Ford Explorer. While the Firestone recall was voluntary, news stories reported six months earlier prompted the recall when it was discovered that approximately 50 people had died in rollover accidents caused by tread separation. The number of deaths in the United States associated with this catastrophic tire failure would eventually rise to 160, with over 600 injuries.

In the resulting legal furor over responsibility for the tires, Firestone charged that Ford shouldered a large part of the responsibility as well because the design of the Ford Explorer was flawed. While this issue will undoubtedly be debated in the court for years to come, one critical issue was raised. When did Firestone know about the problem? In legal documents, Firestone was forced to admit that they were first made aware of the tread separation problem in 1997 from a dealership in Saudi Arabia who noticed an unusual number of these failures. One year later, a technical branch manager from this Middle East dealership wrote a memo stating, "I have to state that I believe this situation to be of a safety concern, which could endanger both the vehicle and more importantly the user of the vehicle...Do we have to have a fatality before any action is taken on this subject?"

This tread separation problem was not just limited to Middle Eastern markets. Similar problems surfaced in Thailand, Malaysia, and Venezuela, and the common denominator for these regions was the hot climate. A 1999 internal Firestone memo showed that management was aware of the overseas tread separation problem but resisted consumer notification of overseas customers for fear that North American markets would become aware of the problem and demand action.

Firestone hired an independent expert to determine the cause of the tread separation, Dr. Sanjay Govindjee, who stated in his report, "From a mechanical and materials engineering perspective, the phenomena of belt separation in these tires resulted from a crack that grew in the rubber between the two belts. This cracking is influenced by a number of factors, including climate, design of the tire, manufacturing differences at Firestone's Decatur plant and usage factors."

The revelation that the majority of the problem tires were manufactured at the Firestone Decatur, Illinois plant was not a new one. A Firestone October 1999 report proved that they knew the tires from Decatur plant experienced significantly higher failures than at other plants. In addition, a subsequent Ford investigation revealed that failure rates for the AT and ATX tires manufactured at the Decatur, Illinois plant exceeded 600 tires per million in 1994, 500 per million in 1995, and 300 per million in 1996. These same tires manufactured at other plants failed at the rate of 100 tires per million. When questioned, Tom Baughman, engineering director at Ford North America Truck said the industry standard for tire failure is 5 per million.

1. Who should bear the ultimate responsibility for the deaths resulting from the tread separation: Firestone management, the engineer who designed the tire, or the Decatur plant engineer responsible for the production process?
2. Are design engineers absolved of responsibility when their product enters the manufacturing stage? Are they responsible for what happens in the management and marketing sequence?
3. When do you think would have been an appropriate time to institute a recall? Why?
4. If the industry standard for tire failure is 5 per million, why do you think such gross failure rates were accepted and who was responsible?
5. Is a tire failure rate of 5 per million ethically acceptable and how do you think this number is determined?

ETHICAL ISSUES IN OTHER AREAS OF ENGINEERING DESIGN

While risk and safety considerations in the design process seem to be the focal point of ethical dilemmas, ethical issues also arise in many other areas of engineering and design. Environmental issues, intellectual property concerns, conflicts of interest are just some of the areas engineers may encounter ethical questions. This section will provide a brief overview of these issues, and if you have further questions, consult the recommended reading list at the end of the chapter for additional sources of information.

Engineers And The Environment

The phrase "Green Engineering" has recently become popular in engineering circles, and is a rapidly growing field in engineering industry. The word *green* implies environmentally conscious, and unfortunately in the past, engineering in general has not been seen as a profession that promotes the environment. Chemical plants, dams that flood vast farmlands, and noise and air pollution from automotive engines are just some of the engineering projects that often receive negative publicity. However, even though many engineering designs promote human progress over nature, there has been increasing emphasis on engineering design that minimizes the impact on the environment.

To this end, many branches of engineering have taken an active role in promoting the role of the environment in the life cycle of an engineering process or product. As noted in the earlier professional societies section, the American Society for Civil Engineers is committed to sustainable development, which means that engineering projects should not negatively impact the environment to the extent that it is permanently altered for future use. The ASCE's commitment to green engineering directly influences future opportunities for materials engineers, and provides opportunities for innovative research in materials design. In addition, many universities offer undergraduate green engineering programs that incorporate sustainable development practices in engineering design.

Intellectual Property

Chapter Seventeen focused on definitions and discussions of intellectual property issues, and this is one area that most material engineers will find themselves faced with in their futures. It is critical that all engineers gain a basic understanding of intellectual property issues, especially in regards to employment. Many engineering graduates are surprised and unprepared for the lengthy contracts they are presented in their first jobs. Of particular concern, the contracts will state that the company owns any design created by an employee, and should a new engineer wish to quit, the company can stipulate that the employee cannot work for any competitor for a stated period of time. One intellectual property area that cannot be clearly defined either legally or ethically is that of *tacit knowledge*, or that know-how and experience that a person gains simply by working in a particular industry or culture. Engineers cannot leave behind this tacit knowledge once they decide to change jobs, and it is often difficult to clearly define what knowledge is proprietary and what is experience, and what if any can be translated to a position with another company.

Legally, employers are within their rights to insist on contractual and legal safeguards of both their investments and proprietary information. Ethically, potential employees may disagree and feel that their individual rights are violated by such contractual demands. If an engineer disagrees with such employment practices, he can simply find another employer. However, should a dispute arise concerning copyright, patent, or trade secret issues while employed for a company, the engineer should seek advice from an intellectual property attorney, a professional society, or both.

Loyalty, Conflicts of Interest, and Whistleblowing

The fourth ABET Code of Ethics canon states that an engineer shall "act for each employer or client as faithful agents or trustees, and shall avoid conflicts of interest." This canon essentially states that engineers shall be loyal to their employers. As seen in the intellectual property section, employers can legally demand a certain level of loyalty from their employees through intellectual

property laws, but ethically engineers should be committed to avoiding situations that might result in a conflict of interest.

A professional conflict of interest occurs when an engineer is faced with a situation in which she may have an interest, which may compromise their ability to act as a faithful agent for either a client or employer. For example, suppose Contractor A offers to take an engineer responsible for hiring contractors to dinner? Does a conflict of interest exist? Possibly. If Contractor A is attempting to circumvent proper bidding channels, then a conflict of interest may exist. What if Contractor A offered to give the engineer a week's vacation in Hawaii? Certainly that could be seen as an outright bribe. However, what if Contractor A is a relative and they often have dinner? What if during this dinner, they are seen by Contractor B, who does not know of the familial relationship. Then an apparent conflict of interest exists, and while no improper behavior took place, the perception exists that Contractor A is given preferential treatment.

Conflicts of interest can be very complicated and often require delicate handling. They can occur in every aspect of engineering from employee relations to intellectual property issues. Unfortunately, even the most well-intentioned ethical engineers can find themselves in either a potential or apparent conflict of interest through no overt fault of their own. To avoid finding yourself in this position, it is always best to notify a superior of any personal relationship that could be seen as a professional conflict of interest. In addition, seek advice in situations where you are unsure and remember that your professional society can offer guidance in such confusing situations.

One other area concerning conflicts of interest and company loyalty is whistleblowing. What should engineers do if they find themselves in situations where loyalty to their companies might mean taking unethical actions? This situation is where the concept of whistleblowing comes into play. There is some confusion as to what exactly whistleblowing is. Many people think whistleblowing takes place when an employee goes to the media to essentially tattle tale on unsavory and/or unethical business practices of her employer. While this is one form of whistleblowing, it is the most extreme. Whistleblowing occurs when any employee goes outside normal company channels to air a grievance or dispute. Whistleblowing can essentially take two forms, internal and external.

If you feel your superior is asking you to take an action that could result in harm to the public, you should always attempt to resolve this conflict at the lowest level possible before going outside the company. You can always approach another manager within your area of the company or outside your area but still within the company. If that doesn't solve the problem, you can notify the head of the company, or even the Board of Directors or equivalent governing body. All of the aforementioned solutions are considered internal forms of whistleblowing. When you have exhausted all of these avenues, then it is ethical to violate company loyalty and use an external form of whistleblowing. One option would be to notify a regulatory or governmental body, but the media should be your last resort and this should always be qualified with immediate safety concerns. Your first ethical responsibility is to the public, but loyalty to your company is an important consideration. Any decision to blow the whistle should be weighed carefully, and following the previously mentioned design steps in making this decision will be critical for resolving a potential conflict and protecting both your career and your company.

Case Study #3

The Bjork-Shiley Heart Valve [8, 9]

In 1979, The Federal Drug Administration (FDA) approved the use of an artificial heart valve that was designed to improve blood flow and reduce the risk of blood clots in the heart. Shiley, Inc., a Pfizer subsidiary, manufactured this new artificial heart valve, called the Bjork-Shiley Heart Valve, model C-C. The valve consisted of a convexo-concave disc inside a Teflon coated metal ring with two wire holders mounted on each side of the disk. One of the wires, the inflow strut, was an integral part of the ring while the second wire, the outflow strut, was welded on after the manufacture of the ring.

Failures of the valves were first noted during clinical trials, however, despite these problems, the FDA approved the C-C valves. Soon after the valves were released on the open market, they began to experience more failures, which were sometimes fatal. In subsequent investigations, it was discovered that all of the C-C valve failures were fractures of the outflow strut. Between 1980-1983, Shiley recalled the valves three times and in each instance, replaced the defective valves with purportedly improved versions. The FDA asked Shiley twice between 1982-1984 to remove the valves from the market on a voluntary basis, but Shiley refused. A total of 86,000 valves were implanted worldwide. Of these, 450 fractured resulting in approximately 300 deaths.

Independent investigations revealed discrepancies in both the manufacture of the parts and the quality control process. The Shiley supervisor for the drilling and welding procedures on the C-C valves from January 1982 – September 1983, George Sherry, came forward with disturbing information concerning the engineering and the manufacturing of the valves. Sherry, who left Shiley in protest over the poor manufacturing procedures, stated that it was impossible to

manufacture the valves according to the drawings. The welders, who were not certified he said, used amperage levels that were 3-4 times greater than recommended, causing welded joint temperatures to rise too quickly, which would weaken the joint.

In addition, if cracks were discovered in the valves, which sometimes occurred in the manufacturing process, the valve was to be either rewelded or discarded. Unfortunately, many valve records were falsified, indicating that some of the valve cracks had not been rewelded, and instead merely polished over. Sherry testified that management knew of the problems, and in one internal Shiley memo, management indicated concern over the rapid rate of valve inspections and the fatigue of the quality control inspectors.

Due to mounting lawsuits, Pfizer and Shiley removed the Bjork-Shiley C-C artificial heart valves from the market in 1986. In a class action lawsuit, Pfizer paid $75 million dollars for research on the valve fractures and $300 million dollars for patient and family compensation.

1. How would a utilitarian view this case? How would this position compare to the Respect-for-Persons viewpoint?
2. What were the ethical and professional responsibilities of the engineers who designed the valves? Were they absolved of any responsibilities because the FDA approved the devices?

3. Do you think the engineers adequately tested their design of the artificial heart valve? Should materials and designs of biomedical devices be held to a more rigorous testing protocol that those used in other engineering applications?
4. Of the 86,000 valves implanted, 0.5% failed. Is this an acceptable failure rate and if not, what would be? How do you determine an acceptable failure rate, especially in biomedical engineering?
5. Should engineers anticipate the use of incorrect manufacturing techniques and what can be done to prevent this in the design process?
6. How much responsibility for the valve failures and resulting deaths should the management of Shiley bear?
7. Should George Sherry have come forward with his information sooner? What other options did he have?

CONCLUSION

This chapter merely serves as an introduction into the complex and extensive field of engineering ethics. Ethics in general has been an entire field of study since Socrates in the 5th century B.C., so one chapter in a design textbook can barely even scratch the surface of such a complex topic. There are many more ethical topics that every engineer should consider; employment concerns, global issues, weapons design, just to name a few.

For the future engineer, probably the first source for information relating to professional ethics should be your professional society. If your society does not provide detailed ethical guidance, the National Society of Professional Engineers or the National Institute for Engineering Ethics are also excellent starting points (see website list below.) Just as in technical aspects of engineering, ethical issues are constantly emerging and changing, especially in areas such as intellectual property. Professional societies can help you stay on top of these critical ethical and legal issues. The Recommended Reading section below also contains several engineering ethics books that could prove useful.

Just as preventive maintenance is crucial for efficient and safe operation of machinery, preventative ethics is critical for recognizing and preventing future ethical conundrums. The ability to recognize an existing or potential ethical dilemma is not always straightforward and often requires analytic skills that engineers do not typically hone. The study of ethics in engineering, while relatively new in some universities is only going to become more emphasized as technology advances and profoundly impacts the world in which we live. The professional engineer should consider the ethical issues of his or her design to be just as important as the technical aspects because their impact could conceivably be much more far-reaching.

RECOMMENDED READING AND BIBLIOGRAPHY

1. Harris, C.E., M.S. Pritchard, and M.J. Rabins, *Engineering Ethics: Concepts and Cases*. Second ed. 2000, Stamford, CT: Wadsworth.
2. *China's Three Gorges Dam: Is the "Progress" worth the ecological risk?*, in *Popular Science*. 1996.
3. Nof, D., *China's Development Could Lead to Bottom Water Formation in the Japan/East Sea.* Bulletin of the American Meteorological Society, April 2001. 82(4).
4. Petroski, H., *To Engineer is Human*. 1992, New York: Vintage Books. 251.
5. *Public Citizen Firestone Tire Resource Center, http://www.citizen.org/fireweb/*. 2001.
6. *Firestone Tires. The firestorm continues*, in *Consumer Reports*. 2000. p. 9.
7. Govindjee, D.S., *Firestone Tire Failure Analysis*. January 30, 2001.

8. *Ethical Issues in Biomedical Engineering: The Bjork-Shiley Heart Valve.* IEEE Engineering in Medicine and Biology, (March 1991): p. 76-78.
9. Burkholz, H., *The FDA Follies.* 1994, New York: Basic Books, HarperCollins Publishers, Inc. 228.
10. Schinzinger, R., and Martin, M.W., *Introduction to Engineering Ethics.* 2000, Boston: McGraw-Hill.

USEFUL WEBSITES

1. The American Society for Civil Engineers Code of Ethics: http://www.asce.org/ethics/
2. The National Society for Professional Engineers Code of Ethics: http://www.nspe.org/ethics/
3. The National Institute for Engineering Ethics: http://www.niee.org/
4. The Online Ethics Center for Engineering and Science: http://onlineethics.org/
5. The Accreditation Board of Engineering and Technology Code of Ethics: http://www.abet.org/Code_of_Ethics_of_Engineers.htm
6. Virginia Tech Green Engineering: http://www.eng.vt.edu/eng/green/green.html

PROBLEMS

1) Answer the questions for Case Study #1.
2) Answer the questions for Case Study #2. In addition, safety standards are generally not as stringent in Middle Eastern and Asian markets. Is it ethical for a U.S. company to design and sell a product overseas that would not pass safety standards in the United States? Why or why not?
3) Answer the questions for Case Study #3. How far into the future should engineers track the successes and failures of their designs? Explain your answer.
4) Go to the ethics Internet page on the website for the National Society of Professional Engineers (http://www.nspe.org/ethics/), and take the ethics quiz. Discuss your missed questions.
5) If you work for a ceramic design company and you invent a ceramic device completely on your own time at home with no company materials, do you think you should own the patent? Does the law support you? Research and discuss your answer.
6) Apply the five steps of ethical decision making to Case Study #1 and formulate a decision as to whether or not the dam should be built.

3

Materials Selection Methedology

W.J. Lackey

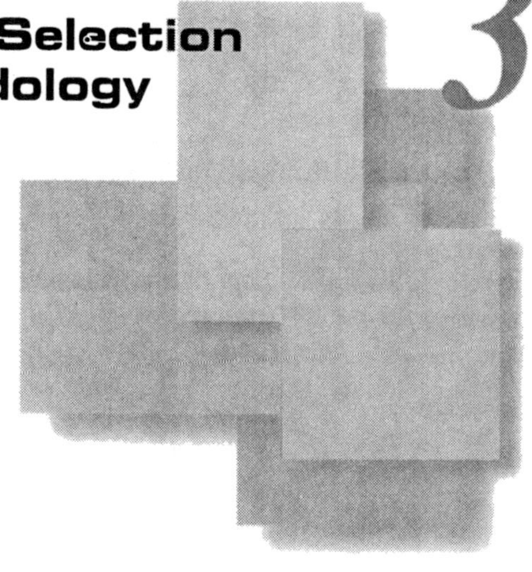

MATERIALS SELECTION METHODOLOGY

W. Jack Lackey
George W. Woodruff School of Mechanical Engineering
Georgia Institute of Technology
Atlanta, Georgia 30332-0405

ABSTRACT

A method is described that permits selection of the best material for a specific structural or thermal component. The initial step typically utilizes knowledge of mechanics or heat transport to permit derivation of a Material Index, i.e. a figure of merit consisting of material properties. As an example, the best material for making a light, strong tie has the largest value for the ratio of strength divided by density. Existing plots of material property data or software are used to identify materials having large values for the Material Index. Methods for considering cost, cross sectional shape, and process are also described.

INTRODUCTION

Selection of the best material for a particular component is best done by an expert who has extensive knowledge and experience. The person needs to be intimately familiar with the function and requirements for the component as well as being equally familiar with the properties and performance of numerous materials. Unfortunately, not many such people exist. Another approach is necessary, particularly in the case where a new product is being designed and thus the experience base is limited. One alternate approach would be an "expert system," that is, a computer program that combines the knowledge of several material selection experts with a vast database and provides an answer after comparing the input requirements with the properties, performance, availability, fabricability, cost, etc. of all ceramics, metals, polymers, and composites. Such computer programs do not exist today. Fortunately, there are other alternatives. The one that the author likes best is a graphical approach developed by Professor Michael F. Ashby. Much of this chapter is based on his book.[1] Other useful books on materials selection are available.[2-4]

Professor Ashby's approach is primarily for selection of materials during the initial stage of mechanical design. It is here, during conceptual design, that one must consider essentially all materials in order to minimize the chance that promising materials are not overlooked. To facilitate this difficult task, graphs

that display material properties are available.[1] The graphs are plots of one property versus another, such as the examples shown in Figures 1 and 2.

Figure 1 plots Young's modulus versus density for many types of materials. Note that the units on density are millions of grams per cubic meter. This gives the same numerical value as the more customary units of grams per cubic centimeter. Each material considered is represented by a "bubble" on the graph. For example, the bubble for alumina shows that the elastic modulus and density vary over a range, accounting for different grades (sources) of alumina. A quick scan of the graph shows that most technical ceramics have higher moduli than most metals. Porous ceramics, such as structural clay products (bricks) and pottery have lower moduli than the technical ceramics. The graph also confirms what the reader likely already knows about polymers. They have low values for

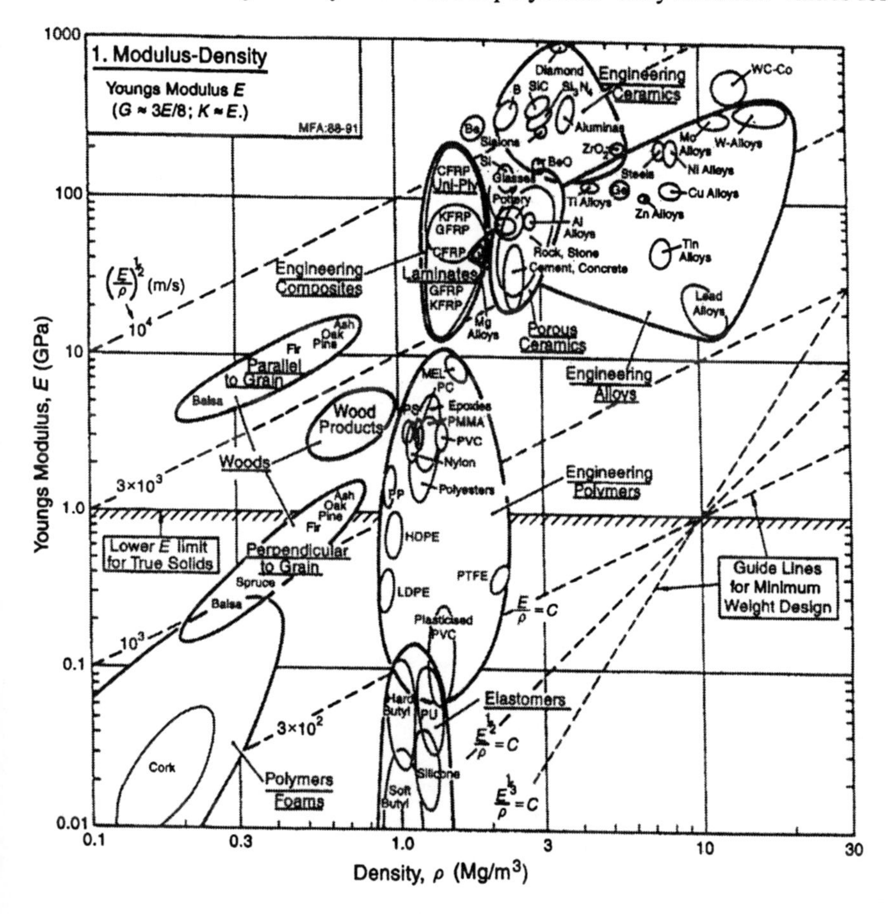

Figure 1. Graph showing Young's modulus and density for many materials.[1]

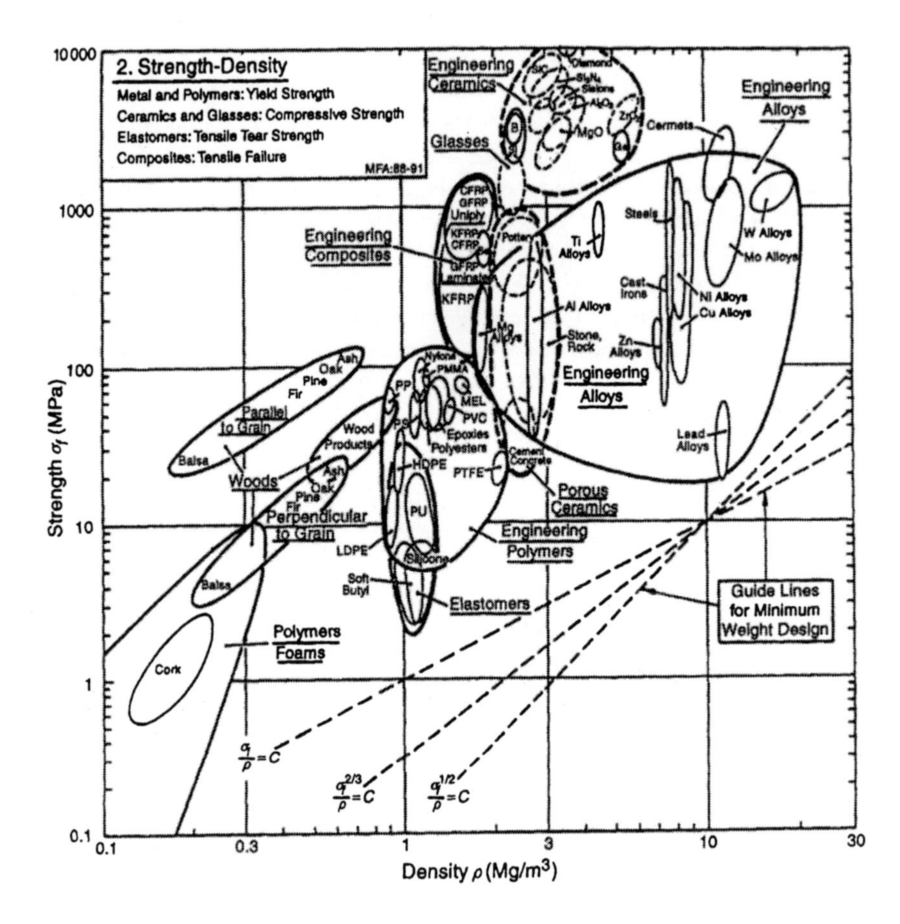

Figure 2. Strength versus density.[1]

Young's modulus, and polymer foams have even lower values. Figure 2 is a graph of strength versus density and is described later.

Briefly, the selection procedure consists of five steps which are presented in detail through case studies described later.

- Step 1: Identify the design goals and constraints.
- Step 2: Derive an expression, depending on the specific design, that permits identification of the property(s) to be optimized.
- Step 3: Use existing property graphs to select a short list of candidate materials.
- Step 4: Compare materials, considering available/practical shapes.
- Step 5: Select preferred materials(s) based on process and cost analyses.

As implied above, the methodology to be described permits quantitative analysis of the influence of component shape and bulk material cost. A graphical approach for selecting the best fabrication process will also be explained. Process selection is based on consideration of component attributes such as size, aspect ratio, minimum section thickness, desired tolerance and surface finish, maximum use temperature, etc.

CASE STUDY: LIGHT, STRONG TIE

A simple example will serve to illustrate the basics of the material selection methodology. Suppose we wish to select a material for a light, strong tie. As shown in Figure 3, the tie of known length, L, must hold a heavy object of known weight without failing. The weight exerts a given force, F, on the tie and places it in tension. Further, we wish to select a material for making the tie so that the tie will be light. We would wish the tie to be light if it were to be a component in an aerospace application.

This is a simple problem and the reader may guess the correct answer. It is perhaps obvious that the best mateiral for a light, strong tie is one that has a low density and large tensile strength. That is, the best material for the tie has the maximum value for the ratio of tensile strength, σ_f, to density, ρ, i.e. maximum value for σ_f / ρ. We call this ratio a Material Index, M. It consists of material properties. Even though we have obtained the Material Index for this problem by an educated guess, let us derive it. The general method we use here will work for more complex problems where the Material Index cannot be obtained by guessing.

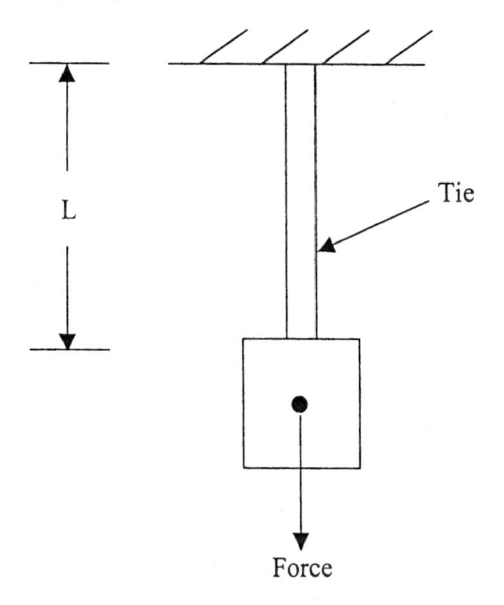

Figure 3. Schematic of tie.

The method for deriving the Material Index consists of first identifying the goal. For a light, strong tie, the goal is to select a material that minimizes the mass of the tie. However, we have a constraint that the tie must not fail. Many material selection problems, at least the easier ones, have one objective and one constraint.

The general approach for deriving a Material Index is to first write an equation for the attribute we are trying to minimize.

For the tie, this equation describes its mass. The mass of the tie is obtained by multiplying its density by its volume as shown in Equation 1.

$$m = \rho A L \tag{1}$$

where: m = mass of tie
ρ = density of tie material
A = cross sectional area of tie
L = length of tie

We obtain a second equation by considering the constraint. The constraint is that the tie must not fail. In other words, the stress, σ, in the tie must be less than the tensile strength, σ_f, of the tie material as expressed in Equation 2.

$$\sigma \leq \sigma_f \tag{2}$$

Since the stress in the tie is equal to the force exerted by the heavy weight divided by the cross sectional area of the tie, we can rewrite Equation 2 as follows:

$$\frac{F}{A} = \sigma_f \tag{3}$$

Next solve Equation 3 for the free variable, A, obtaining:

$$A = \frac{F}{\sigma_f} \tag{4}$$

"Area" is called a free variable in this problem since we are free to vary the size of the tie, i.e. its cross sectional area. Think of it this way: If we selected a high strength-high density material, the cross sectional area of the tie would be less than that for a tie made of a lower strength-lower density material. The mass of the two ties might be equal.

Let us get back to deriving the Material Index. The next step is to substitute the expression for the free variable, i.e. Equation 4, into Equation 1. This yields:

$$m = \rho(F/\sigma_f)L \tag{5}$$

or

$$m = FL(\rho/\sigma_f) \tag{6}$$

This is the desired equation for the mass of the tie. The first two terms, F and L, are fixed and are known for a specific problem. Thus, it is clear that the mass of the tie can be minimized by selecting a material that has a small value for ρ/σ_f. The reciprocal of the term in parentheses is σ_f/ρ. This is the Material Index we defined previously.

$$M = \frac{\sigma_f}{\rho} \tag{7}$$

Obviously, we need to select materials which have maximum values for the Material Index in order to minimize the mass of the tie.

Now that we know that the Material Index for the light, strong tie problem is σ_f/ρ, how do we use this knowledge to select the best material? It seems simple. We "look up" the strength and density of various materials and plug the values into Equation 7 and calculate the Material Index for each material. The material that has the largest value for M is the best choice for the tie. Or is it this simple? We can only perform some limited number of calculations and we might overlook a superior material, i.e. not consider it in our set of calculations. We need a more thorough method that examines all, or at least many, materials. Fortunately, much of the "look up" work has been done for us and is available in easy-to-use graphical form. Many useful graphs are available in Ashby's text,[1] and these and others can be easily drawn using the Cambridge Engineering Selector software.[5] The graphs drawn using the software show more materials than the graphs in the book.

The appropriate graph for the light, strong tie problem was previously presented as Figure 2. Note that both strength and density are plotted on logarithmic scales, the numbers shown are actual rather than logarithmic values. Tensile strengths are plotted for metals, polymers, and composites, while compressive strengths are plotted for ceramics. For our tie problem, we need tensile strength. Thus, for ceramics, the values on the graph should be adjusted by dividing by ~15 to reflect the fact that ceramics are weaker in tension than compression.

The better materials for the light, strong tie appear at the upper left corner of the graph. How do we pick the best two or three materials? If we take the logarithm of both sides of Equation 7, we obtain:

$$\log \sigma_f = \log \rho + \log M \tag{8}$$

For a plot of log σ_f on the y-axis and log ρ on the x-axis, this equation plots as a straight line for a fixed value of the Material Index, M. Materials of equal merit have the same value for the Material Index. Thus, materials of equal merit fall on the same line. This is an important fact. It tells us that on a graph such as Figure 2, or the simplified version shown in Figure 4, we can draw lines that identify materials that are equally good for our light, strong tie. A series of such lines having a slope of unity are shown in Figure 4. For each line, there is a fixed value for M, the Material Index. The next step is to realize that the line that is closest to the upper left corner of the graph has the largest value of the Material Index. That is, materials located near the upper left corner of the graph are best for use in making a light, strong tie. Points A, B, and C in Figure 4 represent three different materials. Materials A and B are on the same line and thus are equally good choices for the tie. They are better choices than material C. The tie made from material C would be inferior in that its mass would be larger than the mass of a tie made from either materials A or B. This is reiterated by Equation 9.

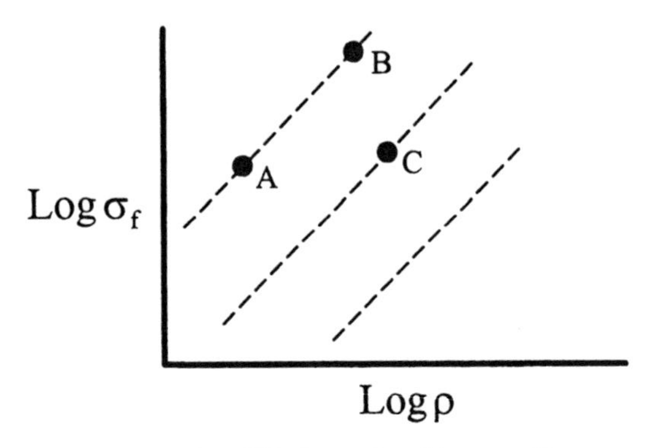

Figure 4. Simplified graph of log tensile strength versus log density, similar to Figure 2.

$$M_A = M_B < M_C \tag{9}$$

Notice that a tie of length, L, capable of withstanding a given force, F, has the same mass regardless of whether material A or B are used. The cross sectional area of the tie made from material A would be larger than that made with material

B because A has a lower density, but the masses of the two ties would be the same. This can be seen from Equation 6, since the two ties have the same value for the Material Index, σ_f / ρ, and F and L are fixed for a specific design. If other aspects of the design restrict the space available for the tie, then the large cross sectional area of the tie made from material A might be a disadvantage compared to the use of mateiral B. Since the tie has the same mass regardless of whether material A or B are used, the final decision would be made by considering other factors including availability and cost.

Based on what we have learned so far, we can pick several promising materials from the upper-left portion of Figure 2 using a design guideline having a slope of unity. Keeping in mind that the "bubbles" for ceramics in Figure 2 have to be lowered by a factor of ~15 since we are interested in tensile strength, the unidirectional fiber reinforced polymers appear to be the best choice for a light, strong tie. In particular, the carbon fiber reinforced polymers (CFRP) and glass fiber reinforced polymers (GFRP) are attractive. Wood and the stronger of the titanium and steel alloys are only slightly less satisfactory. At this point, cost and other factors should be investigated before making a final selection.

A more general method for considering cost is now considered. The graph in Figure 5 shows a plot of strength versus relative cost per unit volume of material. Relative cost, C_R, is the cost of the material per unit mass divided by the cost of mild steel reinforcing rod. Steel rod currently costs about $0.3/kg. Relative cost is used in an attempt to keep the graph from being made obsolete by inflation. The x-axis in Figure 5 is the product of relative cost, which is dimensionless, and density in Mg/m^3. Figure 5 can be used to select the lowest cost material for a strong tie using the same procedure described above for Figures 2 and 4. That is, to select the least expensive material, the guideline of unity slope is moved toward the upper left corner of Figure 5. Materials in this region are the best for an inexpensive, strong tie. Again, materials located on a given line are equally good.

CASE STUDY: LIGHT, STIFF BEAM

We now consider a light, stiff beam. A second case study will reinforce our understanding of the selection procedure. Consider that we wish to select the best material for a light, stiff beam. That is, we wish a beam of minimum mass that will elastically deflect no more than some specified amount, δ, when loaded with a force, F. An example beam is shown schematically in Figure 6.

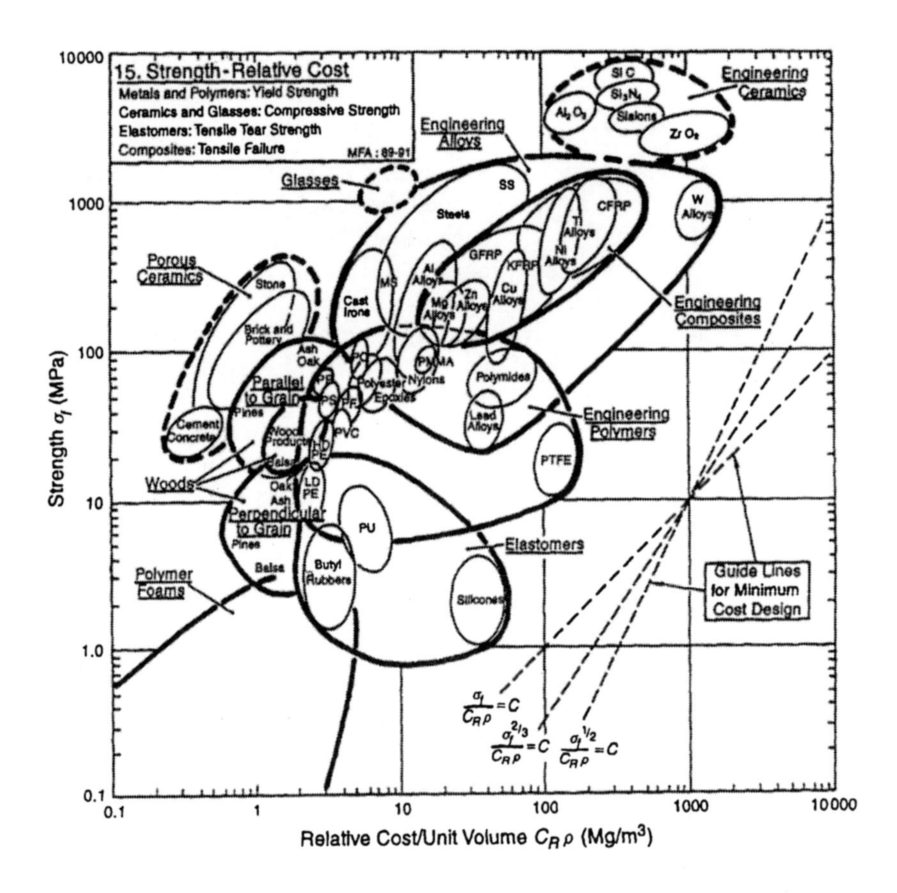

Figure 5. Graph for selecting low cost, strong tie.[1]

The goal is to minimize the mass of the beam and the constraint is to minimize deflection to less than or equal to δ. As before, the following steps will be followed to permit derivation of the appropriate Material Index:

1 - Write an equation using the objective,
2 - Write a second equation using the constraint,
3 - Solve the second equation for the free variable,
4 - Substitute the expression for the free variable into the first equation.

The first step is to write an equation for the mass of the beam. The equation for the mass of a beam having a cross sectional area, length, and density of A, L, and ρ, respectively, is:

$$m = \rho AL \tag{10}$$

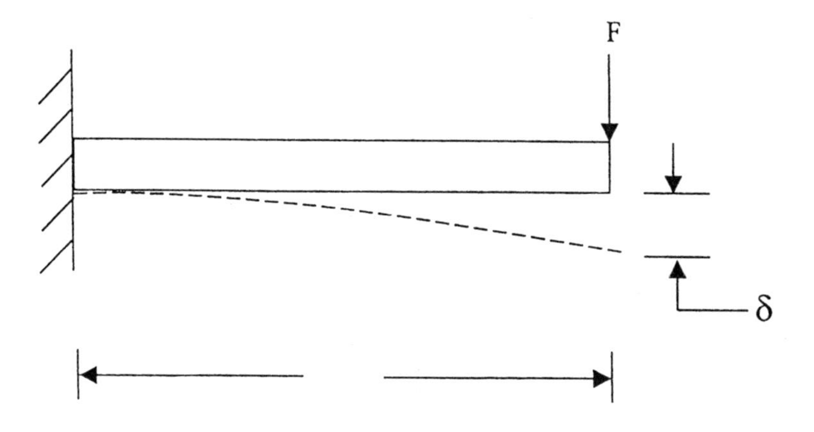

Figure 6. Beam carrying a load F with deflection of δ.

The second step is to write an equation that relates the elatic deflection of the beam to the applied load and beam geometry. The required equation is retrievable from any text on the mechanics of materials. A convenient compilation of equations useful for solution of a wide range of problems is provided as an appendix to Professor Ashby's text.[1] The required equation is:

$$\delta = \frac{FL^3}{C_1 EI} \tag{11}$$

where C_1 is an integer, E is Young's modulus, and I is the second moment of area. The value of I depends on the shape of the cross section of the beam.[1] Assuming the beam is circular, I equals $\pi r^4 / 4$ or $A^2 / 4\pi$. Substitute this for I into Equation 11 obtaining:

$$\delta = \frac{4\pi FL^3}{C_1 EA^2} \tag{12}$$

The third step is to solve the equation involving the constraint, i.e. Equation 12, for the free variable, A.

$$A = \left(\frac{4\pi FL^3}{C_1 \delta E} \right)^{1/2} \tag{13}$$

Finally, substitute this expression of the free variable, A, into Equation 10 obtaining:

$$m = \rho L \left(\frac{4\pi F L^3}{C_1 \delta E} \right)^{1/2} \qquad (14)$$

$$= \frac{\rho}{E^{1/2}} \left(\frac{4\pi F L^{5}}{C_1 \delta} \right)^{1/2} \qquad (15)$$

For a specific beam problem, all of the quantities inside the parentheses of Equation 15 will be constant and thus only the density, ρ, and Young's modulus, E, of the beam material will influence the mass of the beam. To minimize the mass of the beam, we desire a material having a low density and high Young's modulus. Therefore, the Material Index, M, is:

$$M = \frac{E^{1/2}}{\rho} \qquad (16)$$

Note that since we always wish to maximize the Material Index, M, we have written in Equation 16 the reciprocal of the expression that appeared in the mass equation.

As before, it is instructive to take the logarithm of both sides of Equation 16 obtaining:

$$\log E = 2 \log \rho + 2 \log M \qquad (17)$$

This is the equation of a straight line for a plot of log E on the y-axis and log ρ on the x-axis. Note that the slope of the line is 2. Using the same logic as beore, materials having equal values for the Material Index, M will be of equal merit for a light, stiff beam. Figure 1 is the appropriate graph for selecting the best materials. The guideline, with slope equal to 2, is moved toward the upper left corner of the graph to identify the best materials. Several promising materials are balsa wood, carbon fiber reinforced polymer (CFRP), beryllium, boron, silicon carbide, and diamond. The presence of balsa wood in the list reminds us that even though the mass of beams made from each of these materials will be about the same, since each of the materials is on or very near a given line, their cross sectional areas will be very different. The presence of diamond in the list also illustrates that the cost will vary. Cost may be factored into the selection process by using the graph shown in Figure 7. Use of a guideline of slope equal to 2 on

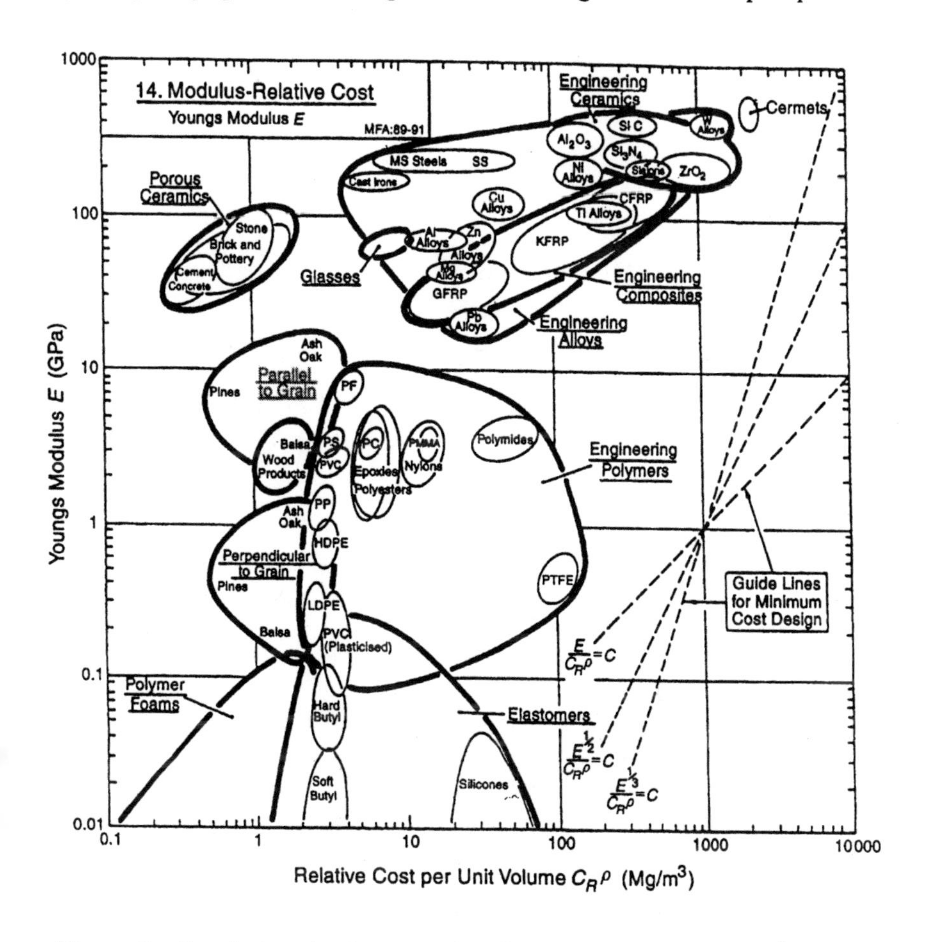

Figure 7. Graph for selecting materials for a low-cost, stiff beam.[1]

Figure 7 leads to the identification of concrete, brick, and stone as the least expensive materials for beams. Not coincidentally, these are the materials we typically use for construction of large structures such as buildings where the use of large quantities of materials dictates that we be cost conscious. Note that cast irons and mild steels are also possible choices. Clearly, steel beams are often used. Steel has superior tensile strength and fracture toughness, which argues for their use when these properties are also considered.

CASE STUDY: LIGHT, STRONG BEAM

When resistance to failure is the constraint of interest, as would be the case for a light, strong beam, the appropriate Material Index is:

$$M = \frac{\sigma_f^{2/3}}{\rho} \tag{18}$$

where σ_f is the yield strength for ductile metals and polymers or tensile strength for ceramics and brittle polymers. The reader may wish to derive this equation using the procedure used in the prior two derivations. The constraint is that we stipulate that the stress in the beam must be less than or equal to the stress at failure, i.e. the strength of the material. You will need to know the equation involving the constraint. It is as follows:[1]

$$\sigma_f = \frac{FLY_m}{CI} \tag{19}$$

where C is an integer, I is the second moment of area, and Y_m is the distance from the centerline of the beam to the surface of the beam. This is, of course, equal to the radius for a circular beam. You may also wish to show that the same Material Index, i.e. Equation 18, is obtained regardless of whether the beam cross section is assumed to be circular, square, rectangular, etc. By taking the logarithm of both sides of Equation 18, you can show that the appropriate graph to use is a plot of log strength (y-axis) versus log ρ (x-axis), and the guideline will have a slope of 1.5.

CASE STUDY: ENERGY EFFICIENT FURNACE WALLS

An important materials selection problem for ceramic engineers is choosing the best material for use in constructing the walls of furnaces, i.e. kilns. Assume that the goal is to minimize the energy, i.e. cost, of heating the furnace from the ambient temperature, T_i, to the operating temperature, T_f, where it is held for a time, t. There are two sources of energy loss per unit area: 1) loss by steady state conduction, Q_1, through the walls of the furnace, and 2) the energy required to raise the temperature of the walls, Q_2. These losses are given by the following two equations:

$$Q_1 = \frac{k(T_f - T_o)t}{w} \tag{20}$$

$$Q_2 = C_p \rho w \left(\frac{T_f - T_o}{2} \right) \tag{21}$$

where: k = thermal conductivity
C_p = specific heat
ρ = density
w = wall thickness

If we were only trying to minimize the energy loss via conduction, we would pick a material that had the smallest value for the thermal conductivity and we would make the walls very thick. If we only considered the energy lost in raising the temperature of the walls, we would select a material having a low value for the volumetric specific heat, $C_p\rho$, plus we would make the walls very thin. These are conflicting objectives; how are we to pick the best material and wall thickness?

To minimize the total energy loss, we must minimize the sum of Q_1 and Q_2.

$$Q = Q_1 + Q_2 = \frac{k(T_f - T_o)t}{w} + \frac{C_p \rho w (T_f - T_o)}{2} \tag{22}$$

To determine the optimum thickness, we differentiate Equation 22 with respect to the wall thickness, w, and equate the result to zero, obtaining:

$$w = \left(\frac{2kt}{C_p \rho} \right)^{1/2} \tag{23}$$

Substituting Equation 23 into Equation 22 in order to eliminate the wall thickness, w, gives:

$$Q = (T_f - T_o)(2t)^{1/2}(kC_p\rho)^{1/2} \tag{24}$$

Since for a given process, T_f, T_o, and t are fixed, Q can be minimized by minimizing $(kC_p\rho)^{1/2}$, i.e. by maximizing the Material Index, M,

$$M = (kC_p\rho)^{-1/2} \tag{25}$$

By using the relation between thermal conductivity and thermal diffusivity,

$$\alpha = \frac{k}{C_p\rho} \tag{26}$$

where α is the thermal diffusivity, it is possible to rewrite Equation 25 as

$$M = \frac{\alpha^{1/2}}{k} \tag{27}$$

Taking the logarithm of both sides of this equation yields

$$\log k = 1/2\ \log\alpha - \log M \tag{28}$$

This shows that Figure 8 can be used to select the best material(s) for the walls of the furnace. The design guideline has a slope of ½. The Material Index, M, is maximized by moving the design guideline toward the lower right corner of Figure 8. We look toward the lower right corner this time because we desire to pick materials having low conductivities. Actually, our goal is to maximize the Material Index, given by Equation 27. In other words, the best materials are located below the design guideline shown in Figure 8. This shows the best materials to be polymer foams, cork, plaster, wood, and polymers. These materials work only as long as low temperatures (~150°C) are involved.
For higher temperatures, porous ceramics are, of course, the appropriate choice. Either Equation 25 or 27 can be used to calculate values of the Material Index for specific porous ceramics. The material with the largest value is best. Equation 23 can be used to calculate the optimum wall thickness. It varies from one material to another.

For some applications, there may not be sufficient space for very thick walls. By using Equation 26, it is possible to rewrite Equation 23 as

$$w = \left(\frac{2kt}{C_p\rho}\right)^{1/2} = (2\alpha t)^{1/2} \tag{29}$$

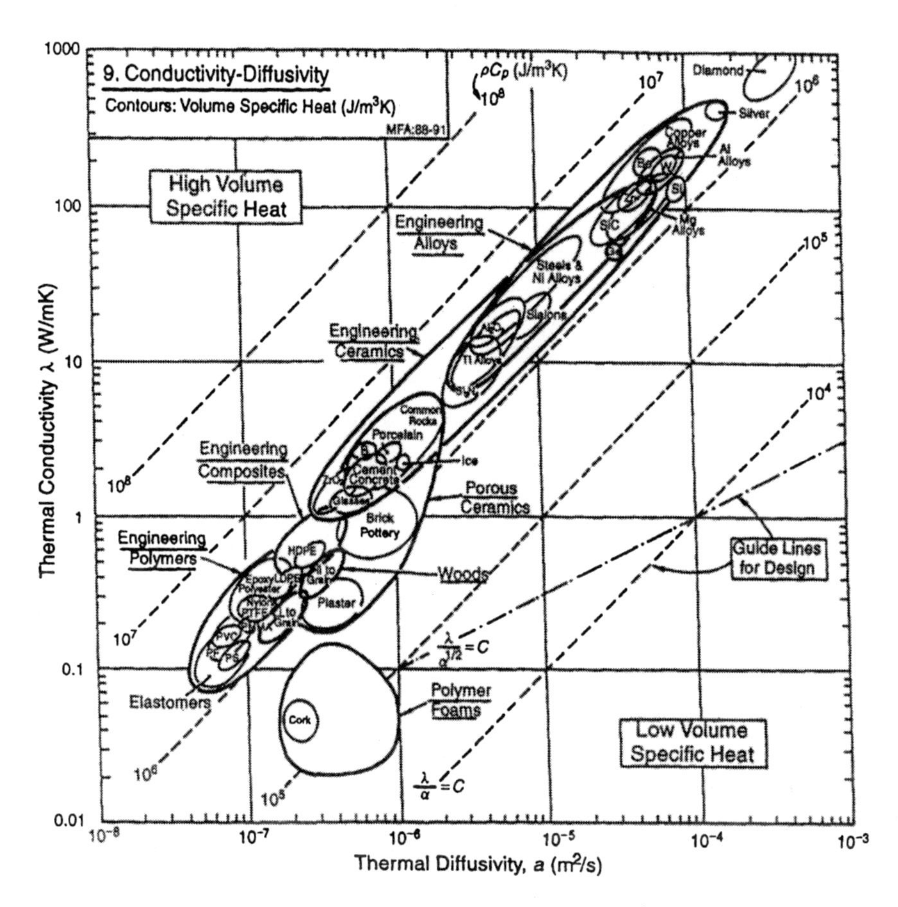

Figure 8. Thermal conductivity versus thermal diffusivity.[1]

This shows that materials having large thermal diffusivities require thick walls for a fixed value of t. If limited space is available and we must also fix the wall thickness, then Equation 29 shows that only materials with thermal diffusivities less than $w^2/2t$ are permissible. We still use Equations 25 or 27 to select the most energy efficient material from this restricted set of materials. Clearly, mechanical and chemical properties as well as cost should be considered before making a final selection.

Ashby tabulates Material Indices for a large number of other design problems.[1] These include deflection, strength, vibration, and fracture of ties, shafts, beams, columns, plates, flywheels, and internally pressurized tubes and spheres. Several thermal, thermo-mechanical, and electro-mechanical problems are also included.

INFLUENCE OF COMPONENT SHAPE ON MATERIAL SELECTION

So far, we have compared materials of the same cross sectional shape. The cross section could have any shape, but it had to be the same for the different materials being considered. This is not "fair" since some materials are better suited than others for being shaped into efficient cross sections. For example, a steel beam is most efficient when shaped like an I, i.e. an I-beam. On the other hand, a beam made of rubber (an unlikely choice) would likely buckle if it had the same cross section.

We treat the fact that different materials have different practical limits for the extent to which they can be shaped by introducing a quantity called the shape factor, ϕ, into the Material Index. Without going through the derivation,[1] the reader is asked to accept that for the case of the light, stiff beam considered previously, the Material Index including shape is:

$$M = \frac{\left(E\phi_B^e\right)^{1/2}}{\rho} \tag{30}$$

where ϕ_B^e is the shape factor for the elastic deflection of a beam. Compare this equation to Equation 16 where shape was not considered.

To get a "feel" for the shape factor, consider two beams both having rectangular corss sections. Suppose one beam has a cross section of 2x4 while the other has a cross section of 1x8. They both have the same cross sectional area, but the 1x8 has a larger value for ϕ_B^e, and, not coincidentally, is also more susceptible to buckling. The general formula for the shape factor ϕ_B^e is:

$$\phi_B^e = \frac{4\pi I}{A^2} \tag{31}$$

where I is the second moment of area as before and A is the cross sectional area. Using this equation and Equation 32, which gives I for a rectangular cross section of height, h, and width, b, we can calculate ϕ_B^e for the two beams.

$$I = \frac{bh^3}{12} \tag{32}$$

These calculations show that the 2x4 has a shape factor of 2.1, while the shape factor is 8.4 for the 1x8. The shape factor for the 1x8 is 4 times larger than that for the 2x4. This means that the 1x8 is 4 times stiffer than the 2x4, where stiffness is defined as force divided by deflection. That is, the load on the 1x8 could be 4 times that on the 2x4 for the same amount of deflection. Alternatively, for the same load, the 1x8 would deflect a fourth as much as the 2x4.

Let us think about how the shape factor can be incorporated into the selection of materials using a light, stiff beam as an example. First, using Figure 1 as we did before when we were not considering shape, a few promising materials are selected. For example, assume we have narrowed our selection to two materials, A and B, shown in Figure 9. Material B appears superior to material A since it is closer to the guideline shown in the figure. But we have not yet considered shape.

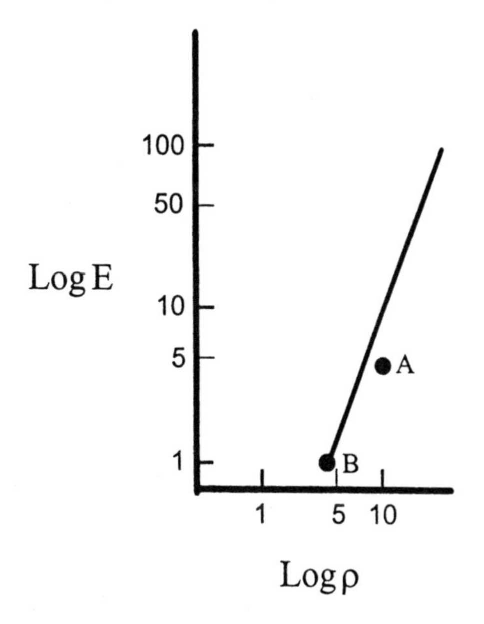

Figure 9. The materials A and B being considered for a light, stiff beam.

If we take the logarithm of both sides of Equation 30 we obtain

$$\log E\phi_B^e = 2\log\rho + 2\log M \qquad (33)$$

This equation suggests that we can use a log E versus log ρ plot, with guidelines having a slope of 2, to select materials if we multiply the shape factor, ϕ_B^e, times the Young's modulus, E. To show the result graphically, we simply move the points up vertically on the graph. Suppose material A is capable of a shape factor of 20 while material B is capable of only a shape factor of 2. We multiply the value of E for material A by 20 obtaining 5x20 = 100. We show the new position of material A after accounting for its shape factor as point A* in Figure 10. Similarly, we multiply the modulus of B by 2, obtaining 1x2 = 2. The new position, B*, is shown in Figure 10. Now, after considering shape, material A is

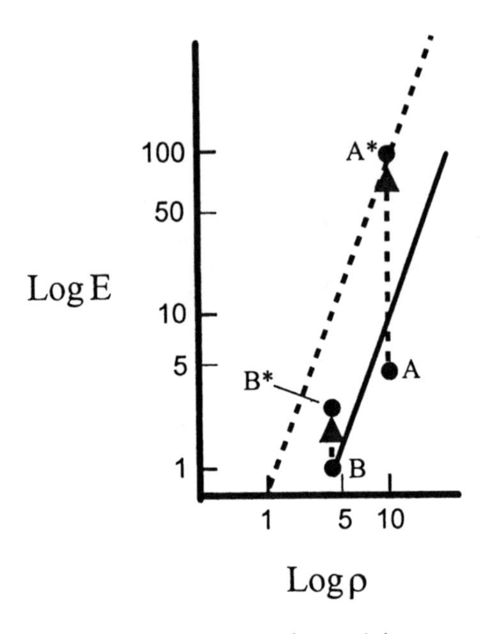

Figure 10. Positions of materials A and B after considering shape.

found to be a better material for the beam. Point A* is on a new guideline that is located nearer the upper left corner of the graph. Point B* is located below this new line.

An alternative non-graphical procedure is to first pick several promising materials and then calculate the Material Index. For the light, stiff beam example we have just considered, the procedure would be to plug E, ϕ_B^e, and ρ values into Equation 30. The best material would be the one with the largest calculated value of M. For the materials A and B considered above, the calculated value of M are $(5x20)^{1/2}/10 = 1$ and $(1x2)^{1/2}/5 = 0.28$, respectively. Once again, material A is shown to be the best material when shape is considered.

The reader is warned that the graphical approach requires the use of the Cambridge Engineering Selector software[5] (CES) for some design problems, while the approach of calculating M by plugging in values will always work.

The reader should also be aware that there are four types of shape factors. We have considered only ϕ_B^e which is appropriate for the elastic deflection of beams. For design of a light, strong beam, the appropriate shape factor is ϕ_B^f. For torsional loading, other shape factors are required as described by Ashby.[1]

MULTIPLE OBJECTIVES AND CONSTRAINTS

The design problems we have considered so far have been oversimplified. Real design problems often involve more than one objective and constraint. For example, we may wish to select a material for an aerospace application, where we have the goals of minimizing both mass and cost and we have the constraints that the component must not deflect more than some fixed amount and also must not fail when subjected to a given load. If it is a part for an engine, we also are constrained by the material's melting point, creep and oxidation rates, etc. Such problems typically do not have a unique solution. Nevertheless, methods have been developed for attacking such problems, two of which are explained below.[1,6]

Successive Treatment

For illustrative purposes, consider a fairly simple design problem where we wish a light, stiff, strong beam. We could first select several potential materials for a light, stiff beam using the Material Index $E^{1/2}/\rho$ as before. We could then select a set of materials for a light, strong beam using the Material Index $\sigma^{2/3}/\rho$. Materials that are common to both sets would likely be good choices. This general procedure could even be applied if we were trying to solve a design problem that had more than two constraints. We would just identify multiple sets of materials and again look for a material that appeared in each set. We may have to enlarge our initial sets if a common material does not occur in each set. This method is far from perfect, and two people may not select the same material.

Weight-Factors

Another method, which also involves some subjectivity, utilizes weight-factors. The key properties or Material Indices are first identified and then assigned weighting factors. The most important properties are assigned higher weight-factors based on the judgement of the designer or design team. Clearly, this step is subjective. The weight-factors are selected so that the sum of all the weight-factors is 1. A list of materials of interest is generated. The next step is to normalize each property or Material Index. For example, suppose fracture toughness is a property of interest. A normalized value for fracture toughness is obtained for each material of interest by dividing each fracture toughness value by the value for the material having the largest fracture toughness. The normalized values for each property are then multiplied by the appropriate weight-factor and the sum calculated for each material. The material having the largest value for the sum is selected as the best material. Clearly, the "best" material depends on the values selected for the weight-factors. For example, one team may believe that strength is more important than density and would therefore assign a higher value to strength, while another team assigns a higher value to density. The two teams likely will not select the same material. Skill, i.e. experience, in assigning weight-factors is very important. Ashby[1] and Bourell[6] provide good descriptions of several other useful methods for the selection of materials when multiple goals and constraints are present.

CAMBRIDGE ENGINEERING SELECTOR SOFTWARE

Computer software called the Cambridge Engineering Selector[5] has been developed specifically for use in graphical selection of materials. It readily permits construction of numerous graphs such as those shown previously where one property is plotted against another. Any property that appears in the database can be plotted against any other property. Thus, the software can be used to plot unusual but useful graphs such as that shown in Figure 11, where the price of various materials is plotted versus their hardness. The software also provides tabulations of physical, mechanical, thermal, and electrical properties for important ceramics, metals, polymers, and composites. Some unusual materials are included such as ice, leather, bambo, etc. The software also provides addresses of vendors for specific materials and shapes. In addition, the software describes many material fabrication processes.

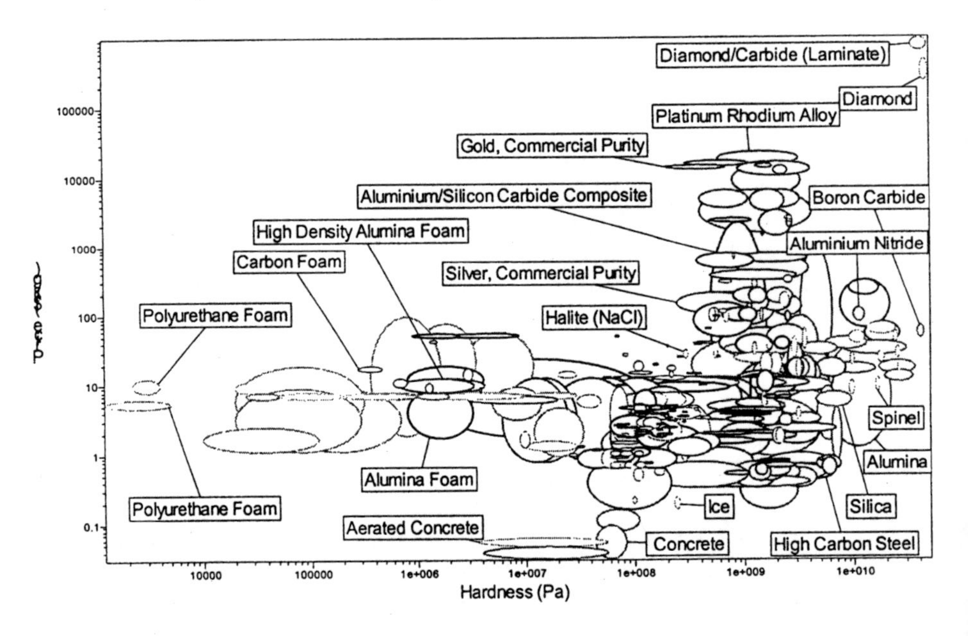

Figure 11. Price versus hardness graph created using the Cambridge Engineering Selector software.

One of the disadvantages of the CES software is that the database does not contain as many specific grades of materials as desired. This has been recognized by the vendor, Granta Limited, and many more materials have been added over the past couple of years. Recently, ASM International has agreed to provide funding and make their material database available for inclusion in the CES software. This will significantly advance this user-friendly software and will likely result in it being widely used.

PROCESS SELECTION

Ashby also describes a graphical approach for process selection.[1] Figure 12 shows one of several graphs that are used. In this example, it is assumed that the tolerance and surface finish of the part to be fabricated are known. Tolerance and surface finish are termed process attributes. Their values for a specific component define a point or small region on the graph. Processes capable of meeeting these requirements contain, i.e. enclose, this region. For example, Figure 12 shows that investment casting, as well as many other processes, have the potential for making a part having a surface finish of 1 μm and a tolerance of 0.1 mm. The designer then sequentially utilizes several other graphs to narrow the selection. A second graph that might be used is shown in Figure 13. Here the process attributes are hardness and melting point. Note that most of the casting processes are not feasible for materials such as ceramics that have very high melting points. For these materials, possible processes include powder methods, chemical vapor deposition, etc. Paper copies of many of the graphs required for process selection are available,[1] but the multiple steps required to select the preferred process are simplified by use of the CES software.[5]

Others have described alternative methods for process selection with emphasis on the interaction between manufacturing and design.[6-11]

It is important to realize that the function of a component, its shape, materials selection, and process selection are certainly interdependent and should be performed concurrently in order to avoid costly iterations. It is not productive to select a material and, some time later, realize that an economical process for fabricating that material into the desired shape is not available. That is, function, shape, material, process, and cost must be considered together throughout the design process.

SOURCES OF PROPERTY DATA

A very large number of databases for material properties and specifications performance are available. Westbrook[7] provides an extensive compilation subdivided by material types. More than 30 sources are given for ceramics, glasses, and cements. Many more sources are provided for metals, polymers, and composites. Sources of materials purchasing information are also provided.

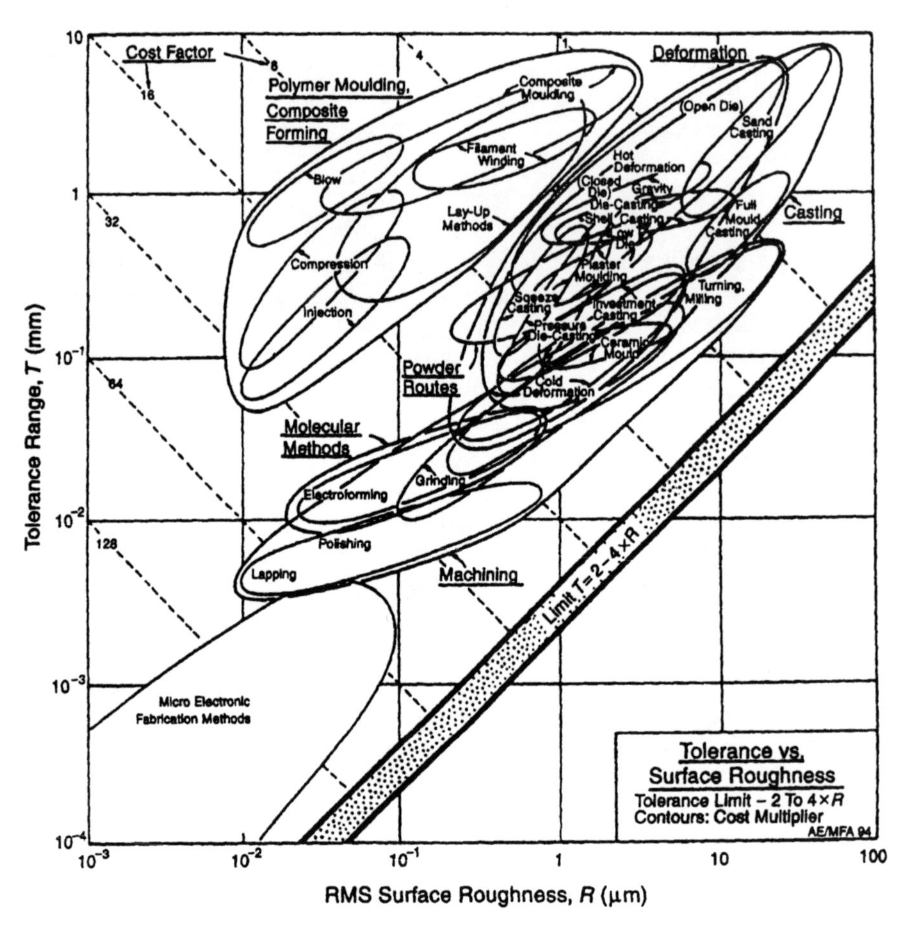

Figure 12. Example of a process selection graph.[1]

Guidance in evaluation, interpretation, and estimation data is available from a number of sources.[1,5] Clearly, since many properties of ceramics are very sensitive to microstructure, great care must be exercised in choosing property values. For conceptual design, handbook and database values may be acceptable. For detailed design, data from vendors or in-house test data are preferable.

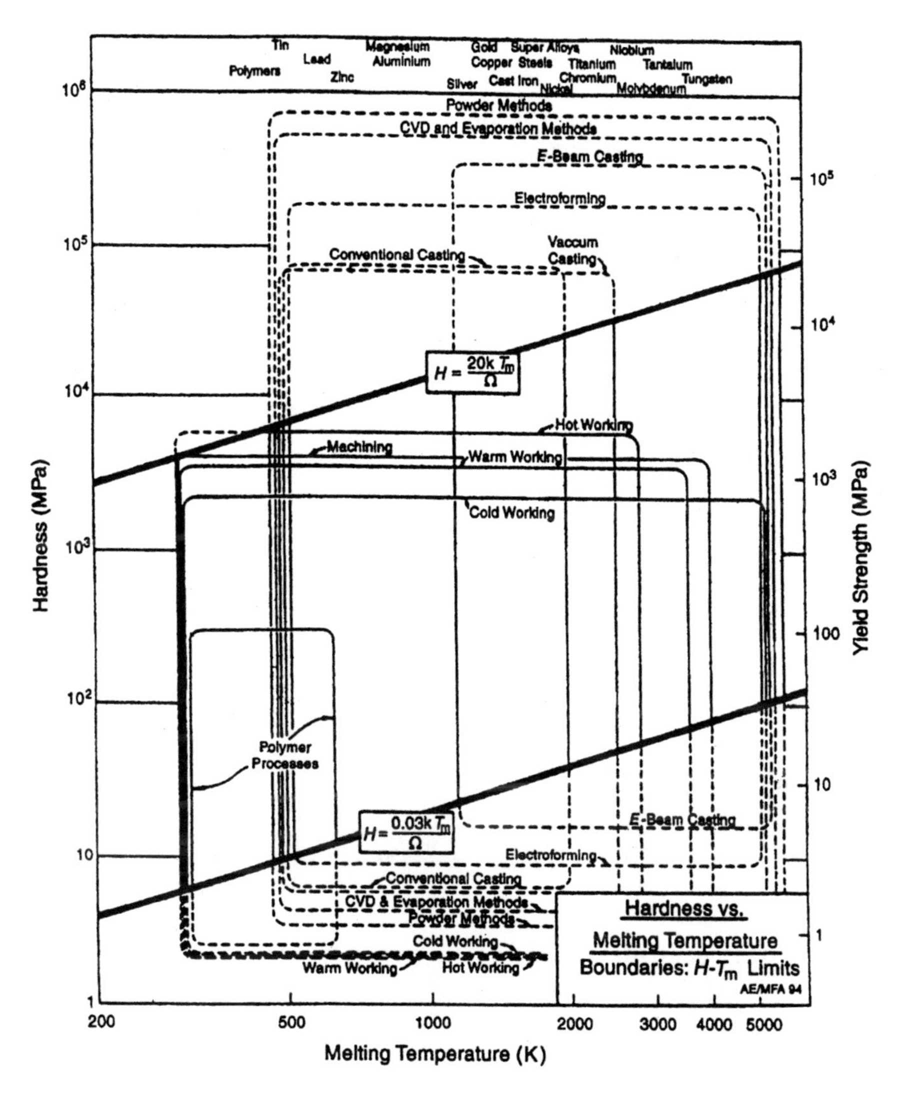

Figure 13. Hardness versus melting point graph used for process selection.[1]

Acknowledgment:
Permission of Michael F. Ashby to use graphs from his text is appreciated.

REFERENCES

1. Michael F. Ashby, Materials Selection in Mechanical Design, Second Edition, Butterworth Heinemann, Boston, 1999.

2. J. A. Charles and F. A. A. Crane, Selection and Use of Engineering Materials, Second Edition, Butterworth Heinemann, Boston, 1989.

3. Materials Selection and Design, ASM Handbook, Volume 20, ASM International, Materials Park, Ohio, 1997.

4. G. T. Murray, Introduction to Engineering Materials: Behavior, Properties, and Selection, Marcel Dekker, Inc., New York, 1993.

5. Cambridge Engineering Selector, Software available from: Granta Design Limited, Rustat House, 62 Clifton Road, Cambridge CB1 7EG, United Kingdom. See www.grantadesign.com. Also available from ASM International: See www.asminternational.org.

6. David L. Bourell, Decision Matrices in Materials Selection, pp. 291-296 in Reference 3.

7. Jack H. Westbrook, Sources of Materials Property Data and Information, pp. 491-506 in Reference 3.

8. Henry W. Stoll, Introduction to Manufacturing and Design, pp. 669-675 in Reference 3.

9. Geoffrey Boothroyd, Design for Manufacture and Assembly, pp. 676-686 in Reference 3.

10. John A. Schey, Manufacturing Processes and Their Selection, pp. 687-704 in Reference 3.

11. David P. Hoult and Lawrence Meador, Manufacturing Cost Estimating, pp. 716-722 in Reference 3.

PROBLEMS

1. Describe, using your own words, the general procedure for deriving a Material Index.

2. Describe how graphs, such as Figure 1, are used to select materials.

3. Use Figure 5 to select three promising materials for use in fabricating a low cost, strong tie.

4. Using Figure 5, calculate the cost of teflon (PTFE) in dollars per kilogram. You can obtain a value for the density of teflon from Figure 2.

5. Derive the Material Index for a light, stiff tie, i.e. a tie that has minimum mass and elongates no more than some given, fixed value. Assume that the length of the tie and the applied force are fixed and known.

Statistical Design

4

H. El-Shall

K.G. Christmas

STATISTICAL DESIGN

Hassan El-Shall, and Kimberly G. Christmas
Department of Materials Science and Engineering
University of Florida
Gainesville, Fl. 32611

ABSTRACT

Industrial products are the results of several studies including: bench scale investigations in academic and industrial research laboratories, in pilot plant studies, and quality control to check design specifications through out production. All of these activities require experimentation and collection of data. Statistics should play a role in every facet of data collection and analysis, from initial problem formulation to the drawing of final conclusions.

Statistical experimental design is an effective tool that ensures precise information about the responses of interest, and guarantees economic use of resources. In addition, statistical data analysis techniques aids in clearly and concisely summarizing salient features of experimental data.

In this paper, various statistical designs including screening, full factorial and optimization designs are explained. Analysis of variance "ANOVA" is introduced and utilized together with practical examples to illustrate the effectiveness of these methodologies in better designs of ceramics and other materials.

INTRODUCTION

Engineers and scientists are usually well trained in the mechanics of experimentation. In most of the cases, however, the strategy typically used is the one-variable-at-a-time strategy. This is not only inefficient strategy; it suffers from several disadvantages as discussed later.

The best experimental designs result from the combined knowledge of science, engineering, technology, and statistics in the area of interest. Statistics help the experimenter to perceive and understand the real world with far greater precision than would otherwise be possible.

The following is a comparison between the statistical design methods and one- variable-at-a time strategy of experimentation.

One-Variable-at-a-Time

1. The number of tests could be prohibitively large. Thus, many variables might remain unexamined and hence, undiscovered, because of the lack of time and money.
2. Analysis of variance is not performed. Therefore, results do not have an adequate level of confidence.
3. Interactions between variables may not be detected or determined. Thus, optimum results may not be obtained

Statistical Methods

1. A low number of tests can be used to obtain meaningful results about main effects of several variables at the same time. In addition, interaction effects can be evaluated.
2. A valid analysis of variance can be performed, so that it is possible to know the significance of the results.
3. A systematic screening of the less important variables and the discovery of variables that are important is a major advantage of these designs.
4. A mathematical model could be obtained to predict the effects of changes in factor levels within the design region. The models can be used also to generate multidimensional surface responses. Consequently, optimum results can be obtained.

In order to give the reader a fair idea about the use of statistical designs in materials research, a review of certain statistical concepts and tools is necessary. In addition a short review of the different designs will be presented. It should be kept in mind, however, due to limited pages in this chapter, the reader should consult other excellent texts and references listed below for detailed information.

IMPORTANT STATISTICAL TOOLS

(ANOVA) Analysis of Variance

The analysis of variance is a rapid estimation method for separating the variability possessed by a collection of data into components or sources. A mathematical model is usually used with it. In addition, it is always connected with the type of experimental design employed. In the following example, the ANOVA is used for the case of comparing two means. The ANOVA table is also used to calculate s^2, the estimate of the variance of ε_i (the experimental error associated with observation i). It is coupled with the model $y_{ij} = \eta + \tau_j + \varepsilon_{ij}$, in which y_{ij} is the observation i in treatment j, η is the grand mean of all observations, τ_j represent effects of treatment j. Finally, the statistic F_{ν_1, ν_2}, is used to test hypotheses concerning the treatment effects $\hat{\tau}_j$. The F statistic is the ratio of two estimates of variance. It has a distribution whose critical values are tabulated. The F statistic (or ratio) has a two-parameter distribution, the parameters being v_1 *and* v_2, the degrees of freedom in the two estimates of variance employed in the ratio, and n_j is the number of observation [S. Hunter (1977), G.E.P. Box, W.G. Hunter, and J.S. Hunter (1978)].

Example: Two different types of additives to a ceramic material are compared with respect to a measured response of the ceramic product and the following data are obtained:

Treatment # 1 (11,7,10,8) Total= 36 ; n_j : (j = 1) = 4
Treatment # 2 (3,5,7,1,5,3) Total= 24 ; n_j : (j = 2)= 6

Averages: $y_{av1} = 9$ $y_{av2} = 4$

y_{gav} (Grand Average) = $\frac{36+24}{10} = 6$

Construct the analysis of variance table for the above model and data and test the hypothesis that $\eta_1 - \eta_2 = 0$

Answer: The ANOVA table can be written as follows:

	Sum of Squares	Degrees of Freedom	Mean Square
Crude Sum of Squares: Σy_{ij}^2	452	10	
Correction Factor: $N\,\mathbf{Y}^2_{gav} = (\Sigma y_{ij})^2/n$	360	1	
Treatment Sum of Squares: $\sum n_j \tau_j^2$	60	1	
Residual Sum of Squares	32	8	$S^2 = 32/8 = 4.00$

The degrees of freedom may be explained as follows:

- 10 observations lead to 10 degrees of freedom. The estimate of the grand mean η(y_{gav} = 6) consumes one df leaving nine degrees of freedom. The estimates of treatment effects (τ_1 and τ_2) only consume one degree of freedom. Thus, eight degrees of freedom are left for residuals.

- Treatment SSq can be estimated using the equation $\frac{T_1^2}{n_1} + \frac{T_2^2}{n_2}$ – CF; or generally, $(\sum \frac{T_j^2}{n_j}$ – CF), where T_1 and T_2 are the sums of the observations for treatments 1 and 2, and where n_1 and n_2 are the corresponding numbers of observations. The quantity CF is the correction factor.
- The pooled estimate of variance s^2 is obtained from the residual sum of squares divided by its degrees of freedom.

One purpose of the ANOVA is to help determine whether or not real differences exist between treatment means. The null hypothesis to be tested is that there is no difference between the treatments, i.e. $\eta_1 = \eta_2$ is true. Thus, we need to compare the difference between treatments and the experimental error. Therefore, two independent estimates of σ^2 are composed, one that is based on the experimental errors only, and a second that is based on the difference between the treatments' averages.

In the example, the clear estimate of σ^2 is the residual variance with 8 degrees of freedom obtained from the ANOVA table. The second estimate of σ^2 may be computed on the basis of the assumption that there is no difference in treatments. The two estimates are $\sigma^2 = 60.0$ with 1 degree of freedom and $s^2 = 4.00$ with 8 degrees of freedom, and use the F ratio.

	Sum of Squares	Degrees of Freedom	Mean Squares
Crude Sum of Squares	452	10	
Correction Factor	360	1	
Treatment Sum of Squares	60	1	$60.0 = s^2 \rightarrow \sigma^2 \mid \tau_j = 0$
Residual Sum of Squares	32	8	$4.00 = s^2 \rightarrow \sigma^2$

Taking the ratio of the two estimates gives:

$$F_{1,8} = 60.00/4.00 = 15.00$$

Referring to the F_{ν_1, ν_2} tables in statistics books, the critical value of $F_{1,8}$ = 5.32, that is $\text{Prob}(F_{1,8} \geq 5.32) = 0.05$. Since the observed $F_{1,8}$ is larger than the critical value, then it is a rare event and therefore we reject the null hypothesis that $\eta_1 = \eta_2$. In other words, we accept the alternative hypothesis that there is a difference between treatment effects with 95% confidence.

VARIOUS STATISTICAL DESIGNS

There are several statistical designs as explained by several authors [R.M. Bethea, B.S. Duran, and T.L. Boullion (1985); G.E.P. Box, and J.S. Hunter (1961); G.E.P. Box, W.G. Hunter, and J.S. Hunter (1978); R. Caulcutt (1991); W.J. Diamond(1981); A. Garcia-Diaz and D. Phillips(1995); J.S. Hunter(1977); R.L. Mason, R.F. Gunst, and J.L. Hess(1989); and D.C. Montgomery and E.C. Peck(1982)].These designs can be used by researchers, regardless of their field of investigation, including:

1. Randomized Block Designs

Controlling the experiments' environment should be a major objective of every researcher. However, in many cases, experiments have to be performed in environments that contain sources of variability, which exceed the variability due only to measurement errors. Examples may include variances due to running experiments while using different shifts of workers, batches of raw materials, and/or different pieces of machinery.

Fortunately, through the simple methodology of blocking we can eliminate many sources of variability and even estimate their contribution to the variance of the data. The ANOVA table is very useful since it allows us to obtain an appropriate estimate of the error variance affecting the treatment comparisons after the elimination of the variation that could be removed by blocking. It should be remembered that, in making comparisons between several treatments we will be able to detect small differences between the treatment means only when the variance of the observations is correspondingly small.

In comparing different ceramic products, for example, each must first be fired in a furnace, but the furnace may not be large enough to handle more than a few samples at a time. Care must therefore be taken to have all treatments represented in each furnace, since we know that differences among furnaces represent an important source of variability. Here furnaces are the blocking variable. The variables under study are the treatments. The treatments are randomly assigned to the furnaces (blocks) so that treatment comparisons may be free of other systematic effects. Thus, the experimental design is called a randomized block design [S. Hunter (1977)].

The ANOVA table should be used with the model $yij = \eta + \beta i + \tau j + \varepsilon_{ij}$; where:

η = grand mean; βi = block effects

τj = treatment effects; ε_{ij}= errors

In randomized block designs, it is assumed that the block effects do not interact with the treatment effects.

2. Factorial Designs

These designs are simple and very useful in many practical applications. Through the use of these designs it is possible to investigate the effects of several variables upon a response with a great economy in the number of experimental runs. Using these designs, we can investigate a response, which is a function of several controlled variables, or factors, some of which may be qualitative, and others quantitative. Using the two-level (or two-version) factorial designs, researchers can estimate the major effects of the controlled variables, as well as their interactions.

For k variables, the two-level factorial designs require $N=2^k$ runs. Examples can be given by the simple design for two variables A and B shown below.

The 2^2 Factorial Design: These designs are used to investigate effect of two variables at two levels.

DESIGN MATRIX		OBSERVATIONS
A	**B**	**Y**
-	-	Y_1
+	-	Y_2
-	+	Y_3
+	+	Y_4

The "– " and "+" corresponds to the low and high levels of the variable and Y is the response.

Estimation of effects:

$Y_2 - Y_1$ = main effect of A independent of B (at lower B)

$Y_4 - Y_3$ = main effect of A independent of B (at higher B)

$\frac{1}{2}[(Y_2 - Y_1) + (Y_4 - Y_3)]$ = effect of variable A

This last expression can be written as $\frac{1}{2}[Y_2 + Y_4) - (Y_1 + Y_3)] = Y_+ - Y_-$ where Y_+ is the average at (+) high level and Y_- at (–) low level of A

The main effect of B = $\frac{1}{2}[(Y_4 - Y_2) + (Y_3 - Y_1)] = Y_+ - Y_-$; the subscript on the two averages refers to the levels of B.

The interaction effect = $\frac{1}{2}[(Y_4 - Y_3) - (Y_2 - Y_1)] = \dfrac{Y_4 + Y_1}{2} - \dfrac{Y_3 + Y_2}{2}$ (1)

Thus, $Y_4 - Y_3 = Y_2 - Y_1$ i.e., if the effect of A at high B = effect of A at low B then the interaction = 0.

To identify the observations entering Y_+ and Y_- for the main effects and the interaction effect, use the design matrix to generate the effects matrix as shown below.

DESIGN MATRIX		OBSERVATIONS	EFFECTS MATRIX		
A	B	Y	A	B	AB
-	-	Y_1	-	-	+
+	-	Y_2	+	-	-
-	+	Y_3	-	+	-
+	+	Y_4	+	+	+

In the effects matrix for A & B the + & - are the same as in the design matrix, for AB multiply A by B. For estimation of effects multiply the Y column by the corresponding effects column. Divide their algebraic sum by 2. In this case, 2 is used because we have two runs at each level of the variables.

Example: Effect of Factor (Variable) A = ½ $[-Y_1 + Y_2 - Y_3 + Y_4]$, and so on, for each effect. You can see that each effect is estimated by the difference in the two averages $\bar{Y}_+ - \bar{Y}_-$. Another important parameter we need for analysis is the variance:

$$V(\bar{Y}_+ - \bar{Y}_-) = \left(1/n_+ + 1/n_-\right)\sigma^2 = 4\,\sigma^2/N \qquad (2)$$

where, σ = Standard Deviation
N = Total number of observations
$n_+ = n_- = N/2$

Thus the ANOVA table is used as described earlier and detailed below for the following example.

Example: In a laboratory experiment, a 2^2 factorial design was run to determine the effect of two temperatures and two additives on the strength of a ceramic product. The following data were obtained after firing the samples in four different furnaces.

DESIGN MATRICES				Strength, Y			
Temp.	**Additive***	**X_1**	**X_2**	**I**	**II**	**III**	**IV**
100	A	-	-	4	6	3	4
200	A	+	-	13	10	13	10
100	B	-	+	6	5	5	44
200	B	+	+	14	11	11	12

* The letters A or B are arbitrarily assigned to the type of additive

Analysis of Data

Solution:

DESIGN MATRIX		Y=Average of	EFFECT MATRIX		
X_1	X_2	Strength	X_1	X_2	X_1X_2
-	-	4.25	-4.25	-4.25	+4.25
+	-	11.5	+11.5	-11.5	-11.5
-	+	5	-5	+5	-5
+	+	12	+12	+12	+12
			+14.25	+1.25	-0.25

Effect of X_1 (Temperature) $= +14.25/2 = +7.125$

Effect of X_2 (Additive) $= +1.25/2 = +0.625$

Effect of X_1X_2 $= -0.25/2 = -0.125$ (Temperature x Additive interaction)

Next, it is important to analyze the variability or variance of the data using the ANOVA table.

Source	SSq*	df**	Mean Square
ΣY^2	1299.0000	16	
CF	1072.5625	1	
SY^2	226.4375	15	
Furnaces	6.6875	3	
Treatments	204.6875	3	$204.6875/3 = 68.229 \rightarrow \sigma^2 \mid \tau j = 0$
Residual	15.0625	9	$S^2 = 15.0625/9 = 1.6736 \rightarrow \sigma^2$

* SSq = Sum of Squares ** df = Degrees of Freedom

Using the "F" test:

$$\frac{\sigma^2 \mid \tau j = 0}{S^2} = \frac{68.229}{1.6736} = 40.7678 \quad (3)$$

Compare this value with $F_{3,9}$ from the "F" tables corresponding to 95% confidence level. Probability ($F_{3,9} \geq 3.86$) = 0.05. Since the experimental value $F_{3,9} = 40.7678$ is greater than theoretical value of 3.86. Then, treatment effect is not a rare event i.e., it is significant and there is a real difference among the treatment.

The SSq of treatments can be separated into individual ones e.g. for temperature, additive, and the interaction between temperature and the additive. This can be done as follows:

$$\text{Sum of squares of an effect} = \frac{N(\text{effect})^2}{4}$$

Where:	N = total number of observations			
	df = 1			
For temperature (T)	SSq =	$16(7.125)^2/4 =$	203.0625	df = 1
For Additive	SSq =	$16(0.625)^2/4 =$	1.5625	df = 1
For T x Additive	SSq =	$16(0.125)^2/4 =$	0.0625	df = 1
Total of treatment	SSq	=	204.6875	df = 3

We can write the ANOVA table as:

Source	**SSq**	**df**	**Mean Square**
SY^2	226.4375	15	
Furnaces	6.6875	3	
Treatment	204.6875	3	$68.229 \rightarrow \sigma^2 \mid \tau j = 0$
Temperature	203.0625	1	$203.0625/1 = 203.0625 \rightarrow \sigma^2 \mid$ temperature effect = 0
Additive	1.5625	1	$1.5625/1 = 1.5625 \rightarrow \sigma^2 \mid$ additive effect = 0
Interaction	0.0625	1	$0.0625/1 = 0.0625 \rightarrow \sigma^2 \mid$ interaction effect = 0
Residual	15.063	9	$S^2 = 1.6736 \rightarrow \sigma^2$

For temperature $F_{1,9} = 203.062/1.6736 = 121.3325 > 3.86$ (Critical value) i.e. temperature effect is significant.

For additive $F_{1,9} = 1.5625/1.6736 = 0.9336 < 3.86$ i.e. additive effect is not significant.

For interaction effect $F_{1,9} = 0.0625/1.6736 = 0.03734 < 3.86$ i.e. (not significant).

The 2^3 Factorial Design: This design provides an excellent experimental arrangement for initially exploring a response as a function of three variables. Each variable is held at two levels and all combinations are employed to produce eight experimental conditions. The data can be analyzed as shown in the above paragraphs. However, there is another simple methodology called "Yates Algorithm" that can be used to estimate the main and interaction effects of all variables. This technique is well described in some of the bibliography provided at the end of this paper.

The 2^k Factorial Designs: The two-level factorial designs can be constructed for any number of variables k. Whenever k is very large, fractions of these full factorial designs are used to screen the most significant variables. The Fractional factorial designs are explained briefly in following sections.

Frequently it is not possible to perform all the 2^k runs within a homogeneous environment. For example, we may wish to run a 2^3 factorial on a process that we know is affected by the day of the week. Probably, the 8 experiments could not be run all on the same day. Suppose, however, it is possible to run only four experiments on one day and the other four on another day. It is possible to block the day-to-day variability with high order interaction (three factor interaction), which is mostly not of interest or small. It can be done by arranging the contrast between the two sets of four runs exactly to correspond to the three-factor interaction contrast. Thus, the estimates of main effects and two-factor interactions of the variables are unaffected by the difference between days of operation. The three-factor interaction in the above example is confounded with the difference between the two blocks (days). An example showing such blocking technique is given below.

The following table displays the plus and minus signs for the effects estimated using a 2^3 factorial design listed in standard order. The experiments are then rearranged so that all those carrying a minus sign in the three-factor interaction column are performed in day I, while those with a plus sign are performed in day II.

Thus:

Contrast for Estimating Effects

Days	A	B	C	AB	AC	BC	ABC
I	-	-	-	+	+	+	-
II	+	-	-	-	-	+	+
II	-	+	-	-	+	-	+
I	+	+	-	+	-	-	-
II	-	-	+	+	-	-	+
I	+	-	+	-	+	-	-
I	-	+	+	-	-	+	-
II	+	+	+	+	+	+	+

2^3 in Two Blocks

Block	A	B	C	AB	AC	BC	ABC
(Day I)	-	-	-	+	+	+	-
	+	+	-	+	-	-	-
	+	-	+	-	+	-	-
	-	+	+	-	-	+	-
(Day II)	+	-	-	-	-	+	+
	-	+	-	-	+	-	+
	-	-	+	+	-	-	+
	+	+	+	+	+	+	+

Any effect of days would be merged (confounded) with the three-factor interaction. The statistic (y_+ - y_-) used to estimate the three-factor interaction effect would be exactly the same as the statistic used to estimate the difference between the two days. Since the ABC interaction and block effect are confounded, the main effects and two-factor interactions are now also clear of the block effect.

3. Fractional Factorial Designs

As mentioned above, complete factorial designs have the obvious disadvantage of requiring a large number of runs, i.e., 2^k. Thus, fractional factorial designs are used to obtain useful information and using smaller number of tests. However, using smaller number of runs will not allow us to separately estimate individual effects. To overcome this problem, in general, fractional factorial designs are chosen so that estimates of a greatest interest will be, as much as possible, confounded with higher–order effects, which are expected to be negligible.

For example, in the 2^{k-1}, fractional factorial designs, only one-half of the 2^k factorial design is performed. An example would be a two-level factorial design for four-variables in eight runs, that is, the 2^{4-1} fractional factorial design. By choosing the half-replicate design appropriately we can ensure that main effects are confounded only with three-factor interactions, which are expected to be negligible. Thus, the design provides useful estimates of the main effects with only half the number runs of the full factorial.

The usual way to construct a 2^{k-1} design is to start with the two-level design for (k-1) variables and then use the + and – signs of the highest-order interactions, the (k-1) factor interaction, to determine the high and low versions of the k^{the} variable.

Example: Write down a 2^{4-1} fractional factorial design in variables A, B, C, and D.

Solution: First, write down a 2^3 factorial and then the vector of + and – signs associated with the ABC interactions. These signs are then used to define the versions if variable D.

A	B	C	ABC
-	-	-	-
+	-	-	+
-	+	-	+
+	+	-	-
-	-	+	+
+	-	+	-
-	+	+	-
+	+	+	+

Thus the 2^{4-1} =

A	B	C	D
-	-	-	-
+	-	-	+
-	+	-	+
+	+	-	-
-	-	+	+
+	-	+	-
-	+	+	-
+	+	+	+

For more details about the fractional factorial designs including the generators for different designs and, defining relation and resolution of the design, the reader is advised to consult listed bibliography.

4. Screening Designs

In some cases, the number of variables might be so large that a full factorial, a half replicate, or even a quarter replicate would be too costly. If this is so, screening designs can be used to gather information about these variables. Actually, screening designs help us to decide which of several independent variables are worthy of further study. These important designs are explained in details in the literature [R. Caulcutt (1991); R.L. Mason, R.F. Gunst, and J.L. Hess (1989); and Plackett and Burman (1946)].

Probably the most widely used screening designs are the Plackett-Burman designs, which are used to study n = (N – 1) independent variables in N runs or trials. N must be a multiple of four. Thus, three variables could be investigated by conducting only four experiments, and we could study seven variables in eight runs, etc. To obtain the design matrix for a Plackett-Burman design we first select the appropriate design vector from:

N=8 +1+1+1-1+1-1-1
N=12 +1+1-1+1+1+1-1-1-1+1-1
N=16 +1+1+1+1-1+1-1+1+1-1-1+1-1-1-1
N=20 +1+1-1-1+1+1+1+1-1+1-1+1-1-1-1-1+1+1-1

The selection of the variables depends upon the experimenter, but it is customary to leave at least three dummy variables from which the error is estimated, as explained later. The dummy variables could be arbitrarily assigned to any column in the matrix.

Example: Some basic variables that affect a process response are listed below together with three dummy variables. The total number is 15 variables, including:

1. A = Conditioning time
2. B = pH in conditioner
3. C = % Solids in the conditioner
4. D = Type of feed preparation
5. E = Dummy Variable
6. F = pH at the separation reactor
7. G = Mineral composition of the feed
8. H = Amount of Reagents
9. I = Particle Size Distribution

10. J = Dummy Variable
11. K = Speed of Agitation in separation reactor
12. L = Separation Time
13. M = Conditioner Temperature
14. N = Reactor Temperature
15. O = Dummy Variable

Suppose we wish to assess the effects of the 15 independent variables in 16 runs. From the left-hand column of the Plackett-Burman design matrix above, the vector for $N = 16$ should be written as a column as shown in Table I. To generate the second column of the design matrix we copy the first column but move each entry down one row and put the bottom sign up to the top. To generate the third column we copy the second column, again shifting down one row. This process is repeated until we have seven columns, then the final step is to add a row of minus signs.

Plackett-Burman designs are very useful if, in the early stages of a series of experiments, we wish to assess the effects of a large number of independent variables. The sequential use of Plackett-Burman designs is illustrated in the case study described later in this paper.

Table I. Experimental Plackett-Burman Design for 15 Factors

Run No.	Factor Levels														
	1	2	3	4	5	6	7	8	9	10	11	12	13	14	15
1	+	-	-	-	+	-	-	+	+	-	+	-	+	+	+
2	+	+	-	-	-	+	-	-	+	+	-	+	-	+	+
3	+	+	+	-	-	-	+	-	-	+	+	-	+	-	+
4	+	+	+	+	-	-	-	+	-	-	+	+	-	+	-
5	-	+	+	+	+	-	-	-	+	-	-	+	+	-	+
6	+	-	+	+	+	+	-	-	-	+	-	-	+	+	-
7	-	+	-	+	+	+	+	-	-	-	+	-	-	+	+
8	+	-	+	-	+	+	+	+	-	-	-	+	-	-	+
9	+	+	-	+	-	+	+	+	+	-	-	-	+	-	-
10	-	+	+	-	+	-	+	+	+	+	-	-	-	+	-
11	-	-	+	+	-	+	-	+	+	+	+	-	-	-	+
12	+	-	-	+	+	-	+	-	+	+	+	+	-	-	-
13	-	+	-	-	+	+	-	+	-	+	+	+	+	-	-
14	-	-	+	-	-	+	+	-	+	-	+	+	+	+	-
15	-	-	-	+	-	-	+	+	-	+	-	+	+	+	+
16	-	-	-	-	-	-	-	-	-	-	-	-	-	-	-

5. Design for Fitting Response Surfaces:

Several different types of designs are useful for fitting a response surface as explained by several authors including [G.E.P. Box, W.G. Hunter, and J.S. Hunter (1978); R.L. Mason, R.F. Gunst, and J.L. Hess (1989)]. Complete and fractional factorial experiments in completely randomized designs are extremely useful for the purpose of exploring the factor space in order to identify the region where the optimum response is located. The two-level factorials are highly efficient but they are used only for fitting and cannot detect curvature. Best designs should give equal precision for fitted responses at points that are at equal distances from the center of the design (factor space), i.e., they are rotatable designs. For two factors, several rotatable designs are given in [R.L. Mason, R.F. Gunst, and J.L. Hess (1989)]. For more than two factors, several designs are used such as central composite designs as explained below.

Central Composite Designs: The central composite design is constructed by choosing a complete or fractional 2^k factorial layout, then adding 2^k axial, or star, points along the coordinates axes. Each pair of star points is mostly chosen to make the design rotatable. Several runs should be repeated at the central point of the design. A picture of central composite design in three factors is shown in Figure 1.

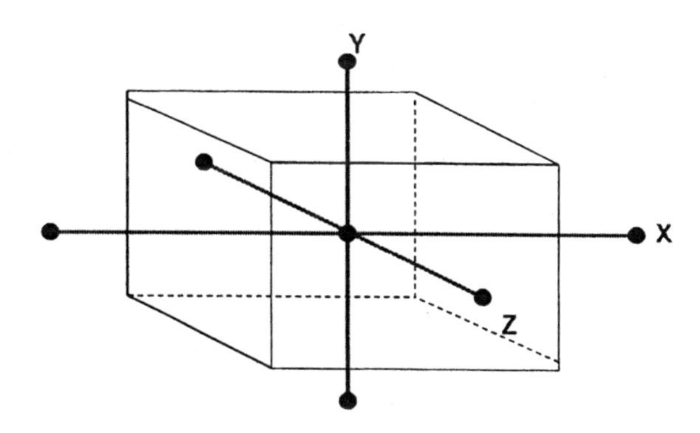

Figure 1. A Schematic Drawing Showing Central Composite Design in three Variables

The total number of test runs in a central composite design based on a complete 2^k factorial is $n = 2^k + 2k + m$. Where m is the number of repeated tests at the central point. An example of a central composite design is presented in the case study described below. The layout of the used design for three factors is shown in Table II.

It can be seen that the three variables (factors) will be tested at five levels each. The −1, 0, and +1 indicate the coded factor at its low level, midpoint (central point), and the high level respectively. In order to obtain the levels to be used corresponding to these coded levels use the following:

- Level at Central Point = (High level- Low level)/2
- Scaling factor = incremental step to be used = High level – Central Point Level
- Code at Low level = -1 = (Low level – Central Point Level)/ Scaling factor
- Code at High level = +1= (High Level- Central Point level)/ Scaling factor
- Code at Central Point= 0 =(Central Point level - Central Point level)/ Scaling factor
- Code at Star point = 1.680 = (Star Point level - Central Point level)/ Scaling factor

Table II. Rotatable Central Composite Design using Three Factors

Run No.*	Coded Factor Levels		
	Factor 1	Factor 2	Factor 3
1	-1	-1	-1
2	-1	-1	+1
3	-1	+1	-1
4	-1	+1	+1
5	+1	-1	-1
6	+1	-1	+1
7	+1	+1	-1
8	+1	+1	+1
9	-1.68	0	0
10	+1.68	0	0
11	0	-1.68	0
12	0	+1.68	0
13	0	0	-1.68
14	0	0	+1.68
15	0	0	0
16	0	0	0
17	0	0	0

*Nonrandomized

CASE STUDY

Aggregation/Dispersion Characteristics of Calcium Oxalate Monohydrate Crystals in the Presence of Protein and Urinary Ions

Introduction: The mechanisms for the formation of kidney stones are not well understood. One possible mechanism is the formation of aggregates of calcium oxalate crystals. However, it is not clearly known the effect of the urinary environment on the aggregation / dispersion characteristics of these crystals. This is attributed to the complex nature of this environment including a large number of variables that need to be studied simultaneously. In this study, screening and central composite statistical experimental designs are used to determine the effect of various factors on the aggregation and dispersion characteristics of previously grown calcium oxalate monohydrate (COM) crystals in different urinary environments.

Variables: Eight variables were examined in the screening stage of this study including calcium, oxalate, pyrophosphate, citrate, pH, protein concentration, order of addition of calcium, and the medium such as ultrapure water vs. artificial urine. The response variable termed as aggregation coefficient (AC) was

calculated from Optical density measurements obtained in slurries prepared under various experimental conditions (runs). Plackett-Burman design was used for screening the most significant variables. The layout of this design is previously presented in Table II. In that design, the following variables were used including:

Plackett-Burman Variables used in the Dispersion study:

i.	pH	ix	protein
ii.	citrate	x	dummy
iii.	dummy*	xi	solution
iv.	dummy	xii	dummy
v.	pyrophosphate	xiii	dummy
vi.	calcium	xiv	order of addition
vii.	oxalate	xv	dummy
viii.	dummy		

* dummy variables are used to calculate the variance of the experimental runs

Results and Discussion of the Screening Design: The response variable termed, aggregation coefficient (AC) was calculated from the optical density data obtained in COM slurries prepared under different conditions corresponding to experimental design runs as given in standard order in Table II. The values of AC are listed below.

Run #	Aggregation Coefficient *
1	42
2	56
3	38
4	42
5	38
6	45
7	41
8	54
9	68
10	49
11	45
12	26
13	16
14	45
15	25
16	55

* The higher the AC, the more the aggregation of crystals

Considering the aggregation coefficient as the response the variable; the effects of variables are calculated as follows:

$$Ei = \frac{\sum Response_{(+)} - \sum Response_{(-)}}{8}$$

Variance is estimated by evaluating effect of dummy variables and using:

$$\sigma^2 = V_{eff} = \frac{\sum E_d^2}{n}$$

Where:

E_d = effect of dummy variable
n = number of dummy variables = 7

Statistical significance of the effects is evaluated using student t-test as calculated using:

$$t = \frac{E_i}{\sqrt{\sigma^2}}$$

At degrees of freedom = 7 and confidence Level = 90%; $t_{7,0.10}$ = 1.415. Comparing t values to 1.415 led to acceptance of variables leading to higher t values and rejecting the variables of lower t values. In other words, variables accepted as significant within 90% confidence level. Such comparison is presented in Table III.

Table III. Calculated t Values for Different Variables and the Decision to Accept or Reject the Significance of Variables on the Bases of Comparing t-Value to Critical Value at 90% Confidence Level

Factors	Effect on Aggregation (t-value)♣	Significant Factor
pH	-0.52	Reject
citrate	-2.43♦	Accept
pyrophosphate	+0.46	Reject
calcium	+1.28	Reject
oxalate	+2.54♦	Accept
protein concentration	-1.83♦	Accept
type of solution	-1.50	Reject
order of addition of calcium	-1.66	Reject

♦ Statistically significant (90% Confidence Level); ♣ (+ ve) sign indicates more aggregation; ♣♣ (- ve) sign indicates a more stable (dispersed) suspension

Conclusions from the Plackett-Burman Design:

- Increasing concentrations of either protein or citrate leads to dispersion of COM.
- Increasing concentration of oxalate leads to aggregation of COM.
- Further research is needed to determine the interactions among the 3 significant factors and to obtain surface response for these variables.

Central Composite Design: A central composite design was used to further investigate the above three most significant factors determined from the screening design. Five levels were used which encompass the extreme concentrations found in urine. The layout of this design is presented earlier in Table 3. In this case also, the aggregation coefficient was determined in every run. The data obtained in the 17 experiments were analyzed using statistical soft-ware program, JMP, by SAS. The results included:

- Main effects
- Interaction effects
- A second order mathematical model
- Profile plots of effect of individual variable
- Interaction plots of combinations of variables
- Contour plots showing surface response as a function of two variables at fixed.

From the central composite design, the data confirms the screening results that protein acts as a dispersant and oxalate promotes aggregation. Interesting interactions between protein and oxalate, along with protein and citrate are observed. These interactions were synergistic or antagonistic, depending on the concentrations of these species. For the oxalate concentration alone, aggregation increases with increases in oxalate concentration. The protein exhibits the opposite effect with the probability of aggregation (aggregation coefficient) is decreased as protein concentration is increased. The change in citrate concentration alone has little affect on aggregation. However, the citrate and oxalate concentrations have a strong interaction. This characteristic is also exhibited between the citrate and protein concentrations displayed in Figure 2a. Protein helps to reduce the aggregating effect of oxalate as shown in Figure 2b. As the protein concentration is increased, aggregation is decreased at all values of oxalate concentrations.

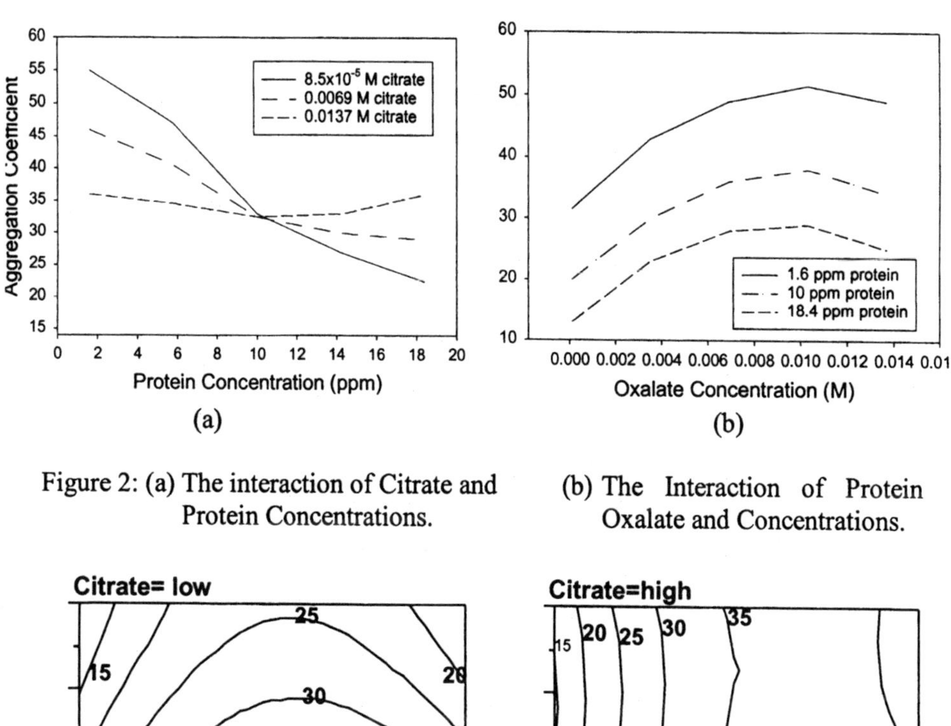

Figure 2: (a) The interaction of Citrate and Protein Concentrations. (b) The Interaction of Protein Oxalate and Concentrations.

Figure 3. A Contour Plot Illustrating How the Aggregation Probability Changes with Respect to Oxalate and Protein Concentration at Constant Citrate Levels)

Surface responses for interactions of oxalate and protein at constant citrate concentrations are illustrated in Figure 3. The contour plots indicate that at low citrate concentration, protein plays an important role in reducing aggregation at all oxalate levels. However, at higher citrate, protein plays a minor role as seen by

the contours, which are parallel to the protein axis. In this case, lower oxalate is desirable for less aggregation.

In summary, statistical designs proved to be useful tools in screening several variables in a complex environment. In addition, interactions between different variables have been identified and quantified. Surface responses illustrate regions in the factor space that may be utilized to control the response variable.

SUMMARY AND CONCLUDING REMARKS

- One-variable-at-a-time research strategy is inefficient and self-defeating, especially if optimum properties/conditions are desired.
- Statistics should play a role in every facet of data collection and analysis, from initial problem formulation to the drawing of final conclusions.
- Statistical experimental design is an effective tool that ensures precise information about the responses of interest, and guarantees economic use of resources. In addition, statistical data analysis techniques aid in clearly and concisely summarizing salient features of experimental data.
- To attain the most benefits from statistical designs, we should choose the design that fulfills the following:
    a. It does not require too many levels of the controlled variables;
    b. It has enough levels so as to permit the estimation of the desired effects or coefficients (e. g., at least two levels if we are planning to fit a straight line);
    c. Enough points should be included to provide degrees of freedom to measure lack of fit;
    d. It must have provisions for replication, or partial replication, so that we may estimate the variance, σ^2 clear of the chosen model;
    e. It must be easy to analyze;
    f. It could be easy to construct from small blocks of runs.
- In this paper, various statistical designs including screening, full factorial and optimization designs are explained. Analysis of variance ANOVA is introduced and utilized together with practical examples to illustrate the effectiveness of these methodologies in better designs of ceramics and other materials.

ACKNOWLEDGEMENT

The patience and dedication of Mrs. Lenny Kennedy during typing this manuscript is gratefully acknowledged. The Engineering Research Center for Particle Science and Technology at University of Florida, the National Science Foundation grant# EEC-94-02989, and the Industrial Partners of the ERC are acknowledged for the partial financial support. The partial finacial support by

NSF grant # NSF Grant # INT-9810982 and Grant # INT-9810983 is acknowledged.

BIBLIOGRAPHY

V.L. Anderson and R.A. McLean, "Design of Experiments: A Realistic Approach", Marcel Dekker, Inc., New York, 1974.

C. Bennett and N. Franklin, "Statistical Analysis in Chemistry and the Chemical Industry", John Wiley and Sons, Inc., pp. 6-16, 76-78 (1954).

R.M. Bethea, B.S. Duran, and T.L. Boullion, "Statistical Methods for Engineers and Scientists", Marcel Dekker, Inc., New York, 1985.

G.E.P. Box, and J.S. Hunter, "The 2^{k-p} Fractional Factorial Designs", *Technometrics*, 3:3, August 1961.

G.E.P. Box, W.G. Hunter, and J.S. Hunter, "Statistics for Experiments", John Wiley & Sons, Inc., USA, 1978.

R. Caulcutt, " Statistics in Research and Development," 2 nd. Edn. Chapman & Hall, New York, 1991

Cochran and Cox, "Experimental Designs", 3rd Ed., John Wiley & Sons, Inc., USA, 1962.

D.R. Cox, "Planning of Experiments, John Wiley and Sons, Inc., USA, 1958.

C. Daniel, "Applications of Statistics to Industrial Experimentation", John Wiley & Sons, Inc., USA, 1976.

O.L. Davies, "Design and Analysis of Industrial Experiments, John Wiley & Sons, Inc., USA, 1971.

W.J. Diamond, "Practical Experimental Designs", Wadsworth, Inc, Belmont, CA, 1981.

W.J. Dixon and F. J. Massey, "Introduction to Statistical Analysis", 2nd ed., McGraw-Hill, pp. 4-27, 1957.

A. Garcia-Diaz and D. Phillips, "Principles of Experimental Design and Analysis", *A Computer Program Used with Text: Interactive Microcomputer Software for the Statistical Analysis of Experiments*, Chapman & Hall, First Edition, 1995.

G.J. Hahn, "Process Improvement Using Evolutionary Operation", *Chemtech,* Vol. 6, pp. 204-206, 1976.

A. Hald, "Statistical Theory with Engineering Applications", John Wiley & Sons, Inc., pp. 44-77,1960.

J.S. Hunter, ' Design of Experiments," Short Course Modules, published by University of Kentucky, College of Engineering 1977.

R.L. Mason, R.F. Gunst, and J.L. Hess, "Statistical Design and Analysis of Experiments with Applications to Engineering and Science", John Wiley & Sons, Inc., USA, 1989.

W. Mendelhall and T. Sincich, "Statistics for the Engineering and Computer Sciences, Dellen Publishing Co., Santa Clara, CA, 1984.

W.Mendenhall, "Introduction to Probability and Statistics", 2nd ed., Wadsworth, pp. 21-45, 1967.

D.C. Montgomery and E.C. Peck, "Introduction to Linear Regression Analysis", John Wiley & Sons, Inc., USA, 1982.

Murphy, "Design and Analysis of Industrial Experiments", Chemical Engineering, June 6, 1977.

Plackett and Burman, "The Design of Optimum Multifactorial Experiments", *Biometrika*, Vol. 33, 1946.

QUESTIONS

1. You are required to conduct an experiment to study heat transfer during manufacturing molded ceramic objects. Three temperatures are to be varied during the experiment: 1100, 1200, and 1300° C. the second variable is cooling time which will vary as 15 and 20 sec. In addition two types of ceramics, type **A** and type **B,** are to be studied. Construct a complete randomized design for this experiment.

2. Design an experiment involving three variables, each at two levels, in which no more than 10 runs can be conducted on each of several test days. Remember that all main effects and interactions are to be estimated. Only 10 days are available to conduct your testing.

3. Design an experiment in which the main effects of the following variables are to be determined: temperature, humidity, speed of mixing, mixing time, concentration of a chemical, solution volume, size of the container, presence and absence of a catalyst, and two laboratory technicians. The experiment should not involve more than 20 runs.

4. Assume response values for the design in question 3 and analyze the data to estimate the main effects, the variance, the t values for each effect, and list the significant variables using 90% significance level.

5. A study of instrument differences was conducted in which ten samples were analyzed using two machines. Are there any significant differences detectable from the following analyses of the ten samples? Mention any assumptions needed to perform your analyses

Sample	Instrument reading	
	Instrument I	Instrument II
1	2.27	2.33
2	2.21	2.69
3	2.45	2.82
4	2.19	2.45
5	2.38	2.48
6	2.36	2.77
7	2.41	2.52
8	2.52	2.66
9	2.43	2.36
10	2.45	2.67

6. Perform ANOVA for the data in question 5.

Use of Computers in Design of Ceramic Bodies and Processes

5

D.R. Dinger

THE USE OF COMPUTERS IN THE DESIGN OF CERAMIC BODIES AND PROCESSES

Dennis R. Dinger
Dinger Ceramic Consulting Services
103 Augusta Rd.
Clemson, SC 29631

USING COMPUTERS TO DESIGN BODIES AND PROCESSES

Introduction

In this 21st Century, computers are expected to be capable of assisting practically every possible task. The capabilities of computers to help design ceramic bodies, or processes to make those bodies, however, are still quite limited. Body and process development remain mostly arts, not easily lending themselves to computerization. One can obviously use CAD software to draw process flow diagrams and to design ceramic wares, but how does one actually use a computer to design a ceramic body?

This paper will describe one approach to ceramic body design that utilizes the capabilities of desktop computers and common software packages. As much as possible, computer-controlled instruments are utilized to measure powder and slip properties. After completion of the measurements, the data are entered directly into spreadsheet databases. Derived parameters can then be calculated to help design the body to achieve target body properties as well as to achieve controllable processes.

Desktop Computers – Some History

Twenty years ago, computers were only just beginning to be used in ceramic labs. At that time, computer usage was limited mostly to research labs, where the computers were added to instruments in supervisory capacities, or as data loggers. There were a wide variety of expensive (by today's standards) computers available which did not easily communicate with one another. Few students and engineers of the day knew much about, nor liked, those computers. Then if you wanted an automated instrument, you had to do the automating, interfacing, and programming yourself. For example, the author spent most of 1979 and 1980 automating three analytical instruments for a research project. As a result of the costs and attitudes of the day, the use of computers in ceramic production facilities was then rather limited.

Now, not only are desktop computers incredibly powerful, fast, and small, but most analytical instruments are shipped with such computers as their controllers. The great advancements made in the last 20 years also allow most instruments today to communicate directly with one another, or through software packages such as the spreadsheet programs that are available with most new computers.

Unlike 20 years ago, today's engineers don't have the excuses that computers aren't available, or that they aren't familiar with their use. Almost everyone, especially in technical fields, knows how to use computers.

But the new questions deal with how these computers can be used in the design of ceramic bodies and processes. Few software packages are available in this area because the ceramics market is small, such specialty programs would be quite esoteric, and the understanding required to design such programs would need to be extensive. Even with such limitations, today's computers should be used to both design bodies, and to control them in the plant.

Predictive Process Control (PPC)

Definition: In 1994, the author and James E. Funk published a textbook on PPC[1], which encourages the use of computers wherever possible throughout the ceramic production process. The modifier *predictive* before *process control* requires that the precise characteristics of all pertinent properties of all ingredient raw materials be measured prior to assembly of production bodies so the characteristics of the total body formulation can be *predicted* (calculated) *prior to* batching.

To accomplish these tasks in the timely manner necessary to control production bodies requires extensive use of computers, software, and automated analytical instruments.

With PPC implemented in a production plant, one should never be surprised by unknown variations in the properties of the final production bodies because all of those same properties are being monitored daily on each raw material.

The PPC Body Formulation: Table I shows a comparison of the traditional body composition versus a body formulation under PPC control.

The traditional body composition descriptors need no explanation. The PPC properties include methylene blue index plus polymer package (MBI+PP), the particle crowding index (PCI) which is the number of particles per true cm^3 of material, interparticle spacing (IPS), the particle size distribution modulus (n), a viscosity from a gelation test ($\mu_{60,10}$), and the total percentage of alkalis in the body.

In traditional ceramics, the clays are used to produce plastic body properties. In electronic, engineering, and other non-traditional ceramics, a polymer package must be used to produce plastic properties in the absence of clays. These two parameters

Table I. Comparison of traditional versus PPC definitions of a body composition

Traditional		PPC	
Clay A	15%	MBI+PP	3.0-3.1
Clay B	15	PCI	0.55-0.58 X 10^{15}
Kaolin	20	IPS	125-128 nm
Quartz	20	n	0.19-0.21
Feldspar	30	$\mu_{60,10}$	1000-1050 mPa·s
	100%	Alkalis	4.0-4.1%

are grouped together as a one property to help characterize a body's plastic forming properties.

The other properties listed define particle size distribution, solids content, viscosity (inherent in which are chemical additive concentrations), and the alkali content of the body. This is a simplified example, of course. A production PPC formulation will include a more extensive list of properties to define the body thoroughly and precisely.

The Goal: The goal of the PPC formulation is to define the body so precisely by its properties that meeting those specifications on a day-to-day basis will produce consistent production bodies. This consistency not only refers to compositional consistency, but performance consistency as well.

Do production problems come and go randomly? ... with the weather? ... or with the phases of the moon? Such problems always seem to exist. They are indications that one or more fundamental properties of the body (and its ingredient raw materials) are not being measured and/or controlled. Such problems should send process engineers searching for the elusive properties so they can be both measured and controlled.

An example comes to mind of a pure oxide component (the main ingredient in a production body) which was defined by quite a few tight specifications, all of which were routinely met by the supplier. These powders usually flowed quite nicely. Nevertheless, every once in a while, a shipment came through that did not flow well at all; mixing of the body was then difficult; and suspension viscosities were high. Testing showed that the angle of repose of the normal powder and that of the occasionally aberrant shipment were quite different, even though all shipments met all listed specifications. No one had ever thought to test this before, so it was not included in the original specifications.

As a result, another specification (and a new test) was added to the list for this material. After pointing out this problem to the supplier, changes were made to their process that eliminated the problem.

Using Computers With PPC: PPC utilizes the speed of computers and computerized analytical instruments to carefully characterize all appropriate properties prior to batch assembly. We could not have proposed PPC or encouraged industry to use PPC in the 1970s or early 1980s because it would have been impossible to perform the required characterizations in the timely manner necessary to implement PPC controls. All of this is quite easily accomplished now at the beginning of the 21st century.

For example, some forms of particle size analysis commonly used 20 years ago took several days to complete. Today, with the available computerized sedimentation and laser scattering instruments, particle size analyses can be performed in a matter of minutes. Similar analysis times apply to specific surface area measurements, X-ray fluorescence, and most other measurements performed by computerized analytical instruments. In many of these cases, the most time-consuming parts of an analysis are the sampling and sample preparation steps.

The implementation of PPC in a plant, therefore, requires extensive use of computers. It requires timely analyses of all candidate batch ingredients using, in most cases, computerized analytical instruments, and then it uses the results of all of those tests to precisely calculate and control the properties of the production body.

A Property Database

The detailed analyses used to implement PPC are necessary requirements when one wants to design a new production body. One must first create a database containing the histories of all properties for each raw material to be used.

Most of today's computer-controlled analytical instruments either work in the Microsoft Windows* operating system environment or they allow data to be easily exported in a format compatible with Microsoft Excel* or other popular spreadsheet formats. These packages allow all pertinent properties to be easily entered and stored in one large data base.

All data for each raw material measured over a period of several weeks should be stored in this database. Even better, if the raw materials are routinely used in other production bodies, historical data from the past several months may be available.

---

*Microsoft Corporation, Seattle, WA

Property data should include several values to define each powder's particle size distribution. For example, when a 4th root of 2 size classification is used, the upper particle size limit (D_L), percentages in several class sizes such as the 21.9-18.4 μm, 10.9-9.2 μm, 5.5-4.6 μm, and 1.15-0.97 μm ranges, as well as the cumulative percent finer than (CPFT) 0.97 μm, should all be included in the database. If submicron particle size classes are routinely measured, data for those size classes should be stored in the database as well.

Regarding particle size analyses, many points of data should be included in the database to characterize the details of the complete particle size distributions. A single particle size characteristic, such as the median particle size of a distribution, is simply insufficient to characterize a powder's particle size distribution.

The database should also contain specific surface area (SSA), elemental percentages (from X-ray fluorescence analyses or other such techniques), mineralogical phase percentages (from X-ray diffraction analyses), suspension characteristics (ion contents in carrier fluids or in feed water), quantitative summaries from DTA, TGA, and dilatometry, etc. The more properties included, the better, more complete, and more useful will be this database.

One should take care to always tabulate and compare apples with apples – never apples with oranges: measured values with measured values, calculated with calculated, a particular instrument's results with other results from the same instrument, etc. For example, specific surface areas can be calculated or measured. The two forms should not be mixed. If both types are used, they should be tabulated separately in the database. Another example is particle size analysis. Many different types of analyzers are available, but results should be stored, and compared, by analyzer type.

The particular properties stored in the database will vary with the type of ceramic to be developed. Percentages of impurity ions, or dopant ions, of particular importance to the process and product should definitely be included.

Raw materials mined by different suppliers will usually contain different, but characteristic, impurity ions. Some such impurities will be important to the body and the process. Some will not. All such values, however, should be included in the database.

Chemically prepared raw materials will also usually contain characteristic impurity ions remaining from their particular chemical preparation processes. Frequently, the number of suppliers of a particular raw material is limited. If the powders' labels are misplaced, each supplier's powders can usually be distinguished by analyzing for their impurities.

Body Design

Average Property Values: With a complete property data base available, the next consideration is that the body formulation must be achievable using the average values of the properties of all candidate materials.

Whether or not PPC is being used to control a process, it is particularly important that the average properties of each of the materials be used to design and define the body composition. With a complete database in hand, it is quite easy to calculate the means and standard deviations of each property of each raw material. If this is ignored, it may be difficult to ever achieve a reproducible set of body properties.

If the mean values of each property for each raw material, when combined in an average batch formulation, cannot produce the desired batch properties, new raw materials should be acquired, or some of the properties of some of the raw materials should be changed.

Most material properties can be expected to vary from batch to batch according to a normal distribution. Means and standard deviations (σ) of properties are routinely calculated and tabulated. But let's remember and consider the definitions applicable to normal distributions. In a normal distribution, 68% of the values lie within $\pm 1\sigma$ of the mean; 95.5% within $\pm 2\sigma$ of the mean; and 99.7% within $\pm 3\sigma$ of the mean.

Described from the opposite point of view, this means that only 4.5% of all values for a normally distributed property will lie more than $\pm 2\sigma$ from the mean, and only 0.3% of values will lie beyond $\pm 3\sigma$ from the mean.

With this in mind, one must never design a body where the raw material properties are required to be greater than $\pm 2\sigma$ from the mean. If the average values of each raw material property cannot be combined to produce the desired body properties, the design should be discarded and the design process should continue until a better solution is found.

One cannot simply look at a control chart for a particular material property, recognize that the desired value is within $\pm 3\sigma$ of the mean, and then continue as if everything is fine and normal. If the required property value is more than $\pm 2\sigma$ from its mean, things are certainly not normal.

For example, if a material has an average volume surface area (VSA) of 10 m^2/cm^3, with a standard deviation of 1 m^2/cm^3 (this is an extreme case, to be sure, but it will better illustrate the point), one should not routinely expect to find surface areas of this material ≤ 8 m^2/cm^3. A value less than 8 m^2/cm^3 would be more than two standard deviations below the mean which, by definition, includes only 2.25% of samples of this material.

Consider the following scenario using this material: A production body is defined by several raw materials with varying properties. The body's VSA is specified to be ≤ 10 m^2/cm^3. On this day, the first 90% of the body composition contributes 9.2 m^2 of the volume surface area, and the final 10% addition to the composition is the material just described. To meet the VSA specification, it must supply the remaining 0.8 m^2 of surface area. Under such conditions, the total body will only rarely meet the 10 m^2/cm^3 specification because the sample of this final material that is added to the body formulation must have a surface area <8 m^2/cm^3, that is, it must be more than 2σ below its mean. Is it possible for this material to have such a surface area? Yes! Will it occur frequently? No!

If the VSA variations of this one component follow a normal distribution and this material is tested on a daily basis, only about two samples out of every hundred will have surface areas in this range (<8 m^2/cm^3). This means that the desired VSA for this powder may only be available on two days of any given quarter (and probably not on the two days they are actually needed).

If a traditional body composition is used and 10% of this material is added to each batch, this body will not only be *out of spec* most of the time, but it will be *out of spec* for unknown reasons. When daily property measurements of all candidate raw materials are not being measured, which material and which property are to blame for any given problem?

If the VSA requirements used in this example had to be met, a different powder source with an average value of 8 m^2/cm^3 should be obtained. Alternatively, the specifications of the formulation could be redesigned to accommodate this particular raw material's surface areas.

Number of Candidate Ingredients: To use PPC, one should have at least one candidate raw material for each of the properties being controlled. Actually, the requirement that the sum of all ingredients equals 100% is also a control specification, so one more raw material is required than the number of control properties.

If properties to be controlled include particle size distribution (3 parameters), specific surface area (1 parameter), dopant level (1 parameter), and any one other property (1 parameter), which is a total of 6 parameters, at least seven raw materials should be available for use. The more properties to be controlled, the more candidate raw materials should be available.

It's possible that six properties can be controlled using only three raw materials, but luck plays a major role in cases like this. Over the long run, routine property variations will frequently prevent the desired body properties from being achieved when fewer candidate materials are available than properties to be controlled.

Candidate raw materials should include variations of each of the ingredients to be used in the body. If the body is to be a barium titanate body, candidate barium titanate powders should include different particle size distributions, different levels of dopant, and different surface areas, for example. At least one of the powders should be close to the average value of each of the properties to be controlled. But some powders should be available that have higher and lower values of each of the properties. A minimum of three candidate powders for each property (one high, one approximately average, and one low) is a good starting point, and even this number does not guarantee that all property requirements can be met in each batch.

In one company's batching process, each batch of their body was produced from one bulk bag of raw material. There is no *control* over particle size distribution, surface area, or any other powder property in such an operation. They simply had to live with the properties inherent in the powder in each bulk bag. They attempted to force the controls on their raw material suppliers. But to control all of the fine production properties affected by the particle physics of the powders, they would have needed specifications that were too tight for their suppliers to meet (or if they could meet them, the powders might then have prohibitively high prices.)

Chemical additives, solids content, etc., could be controlled in that process, but powder properties could not. One powder simply does not allow any such control (no matter how tight the specifications imposed on the raw materials supplier.)

At another company, two bulk bags of powder defined each batch. They, at least, had some control available to them because each bag was labeled with its measured median particle size. But they were not using it. Their procedure was simply to randomly select two bags for each batch. As a result, their particle size distributions and other powder properties varied randomly, and their processing properties varied randomly as well.

How does one achieve the necessary number of candidate raw materials when a minimum number of ingredient materials is now being used? When raw materials are shipped in bulk bags, one can characterize the contents of each bulk bag and carefully select the bags to be used in each batch based on those analyses. Preferably, each batch will be large enough to require the use of several bulk bags. Alternatively, one could slurry the contents of each bag, characterize the slurries, and use those results to assemble each batch.

A better method would be to divide each bulk bag into smaller containers. Then, the contents of each smaller container could be characterized and used as a candidate raw material.

Dividing large samples into smaller ones is a common procedure in industry, but it is not routinely accompanied by the careful characterization of the smaller containers' contents. There is no need to even be careful that each small container be representative of the larger sample, if each container will be analyzed individually. The variations produced may be useful.

However, when each small container will not be individually characterized, great care should be taken when dividing the larger quantity so all smaller samples are fully representative of their larger source. When smaller samples are not representative of the larger sample and careful characterization is not performed on each, major process variations and yield swings will occur. I have seen this particular problem in industry. Reproducibility from batch to batch was nowhere to be found.

One could order materials with different properties from each supplier. Some could be coarser, some finer, some could have higher and lower surface areas, and some could have greater and fewer impurities than the average powders. Each of these powders can then be characterized and made available as ingredients to the final batch.

When powders with widely variable properties are not available from suppliers, powder properties can always be altered in-house. This is especially the case for particle size distribution and specific surface area which can be altered by milling the powders for different lengths of time. These two fundamental properties greatly influence rheological properties, body forming behaviors, particle packing, and shrinkages, and should be high on any list of properties that must be variable among the candidate raw materials.

Computers and Body Design

When a complete data base containing the details of several raw material powders is available, one can use the optimization functions available within spreadsheet programs to calculate batch formulations.

For example, the Solver functions within Excel provide the optimization routines necessary to perform the batch formulation calculations. Optimization routines are common methodologies used to solve large sets of simultaneous equations with multiple unknowns. Within Excel, one can define the exact values of each property desired in the final body, and the Solver functions will calculate the percentages of each of the candidate raw materials that allow those property values to be achieved. Using the optimization functions, each desired body property can be constrained to certain value ranges (for example, the SSA value can be constrained to values >9.8 and <10.2 m^2/g).

If one uses too many constraints to define a body's properties and there are too few candidate raw materials available, the optimization routine may simply respond with "No Solution Possible." Under such circumstances, one must then alter the constraints or change the candidate raw material properties until a solution is possible.

The point to be emphasized is this: when the computer indicates a solution is *not possible*, it will be impossible to use that day's candidate raw materials to produce a body that meets the desired specifications. Knowing this, the engineer in charge will have advance warning that problems are coming and he/she will be able to try different solutions to head off those problems. All of this, of course, happens in the

lab, on a computer, well in advance of the problem(s) actually appearing on the plant floor.

A Production Example

As an example of this, consider what happened in one South American plant that is running under PPC control. The SSA values of new shipments of two different raw materials went high at the same time. As a result, the engineers were unable to achieve their desired batch formulation with the available raw materials. Since this was happening in advance, on the engineer's computer, rather than in the plant, they were able to modify the body formulation in the computer to minimize expected production losses.

They attempted to use other available candidate raw materials (from other storage silos and patios), and to vary the particle size distribution, SSA, and other properties (which they could do by milling and other alternative processing steps) before they finally needed to decide on a course of action. This needs to be emphasized: **All of this took place on the computer** before the raw materials were mixed to form the production bodies for use in the plant.

To do this, of course, the engineers needed to have excellent understandings of the cause-and-effect relationships between the many body properties they were controlling and the resulting forming behaviors they would produce. This is an enormously important requirement. A full discussion of this subject, however, is well beyond the scope of this paper. The PPC textbook[1] is an excellent starting point to learn more about this subject.

As a result of their computer trials, their production yields decreased, but not catastrophically. Had they not been using PPC, their production yields would have dropped off precipitously for some unknown reason, and they would then (when the bad body was delivered to the production floor) have been scrambling to figure out why all the losses and what to do about them. Because they were routinely analyzing raw material properties and calculating body properties in advance, they knew that a problem was coming; they knew what was going to cause the problem; and they were able to plot a best course through the potential mine field that lay ahead.

Other Software to Help Design Bodies

Particle Packing Routines: More than 10 years ago, the author developed particle packing routines to calculate particle size distributions for best packing. These routines[1,2], several[3-7] of which were described in a series of papers in the *Ceramic Bulletin*, have been widely distributed and used since then.

The author has recently been putting the finishing touches on a new set of particle size distribution functions that operate as add-ins to Excel. These functions perform the same calculations as the earlier routines, but they have been broken down into more fundamental steps to allow engineers to design, set up, and perform whatever calculations they deem necessary for their particular problems.

These functions calculate, for example, 4th root of 2 and square root of 2 particle size classifications, surface area and cumulative surface areas, number and cumulative number of particles per true cm^3 of powder, as well as estimated minimum porosities and interparticle spacings for powders with known particle size distributions, surface areas, and solids contents. Although not configured in the same way as the earlier programs, these functions can be used to perform all of the same functions of the earlier programs, but in the Windows and Excel environments.

Glaze Calculations: A spreadsheet to calculate glaze formulations and properties was developed several years ago[8-9]. This routine calculates thermal expansions, melting temperatures, and surface tensions from the oxide formulations of glazes and frits.

This spreadsheet stands as an example of the types of routines that can be developed by engineers to address problems in their particular materials systems. Specifically, this type of routine requires that the properties addressed be additive. This is the identical requirement imposed by PPC when one is selecting properties to be characterized and used for control of a process.

Properties such as specific surface areas, particle size distributions, thermal expansions, and surface tensions, can all be calculated using a simple additive formula:

$$P = a_1x_1 + a_2x_2 + a_3x_3 + \ldots + a_ix_i + \ldots + a_nx_n \quad (1)$$

where P = property value
a_i = property coefficient
x_i = fraction of the ith material in the composition

For the glaze calculations, the property coefficient values are available in the literature[8-11].

Other Additive Properties: Examples of other additive properties include particle size distributions (both histogram and cumulative forms), surface areas, the measurements from differential thermal analysis (DTA), thermogravimetric analysis (TGA), thermal dilatometry, X-ray diffraction, X-ray fluorescence, and quite a few other simple and routinely measured parameters.

For particle size distribution calculations, each property value in Equation 1 simply corresponds to an individual particle size class — and there will be *n* property values (n size classes) needed to characterize the whole particle size distribution. In the case of particle size distributions, each property coefficient will be 1.0.

For surface area calculations, each property value again corresponds to data for an individual particle size class. The property coefficients are the surface areas of a unit volume, or a unit mass, of powders in the particular particle size class (which can be calculated). Optionally, surface area calculations can then be taken a step further by including a shape assumption in the surface area calculation for each particle size class.

Using Equation 1, surface areas can be calculated for each raw material from its particle size distribution, and specific surface areas and all of the other applicable properties can be calculated for each batch from the raw material properties.

These are examples of the types of calculations that can be made for many of the properties that are routinely used in the ceramic industries. If a property is an important control parameter in a particular body formulation, then spreadsheet calculations like these can and should be developed for those properties.

The speed, ease of use, and availability of today's computers suggests that such calculations are not only possible, but that they should be performed routinely in advance of plant assembly processes. There is no reason, except for more complex properties or for unusual circumstances, that any engineer should be surprised by the results of a batching operation.

Firing Curve Design: Jim Funk, published a paper in 1982 describing how to design the optimum firing curve for porcelains[12]. The procedure, which uses reversible and irreversible dilatometry, is applicable to the design of firing curves for any ceramic body.

The author recently developed another spreadsheet program which implements the calculations described in Funk's paper. We tested this program by designing the firing curve for a porcelain body. The production body tested was routinely fired in a tunnel kiln on a 22 hour cycle. Our test design was an 11 hour cycle, implemented in a shuttle kiln in the same plant. Fewer losses were produced in the 11 hour firing during this test than routinely occurred in the 22 hour cycle.

Why? Upon further comparison of the production and test firing curves, it became apparent that even though the overall firing time was significantly reduced, the new firing curve ramped the temperature more slowly through certain trouble temperature ranges than did the production firing curve in the tunnel kiln. The new curve was heating and cooling quickly in the temperature ranges where it was possible to do so without deleterious effects, and it was heating and cooling slowly through the known trouble areas.

This firing curve design procedure can be applied, as mentioned, to any ceramic body. The requirements for use of this procedure are the reversible and irreversible dilatometer curves for the body. The requirements for use of this program are the two dilatometer curves and knowledge of the potential trouble spots on the heating and cooling curves for the body being designed.

Computers and Process Design

General Applications: Not much has been said up to this point concerning the use of computers to help design processes. Obviously, computer programs are available to help lay out plants, but such programs are not applicable to the subject of this paper.

Personally, I don't know of any programs available today to help one design a process. I think such programs can and will be developed as more engineers add pieces of knowledge to the giant processing puzzle. The greatest amount of help available today for those who want to design processes can be found in the types of programs already mentioned.

Comminution: Some software has already been developed for milling operations. When fully developed, such programs should define the changes that occur in individual unit operations. For example, it is possible to predict changes that occur when a powder is fed into a ball mill. The use of such programs will require a full description of the particulars of the mill (its capacities and operational parameters), as well as the full particle size distribution and other pertinent properties of the feed powders. Such programs will predict (calculate) product particle size distributions expected from that mill for that powder. These programs will be especially useful for the development of new bodies.

Mixing: What has just been said for comminution, however, does not apply to mixing. It is relatively easy to write computer programs that assume homogeneously distributed particles (i.e., totally random arrangements). If a good mixer is available that can achieve homogeneity in ceramic bodies, no new computer programs are required.

The problem is that many mixers don't achieve true homogeneity. Chemically prepared materials may be shipped as agglomerates. Many mixers cannot deagglomerate powders. In such cases, one may think that mixers are homogeneously and randomly distributing individual powders throughout the bodies — when they're not.

Programs that assume inhomogeneity of mixing (which is probably more suitable to most process mixers) are more difficult to develop and they will be much more difficult to use. How exactly does one define an inhomogeously distributed powder? That's a hard question. Which particular type of inhomogeneity is characteristic of any particular ceramic production body? Such programs and questions are not beyond today's level of technology, but one must question whether anyone is developing such programs?

Most computer models assume randomness and homogeneity. Inhomogeneities that model real-world situations are much more difficult to define and simulate. Such programs may be 20-50 years away.

Another problem with computer simulations of mixing is that many mixers do not provide the means to maintain the level of mixedness upon completion of the operation. Use of a high-shear mixer, followed by storage in an inadequately agitated tank will allow settling and segregation of the particles by size, and differences in solids content at each level in the tank. Dry mixing, followed by vibration of the particles will also cause segregation of the particles by size — coarse to the top, fines to the bottom. But these are not computer simulation problems. They are process problems that should be addressed in the plant.

Utilization of Programs to Help Design Processes: We should all be on the lookout for programs that model individual pieces of plant equipment, and when we find them, we should acquire and begin to use them. Some of us should be developing such programs. Experience suggests that as greater numbers of raw materials are required to more closely control process batches, companies will simply subject the raw materials they have on hand to varied preparation techniques to achieve the variations required. In such cases, the computer program that can predict the properties resulting from the various preparation techniques will be quite useful to define the specific natures of the preparation techniques.

Summary

The author has been a computer person from the day he was first allowed to use and program computers 34 years ago. Since that time, computers have become smaller, faster, and much, much, much more capable of large-scale calculations in short periods of time. As an educator, the author has seen one relatively consistent characteristic of engineering students: they won't use computers unless all else fails. In the 1970s, computers were relatively new, and most engineers knew little about them for that reason. In the 1980s, computers were no longer new, but programming languages were not particularly user-friendly and programming was rather difficult to perform successfully. In the 1990s, computers were mature, software was simple to use, but capabilities were so great that many were scared away by the sheer magnitude of the capabilities. In this 21st century, let's change this and begin to tap the enormous capabilities available from the computers that reside on most engineers desks and throughout plants.

In my experience, one does not need to be an expert in higher mathematics, calculus, or differential equations to achieve great results on a computer. To the contrary, one has to be able to break down problems into extremely fundamental steps. When this is accomplished, building programs from those simple steps is relatively easy to do.

Today's computers are incredibly fast, performing enormous numbers of calculations in fractions of a second. We should utilize these capabilities to help us predict the outcomes of body assembly and processing methods *prior to* implementation in the plant.

We should utilize the speed and capabilities of today's analytical instruments to the same end. Surface area measurements take but 30 minutes to perform, but several hours for sample preparation. Even several hours to obtain results is sufficiently fast to use those results for body design and process control. The same is true for many other measurements. We need to fully utilize the capabilities from such instruments — and there are many more instruments available than the few mentioned in this paper.

With the data from such analyses in hand, one can then decide whether or not to proceed with batch production. "Yes, I like these results. Assemble the batch!" or "No, these results are unacceptable. Let's try some other variations before we assemble the batch in the plant."

Let's take advantage of these tremendous capabilities and use computers to their fullest extent in the design of bodies and their processes, and in the control of those processes.

ILLUSTRATIVE PROBLEM

Definition of the Problem

Consider an example of a body formulation problem for a whiteware body. A company called WhiteWares, Inc., wants to make a new porcelain body with four raw materials. They believe as a starting point that a composition of 25% ball clay, 25% kaolin, 25% feldspar, and 25% quartz would be good.

When pressed for more details, they disclose that they want this body to have as many colloidal particles as possible, at least 40% plastics in the body, a SSA of 10 m^2/g, an MBI of 4, and an alkali content to be 4.0-4.1%.

How should Whitewares, Inc., proceed?

Database and Spreadsheet Calculation

Although the database should include many more powder properties than these few just mentioned, Table II shows a database applicable to this problem. Data included are the raw materials properties for ingredients available for a single production batch. The table is in the exact form as it appears in the spreadsheet.

The colloidal contents of the materials, defined as the cumulative percentage less than 1 μm, are shown in Row 2. The alkali percentages, defined as the sum of the Na_2O and K_2O oxides in each material, are shown in Row 5. Specific surface area (SSA) values and methylene blue indices (MBI) are shown in Rows 3-4.

The percentages of each candidate ingredient are shown in Row 7. The sum of the whole body composition, which should always be 100% is in Cell J7, and the sum of the plastic ingredients is in Cell E8. Cell J7 contains the formula which

Table II. Raw Material's Properties Stored in a Spreadsheet Database

A	B	C	D	E	F	G	H	I	J
1 Property	Clay 1	Clay 2	Kaol 1	Kaol 2	Feld 1	Feld 2	Qtz 1	Qtz 2	Body
2 CPFT 1μm	40.0	70.0	30.0	20.0	3.0	5.0	0.0	0.50	
3 SSA (m^2/g)	25.0	36.0	22.0	18.0	3.8	8.0	0.2	0.5	
4 MBI	9.0	13.	12.0	7.0	0.2	0.2	0.0	0.0	
5 Alkali %	0.6	0.6	0.3	0.8	8.2	7.3	0.1	0.2	
6									Sum
7 Percentages	25	0	25	0	25	0	25	0	100
8		Sum of the Plastics		50					
9 CPFT 1μm	10.0	0.0	7.5	0.0	0.75	0.0	0.0	0.0	18.25
10 SSA (m^2/g)	6.25	0.0	5.5	0.0	0.95	0.0	0.05	0.0	12.75
11 MBI	2.25	0.0	3.0	0.0	0.05	0.0	0.0	0.0	5.3
12 Alkali %	0.15	0.0	0.075	0.0	2.05	0.0	0.025	0.0	2.3

calculates the sum of all eight percentages, while Cell E8 contains the formula which calculates the sum of the first four.

Rows 9-12 are the contributions to the body from the percentage of each raw material. Each value in Rows 9-12 is the percentage in Row 7 times the appropriate value in Rows 2-5.

Cells J9-J12 contain the calculated body properties for a body of the composition defined by the percentages in Row 7 for each raw material. Each of these cells contains a formula that sums the eight property values to their left.

The Solver Calculation

The Solver function is an add-in to Excel. Other similar optimization routines are available in other spreadsheet programs. If optimizations are not routinely being performed on a computer, the optimization routine may need to be installed from the distribution disk before it can be used. After Solver is installed in Excel, clicking on the "Solver" selection under the "Tools" menu item, will bring up the Solver window.

When using optimization routines, such as Solver, one cell in the spreadsheet must be maximized, minimized, or set equal to a particular value. That is the first choice to be made, and it represents the first entry to be made into the Solver window.

In this case, consistent with WhiteWares, Inc's desire to maximize the colloidal content of their body, the selection of the cell to maximize is an easy task. Cell J9 should be designated as the "Target Cell:" and the "Equal to: Max" button clicked so Solver will maximize the value of the body's colloidal content.

The next Solver text window is labeled "By Changing Cells:" In this example, Solver is going to change the body composition values in Row 7 (in Table II), so the references to the eight percentages in Row 7 should be entered in this window ($B$7:$I$7). As Solver runs, it will change the eight values in these cells in Row 7 as it attempts to meet all of the constraints that must also be entered. Entry of the constraints is next.

The final Solver text window is labeled "Subject to the Constraints:" In this window, using the "Add", "Change", and "Delete" buttons to its right, you can enter and manipulate the constraints for the optimization runs. Table III shows the constraints that were entered, consistent with the company's stated goals for this body.

Table III. Constraints Entered for the Solver Optimization

Constraint	Explanation
$E$8 >= 40	The sum of the plastics should be at least 40%.
$J$10 = 10	The SSA is to be 10 m²/g.
$J$11 = 4	The MBI is to be 4.
$J$12 >= 4	The alkali content is to be in the range 4%
$J$12 <= 4.1	to 4.1%.
$J$7 = 100	The body composition must be 100%.
$B$7 >= 0	
$C$7 >= 0	
$D$7 >= 0	The composition percentages used by Solver
$E$7 >= 0	must be at zero or positive.
$F$7 >= 0	
$G$7 >= 0	
$H$7 >= 0	
$I$7 >= 0	

There are other options that can be selected to run Solver, but the procedures given here are sufficient to run the program and obtain a solution (if a solution is possible).

Note that when the attempt was made to use only the four #1 raw materials, Solver performed its calculation and the results window began as follows: "Solver could not find a feasible solution." The same happened when the four #2 raw materials alone were tried. Only when all eight raw materials listed in Table II were used did the results window begin with: "Solver found a solution."

ANALYSIS OF THE ILLUSTRATIVE PROBLEM

Traditional Body Composition

Note, in Table II which contains the original nominal idea for the body composition (25% of each of four raw materials), that the only target parameter met was the desire for sufficient plastic raw materials. Other than that, SSA was high, MBI was high, and alkali content was low. (See Cells J0-J12 in Table II.)

Obviously, if the company's engineers had done a lot of homework on the available raw materials, it is probable that the four raw materials, in these proportions, might have satisfied the requirements for SSA, MBI, and alkali content. The raw materials chosen for this example, however, did not provide the desired properties.

Optimization Trials Results

Two tests were run using Solver, as mentioned earlier, to determine the compositions required using only the original four raw materials to meet the desired specifications. It could not be done. No combinations of the four #1 raw materials could provide a body that met all desired specifications. When a second sample of each raw material was provided (the #2 samples), and those four alone were run through Solver, it was not possible to meet all desired body specifications with those four alone either.

Only when all eight samples were run using Solver was it possible to calculate a body formulation that met all desired specifications. Tables IV and V show two different results from the Solver operation that met all desired specs. Table IV shows the results when the colloid fraction was maximized. Table V shows the results when the colloid fraction was minimized.

Table IV. Solver Results Maximizing the Colloid Content

	A	B	C	D	E	F	G	H	I	J
1	Property	Clay 1	Clay 2	Kaol 1	Kaol 2	Feld 1	Feld 2	Qtz 1	Qtz 2	Body
6										Sum
7	Percentages	2.51	0	21.15	16.34	46.06	0	13.94	0	100
8			Sum of the Plastics		40					
9	CPFT 1μm	1.00	0	6.35	3.27	1.38	0	0	0	12.00
10	SSA (m^2/g)	0.628	0	4.65	2.94	1.75	0	0.0279	0	10
11	MBI	0.226	0	2.54	1.14	0.0921	0	0	0	4
12	Alkali %	0.0151	0	0.0635	0.131	3.78	0	0.0139	0	4

Raw Materials Selections

Note that Feldspar #2 and Quartz #2 are not included in either solution. This result is an indication that the properties of those raw materials are not useful in making fine adjustments to the body formulation. In such cases, it's advisable to substitute other raw materials in their places, and run the optimization routine again.

It's not necessary that the optimization routine produce approximately equal raw material percentages for each type powder, but a substantial amount of one powder plus a few percent of its variations would indicate that fine adjustments to the properties can be (and are being) made using the whole range of raw materials available.

Table V. Solver Results Minimizing the Colloid Content

A	B	C	D	E	F	G	H	I	J
1 Property	Clay 1	Clay 2	Kaol 1	Kaol 2	Feld 1	Feld 2	Qtz 1	Qtz 2	Body
6									Sum
7 Percentages	0	0	20.53	20.64	45.86	0	12.97	0	100
8		Sum of the Plastics		41.17					
9 CPFT 1μm	0	0	6.16	4.13	1.38	0	0	0	11.66
10 SSA (m^2/g)	0	0	4.52	3.71	1.74	0	0.0259	0	10
11 MBI	0	0	2.46	1.44	0.0917	0	0	0	4
12 Alkali %	0	0	0.0616	0.165	3.76	0	0.0130	0	4

This optimization procedure therefore can be used to help pick and choose raw materials for use in ceramic production bodies. There are many possible raw materials from which to select in traditional ceramics, electronic ceramics, engineering ceramics, etc. Each variation of each material will have slightly different mineralogical and oxide compositions, properties, impurities, and/or dopants that are of importance to the product to be made. The use of this procedure, therefore, can guide in the design of new bodies, and specifically, it can guide in the selection of powders to be used.

Property Selections

Starting Compositions: In the example under discussion, the original goal of 25% of each of four raw materials, or even the more precise property values for SSA, MBI, alkali content, and colloid content, may have been the result of considerable study and engineering decisions. Frequently, such starting compositions were suggested by a knowledgeable engineer, a sales representative, or by a friend. Some such compositions may be in actual use in production, but unless they are accompanied by the suppliers' names and their particular product designations, the general formula alone may not be very useful.

Whether one understands the behavior of a production body in great detail, or one is totally new to a system, body, and/or product, using this optimization procedure can help guide the selection and specification of desired properties of the final body.

Particle Size Distribution and Surface Area: The only particle size distribution (PSD) parameter in the example above is the colloid content. One PSD property is insufficient to completely specify the PSD. Three or more PSD parameters should be used to define a particle size distribution. Specific surface area should also be one of the constrained properties. Surface area analyzers can see and measure all of the surfaces, but particle size analyzers cannot measure the sizes of all of the particles. Size distribution and surface area parameters are complementary to each other — not redundant.

Solids Content and Interparticle Spacing: One of the more important parameters that can be used to define a body is InterParticle Spacing (IPS). This can be calculated using the expected minimum porosity for a dense pack of the powder particles[1,5], the measured surface area of the powder, and the solids content in the production suspension or body. To achieve consistency of day-to-day performance, this is an important parameter to use. This one parameter alone, however, requires the analyses of PSD and SSA, followed by a series of calculations. This may add complexity to the optimization algorithm, but we have found it to be an important control parameter.

Mineralogical and Oxide Compositions: Frequently, one will find different mineralogical types and/or impurities, or different oxide types and/or impurities, predominant in different source raw materials.

For example, clays can be kaolinitic or illitic. The two mineralogical types are quite different, but if the body formulation only calls for a "clay," great variations can be found within that broad category. If such materials are to be used, it may be necessary to analyze and tabulate the kaolinite and illite contents of each "clay."

Then again, clays frequently contain free quartz impurities — and the percentages of free quartz traveling with the clays vary substantially. If a body uses both a "clay" and a quartz as ingredient raw materials, the quartz content of each clay should be analyzed and tabulated.

The amount of "clay" going into each batch can then be increased accordingly to put a consistent amount of actual clay into the body, and the quartz addition to the body can be decreased accordingly. We have known of yields dropping to zero because companies were not paying attention to such details.

If an electronic ceramic ingredient, or a structural ceramic ingredient, contains particularly important impurity or dopant contents, those impurities should be analyzed so they can be closely controlled in the production bodies.

If, for example, the impurity "iron oxide" is an important control parameter in a body, is the "iron oxide" FeO, Fe_2O_3, or Fe_3O_4? Does it matter? Yes! What effects are caused by each of the different forms? Which are helpful? Which are harmful?

Engineers should know the answers to such questions for all of the ingredients in the ceramic bodies for which they are responsible.

Start Small — Add One By One: The properties to be constrained in the optimization routine should be selected carefully, starting with only a few and adding more and more as successful optimization solutions are calculated.

For example, if the percentage of alkalis desired in the body in the example above had been set too high, it would not have been possible to make a body from the available raw materials, even if alkali content was the only major constraint. This would have been the case had the goal for the body's alkali content been set to 9%. No solution would have been possible because none of the raw materials have alkali contents that high.

When one has 50 different properties in a spreadsheet database, 20 of which are going to be constrained in the optimization calculation, it will be difficult (almost impossible) to obtain a successful solution on the first try. The recommended procedure is to select the two or three most important properties; constrain them as necessary; run the optimization procedure; and gain a successful solution before moving on to more complicated sets of constraints.

The rest of the constraints can then be added to the optimization procedure, one by one, until the large, complete set of body formulation constraints are successfully producing solutions.

BIBLIOGRAPHY/SUGGESTED READING

1. Funk, J.E., and Dinger, D.R., *Predictive Process Control of Crowded Particulate Suspensions Applied to Ceramic Manufacturing*, (1994), Kluwer Academic Publishers, Boston.
2. Dinger, D.R., and Funk, J.E., "Predictive Process Control - Computer Programs for Fine Particle Processing Controls," *Ceram. Eng. Sci. Proc.*, **13**[1-2], 296-309 (Jan-Feb, 1992).
3. Dinger, D.R. and Funk, J.E., "Particle-Size Analysis Routines Available on CERABULL," *Ceramic Bulletin*, **68**[8], 1406-1408 (Aug, 1989).
4. Dinger, D.R. and Funk, J.E., "Particle-Size Analysis Routines Available on CERABULL: Overview and MXENTRY Program," *Ceramic Bulletin*, **69** [1], 58-60 (Jan, 1990).
5. Dinger, D.R., and Funk, J.E., "Particle-Size Analysis Routines Available on CERABULL: The MIX10 Program," *Ceramic Bulletin*, **69** [2], 204-206 (Feb, 1990).

6. Dinger, D.R., and Funk, J.E., "Particle-Size Analysis Routines Available on CERABULL: The PCI Program," *Ceramic Bulletin*, **69** [3], 326-329 (Mar, 1990).
7. Dinger, D.R., and J.E. Funk, "Version 2 of Particle Size Analysis Routines Available on CERABULL," *Ceramic Bulletin*, **70** [4], 669-670 (Apr, 1991).
8. Villa, C.E., and Dinger, D.R., "Spreadsheet Calculates Glaze Compositions, I," *Ceramic Bulletin*, **75**[4] 93-96, (April, 1996).
9. Villa, C.E., and Dinger, D.R., "Spreadsheet Calculates Glaze Compositions, II," *Ceramic Bulletin*, **75**[5] 80-83, (May, 1996).
10. West, R., and Gerow, J.V., "Estimation and Optimization of Glaze Properties," Trans. of Brit. Ceram. Soc., **70**:265-268 (1971).
11. McLindon, J.D., "Estimation of Glaze Expansion from Chemical Compositions," Senior Thesis, Alfred University, Alfred, NY, (1965).
12. Funk, J.E., "Designing the Optimum Firing Curve for Porcelains," *Ceramic Bulletin*, **61**[6]632-635 (June, 1982).

PROBLEMS

1. (a) Obtain complete property lists (averages and standard deviations of each) from at least three different suppliers of barium titanate. Properties should include at least three particle size distribution parameters, specific surface area, and oxide composition analyses of the barium titanate as well as any included impurity oxides. Store all of this information across one row of a spreadsheet.
   (b) Using random number generators, calculate the variations for each of these properties expected on a daily basis over a period of a month (add 29 more rows of property data to the spreadsheet.)
   (c) Define a body formulation for a barium titanate body using the original (average) values stored in the first active row of the spreadsheet.
   (d) Using the optimization routine in the spreadsheet, constrain the properties consistent with Step (c), and calculate the new body formulation required for each day of the month as the properties are randomly changing.
   (e) Adjust the body formulation as necessary or find different source raw materials to achieve a successful solution for each day in the month (each of the 30 rows of data in the spreadsheet.)
2. Perform a calculation as defined in #1 above for your company's major product, or for a silicon carbide body, or a silicon nitride body, or an alumina body, or a zirconia body, or etc.

6

Developing a Design Protocol for Load-Bearing Applications

J.J. Mecholsky Jr.

DEVELOPING A DESIGN PROTOCOL FOR LOAD BEARING APPLICATIONS

J.J. Mecholsky, Jr.
Department of Materials Science and Engineering
University of Florida
Gainesville, FL 32611-6400

ABSTRACT

Material selection, characterization, testing and analyses are critical parts of the design process for a structural component. After the requirements for successful application have been provided it is incumbent on the materials engineer to assure proper selection and performance of the ceramic component. Finite element stress analysis, component testing and reliability analyses are important elements of a synergistic approach to the successful design. This paper will address a multi-disciplined, systems approach to designing with structural ceramics. Two examples will be provided to illustrate the process: glass ceramic dental prosthetic components and silicon nitride hybrid bearings.

INTRODUCTION

Tough and strong ceramics are generally desired for structural applications. These applications may include such diverse needs as high temperature components in gas turbine engines to prosthetic devices for the body. There are several ways to obtain strong, tough ceramics: particulate or fiber additions, *in situ* microstructural changes during processing, lamination with either brittle or ductile layers, and/or post fabrication processing. For the purposes of this paper, we assume that the desired ceramic has been developed and that it is available for the application desired. The purpose of this paper is to outline and discuss the approach of designing to prevent failure, i.e., using an improved material, which has been characterized, in a design that has been structurally analyzed. I will first present the general philosophy and approach to material and structural design and then provide two examples to illustrate the techniques.

DESIGN PHILOSOPHY AND APPROACH

The design process is an iterative one[1]. Once the customer provides the general requirements, the search for the correct material begins. The design process is outlined schematically in Figure 1. The requirements include the environment, i.e., the temperature, loads, allowable deflections, geometry of the component and any other special needs, such as, hermeticity, corrosion resistance, and lubricity. A stress analysis shows the maximum level of stress expected for a given application. This stress analysis should be based on experience so that the critical regions can be identified for closer scrutiny. If this experience is not available, then several preliminary tests may be needed to determine the critical regions and properties in a design. The selection of the material is based on a reliability constant, sometimes called a figure of merit (FOM)[1]. After the material(s) has (have) been tentatively selected, testing and characterization of the important properties are conducted. Part of this evaluation should be the use of fractography in which failure of the material can be determined to be due to over stress or fabrication defects[2]. Once the preliminary tests are finished, an evaluation is made to decide if the material(s) is (are) ready for small scale simulated model testing. These tests are usually planned to experience as close as possible the *in-situ*, component, environments. If the material passes these tests then a small number of full-scale tests are usually performed. If successful, then the next step is incorporation of the selected material into a new design and construction begins. However, even after selection, monitoring for the response of the material and component to the environment is warranted and is usually advised. With modern technology, *in situ* monitoring is possible for many applications and is recommended. In fact, this approach should be designed into all new structural applications. All stages of testing should be accompanied by computer analysis and modeling. The testing and modeling should be interactive, i.e., tests should provide information for adjusting the models and the results of the modeling should provide the important parameters to characterize.

PROSTHETIC CROWN

The design and material selection for a tooth prosthesis is based on several important properties to the dentist[3]. One of the key properties is color, or beauty. From a structural viewpoint, the artificial tooth or crown must withstand the corrosive environment of the mouth and must be tough enough to not crack, especially due to point contacts from hard particles such as unpopped corn kernels or peanuts. We cannot select the toughest material available because that material would not fit all of the other requirements. Thus, we want to select the toughest material available or able to be developed that will also satisfy the other requirements for use in the mouth. The prosthesis must match the hardness of enamel because if it is too hard, then the prosthesis will wear down the natural teeth. In addition, there is a soft foundation of dentin upon which the prosthesis rests. Therefore, the elastic modulus cannot be too high because the other natural materials around the prosthesis have relatively lower moduli. It is better in design with biological materials that the moduli are matched as close as possible in most cases. Cyclic fatigue and wear resistance are important parameters that have to be considered as well. In an oral environment, corrosion as well as stress corrosion may become an issue in fail-safe designs and performance. Thus, a composite analysis is required to determine the working stresses and a multi-faceted approach is necessary to design against failure.

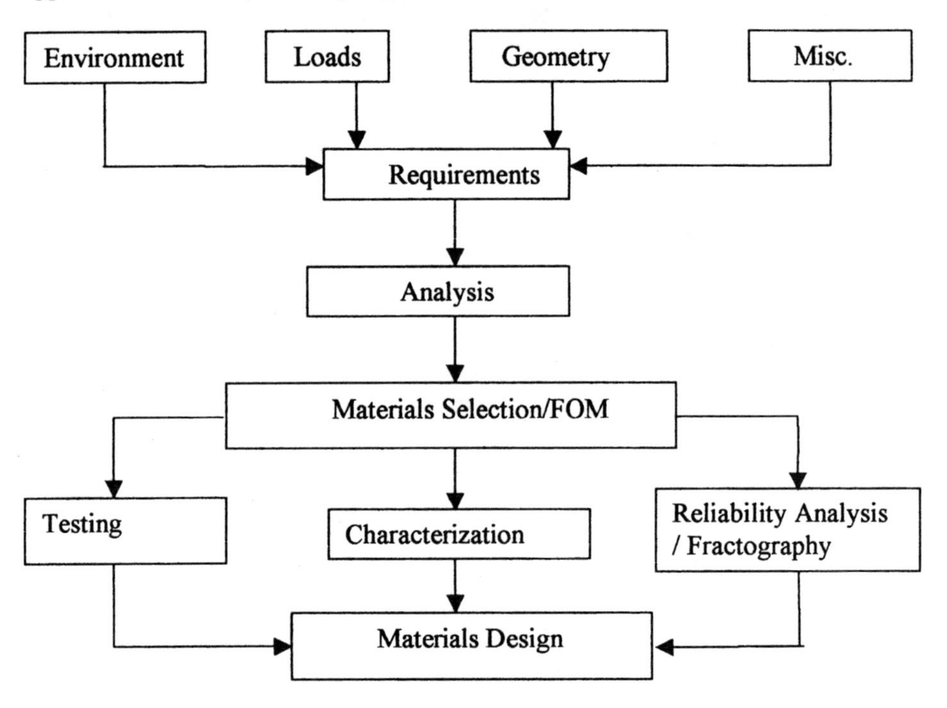

Figure 1. Outline of design methodology.

A summary of the design steps that are listed below can be a guide to the design process. However, each specific material combination can be different, so these "rules" are only a guideline to demonstrate the general approach to selecting new materials for potential use in the oral environment. There are several types of prosthetic devices including crowns, bridges and implants. For the purposes of illustration, we will only discuss a generic crown. The interested reader is referred to an excellent presentation of dental materials.[3]

Summary Of Design Approach For Dental Ceramics

1. List requirements.
2. Select an initial trial structural and material design, i.e., select a shape based on previous knowledge of behavior a material based on laboratory results.
3. Determine the properties of the selected materials.
4. Calculate stresses from model design.
5. Determine regions of maximum stress levels.
6. Decide if materials selected are adequate for the design. Are the stresses expected during use lower than allowable (with a factor of safety) based on the toughness of the materials and the distribution of expected crack sizes?
7. Change structural design if necessary. Change material(s) if necessary. Repeat steps 2-6 until satisfied.
8. Experimentally verify analysis with controlled tests of model.
9. Build prototype for clinical trials.
10. Maintain record of performance during service, at least through the first three years.

A stress analysis based on finite element analysis (FEA) of a crown on dentin can be used for determining the critical stress state in crowns[4] as shown in Figure 2. Notice that a decision has to be made on how to model the cement layer. There are two critical regions associated with a crown loaded as if there is a relatively hard particle contacting the crown as shown in Figure 3. Following well-established contact mechanics, it has been shown that there is a large tensile stress outside the contact area on the contact surface and there is a large tensile stress beneath the contact area on the opposite side of the crown but above the dentin. In general, either can be more critical depending on the thickness of the crown, the relative elastic moduli between the crown material and the dentin and the size of the contact area. Notice that in selecting the mesh size and number for the FEA analysis that there are more and smaller mesh units near the surface and at the boundary between the crown and dentin including the cementitious region. For the purposes of this discussion we are going to ignore the cement layer. The

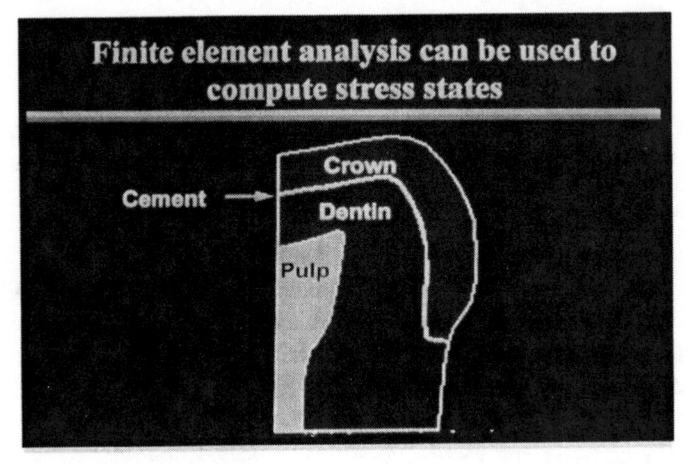

Figure 2. Schematic of prosthetic crown on dentin and pulp in preparation for finite element analysis.

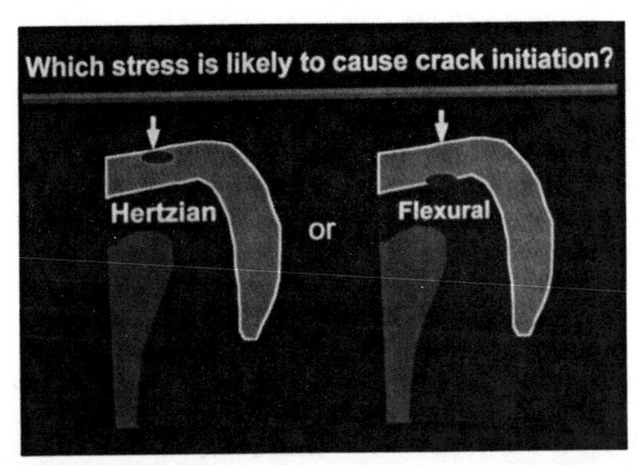

Figure 3. Schematics of loading on crown showing regions of maximum tensile stress (elliptical regions).

inclusion does not change the results discussed here unless a poorly applied cement layer results in very weak bonding. It is possible to provide different moduli and other properties for each element (Figure 4). The results of the analysis in Figure 5 shows that, for the material properties of a typical crown material and dentin, the region of maximum tensile stress is in bending between the two materials.

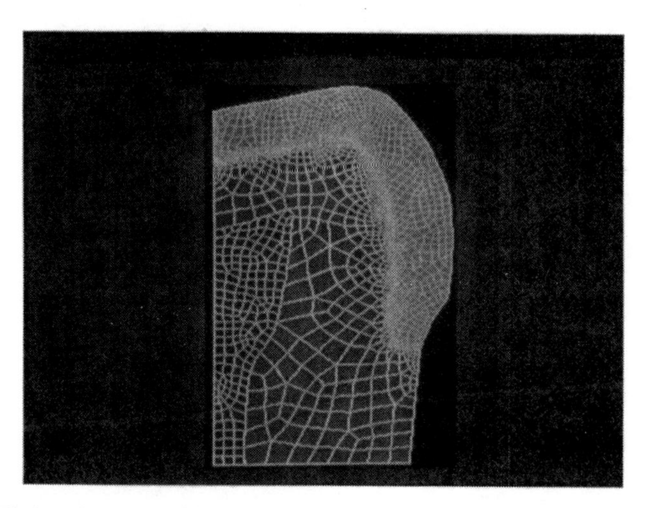

Figure 4. Finite element grid on crown.

To verify that the model in fact reflects reality, several studies were conducted on fractographic analysis of clinically failed crowns. One such study by Thompson *et al.*[5] showed that the failure origin was at the bonded region in the crown material as shown in Figure 6. The stress at fracture, σ_f, of this crown can be determined by measuring the size of the crack, c, and the toughness, K_C, of the material. The equation that relates these quantities is derived from fracture mechanics:

$$\sigma_f = K_c / [Y (c)^{1/2}] \quad (1)$$

where Y = 1.24 for an equivalent semi-circular crack that is small relative to the thickness. In this case, the size of the crack was ~ 140 microns and assuming the toughness was 0.9 $MPam^{1/2}$, the strength is calculated to be 61 MPa. The important aspect of this study was that the location of the origin of failure and stress level was established for *in situ* failures so that future developments can concentrate on these critical regions for improved design considerations. A follow-up laboratory study identified the limits of many of the variables involved in this type of design[4]. This study involved experimental confirmation and FEA analysis of a controlled geometry. With the laboratory study and the clinical analysis along with the finite element analysis, a selection of material, design and application procedure can be made. The actual design and material selection of a crown is more complicated than described here. However, the general approach as described is applicable to the material design process.

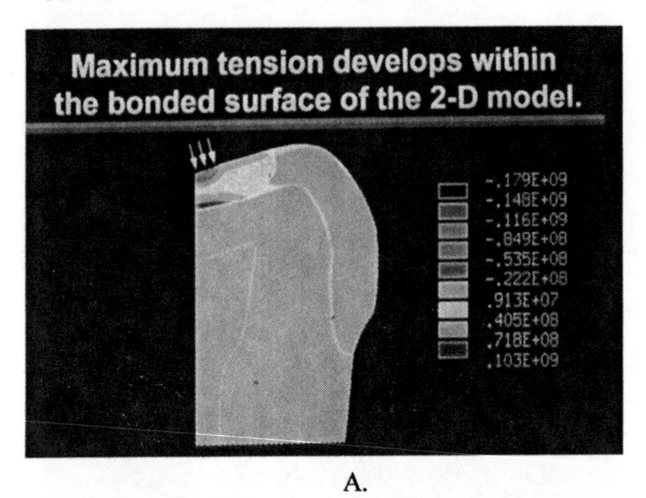

A.

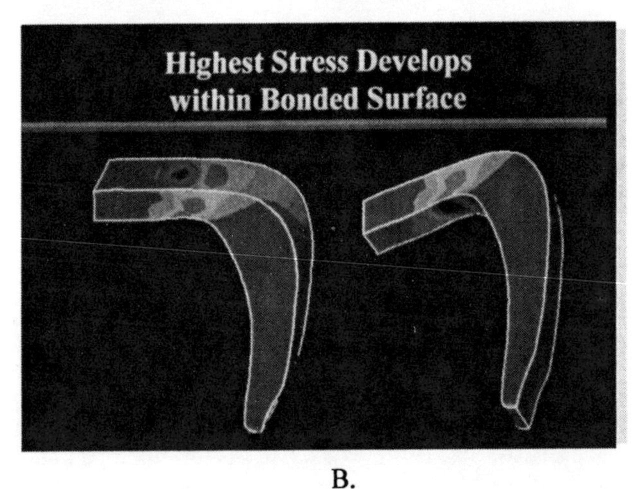

B.

Figure 5. Schematic showing results of finite element analysis: (A) 2-D view, (B) 3-D view of crown. Dark regions show largest tensile stress.

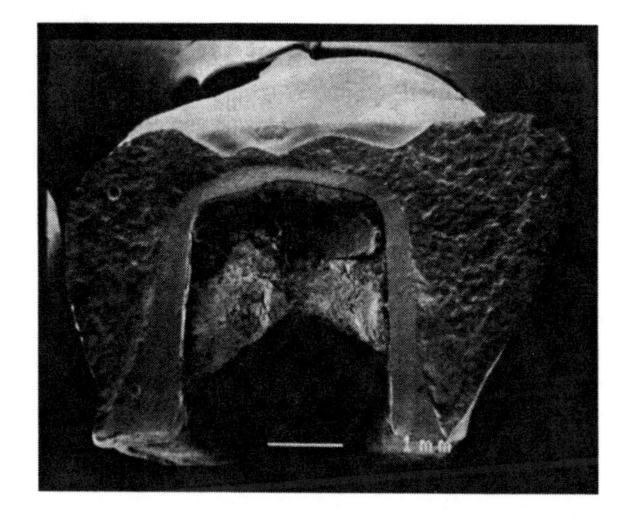

A.

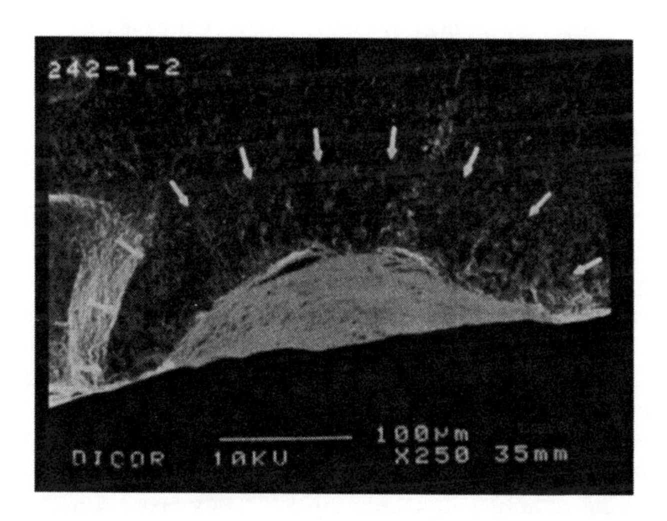

B.

Figure 6. Fracture Surface of Failed Crown: (A) overall view, (B) close-up view of fracture origin.

HYBRID BEARINGS

Because of the great success of Si_3N_4-Chronidur steel hybrid bearings in the space shuttle Discovery, these types of bearings are being considered for use in the next generation of aircraft engines[6]. The hybrid bearing consists of a ceramic ball and tool steel type metal raceways. The space shuttle application was a technical success. Although the hybrid bearing was used successfully in the space shuttle, it was a very specialized application and could not be duplicated for commercial use because of economical reasons. The problem arises from two areas: (1) expense of processing and fabrication of 1.5 inch balls and (2) structural reliability. It is necessary to be able to produce a hybrid bearing at the cost of all steel bearings for commercial applications. Also, it must be a requirement that these bearings do not fail during use. There has been acceptable success for the all-steel bearings. Thus, a joint Defense Research Projects Agency (DARPA)/Air Force program was initiated to produce an affordable silicon nitride ball to be used in a hybrid system. Part of this program involved developing a methodology for determining the maximum allowable flaw size. This methodology is part of the design process that will be discussed here. Once the maximum flaw size is determined then the non-destructive evaluation (NDE) methods available can be evaluated for allowing 100% inspection, and thus assuring 100% reliability. The entire program consists of many parts: (1) processing and fabrication of reduced cost silicon nitride balls; (2) production cost analysis; (3) development of 100% NDE techniques for use on balls and (4) development of a flaw size methodology.

The following is a description of the process of developing a flaw size methodology for silicon nitride balls to be used in hybrid bearings. Obviously this is only one part of a very large design effort; however, it is a good example of the first steps in the development of a new use for an existing material.

The objective for this portion of the program was to develop a critical flaw size methodology for determining the maximum allowable size of cracks or defects for safe operation of a silicon nitride ball in a hybrid bearing system[7]. The technical approach involves three interdependent processes: (1) performing simulated cyclic loading tests to determine levels of degradation as a function of microstructure; (2) performing finite element analysis to determine the location of the maximum tensile stresses during cycling; and (3) developing an equation to predict the maximum tolerable crack size for safe operation during cyclic loading.

The procedure was an attempt to determine what critical size of Hertzian flaw would cause the sample to fracture at that location and not from the flaws intrinsic to the material or processing technique. Conical cracks are formed when a spherical indenter is loaded greater than a critical load. The contact surface usually exhibits a circular crack diameter that is just larger than the contact diameter. The circular and cone crack system is called a Hertzian crack[8,9]. The ball on disc fatigue test was used to simulate loading conditions in Si_3N_4 hybrid bearing systems assuming that external traction forces do not exist on the contact surface. Even though no external traction forces exist, local traction forces were present and caused microcrack formation, Hertzian crack propagation, and eventually material removal. The Hertzian cone crack formation was due to indentation using a blunt indenter and is described in detail elsewhere [8,9].

The set-up for this test can be found in Figure 7. The tests were conducted using a tensile testing machine (Instron, Bedford, MA). Norton Advanced Ceramics (Northboro, MA) supplied the silicon nitride material used in this research. The tests were performed on rectangular bars (4 x 30 x 9 mm) that were polished down to one micron on one side. A WC-Co ball with a 3/16" (4.76 mm) diameter was pressed cyclically into the silicon nitride bar at 10 Hertz for a set number of cycles and with a stress varying from near zero compressive to the chosen maximum compressive stress. The maximum Hertzian contact stress was varied from 2.1 GPa to 15 GPa, and the number of cycles was varied from 10^4 cycles to 10^6 cycles. The equation to determine the maximum amount of contact stress is the following:

$$\sigma_{max} = (6PE^{*2} / \pi^3 R^2)^{1/3} \tag{2}$$

where,

$$\frac{1}{E^*} = \frac{1-\nu_1^2}{E_1} + \frac{1-\nu_2^2}{E_2} \tag{3}$$

and where P is the load applied, R is the radius of the spherical indenter, E is the modulus of elasticity, v is Poisson's ratio, 1 is the designation for the silicon nitride, and 2 is the designation for the (WC-Co) indenter [10].

The samples were then broken in four-point flexure. The schematic of the critical flaw size methodology is shown in Figure 7. The failure stress was recorded, and the size of the flaw that caused failure was determined using optical microscopy. The failure stress, σ, was determined using[11]:

$$\sigma = 3LP/ wt^2 \tag{4}$$

where L is the distance between the inner and outer spans, P is the failure load, w is the sample width, and t is the sample thickness. It was important to determine the source of the failure initiation. Some bars failed from the Hertzian damaged areas, while others failed away from the Hertzian damaged areas.

Finite Element Analysis (FEA)

The role of finite element analysis (FEA) is to provide an estimate of the stress levels around the contact region both with and without crack damage due to the contact process. The stress generated by the bearing contact surfaces is quite complicated. Thus, the stress analysis of this condition is also quite complicated. However, a 2-dimensional analysis can approximate the level of stress that the parts are required to withstand[12]. The contact load is dictated by the design, i.e., 3 GPa for the contact stress. In contact problems, there are tensile stresses generated just outside the contact region[8,13]. We are most interested in tensile stresses because cracks are propagated only if tensile stresses exist somewhere near the crack. A complete finite element analysis was performed to simulate the stress field around a sphere contacting a plate. The presentation of this analysis is beyond the scope of this paper. The results showed that the presence of a crack alters the stress field such that the Hertzian crack that starts to propagate in cyclic loading turns toward the highest tensile region, which has shifted to the surface. Thus, in cyclic loading cracks tend to start into the depth but turn and return to the surface and, as a consequence, a local fracture occurs which appears as spallation.

Summary of the Critical Flaw Size Methodology

1. Simulate cyclic loading by using a WC-Co ball repeatedly loaded on a silicon nitride polished bar.
2. Determine cyclic loading damage using optical and scanning electron microscopy. Develop a quantitative measure of the damage.
3. Determine strength and crack size for rectangular bars subjected to cyclic loading as compared to control group
4. Correlate experimental results with finite element analysis (FEA) of 2-D simulations.
5. Determine degradation levels due to cyclic loading. Establish whether or not a threshold contact stress exists.
6. Determine the critical crack size from measurements of failure from Hertzian damage and the control groups.

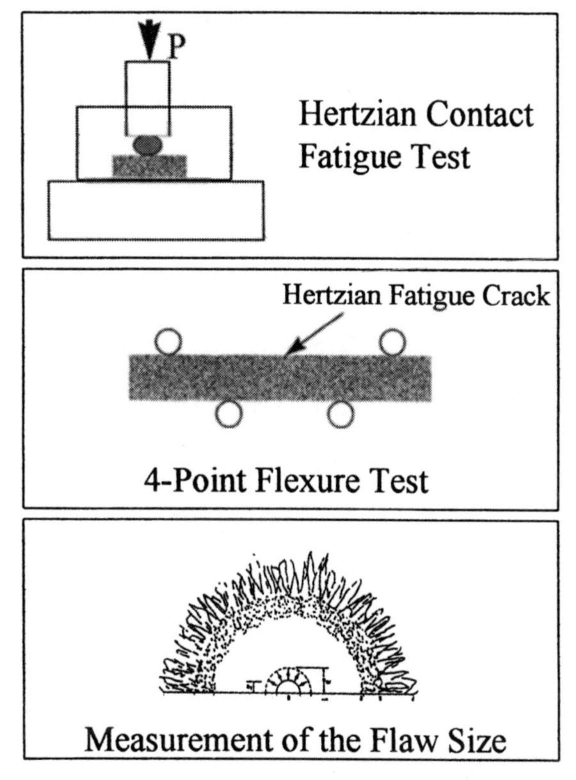

Figure 7. Schematic showing critical flaw size experimental arrangement.

The amount of damage accumulation during the cyclic loading ball-on-plate tests was determined by measuring the area of Stage I damage for fixed load and number of cycles. This approach resulted in a quantitative measurement of damage accumulation based on previous research in which three stages of damage accumulation were determined. For well-polished surfaces, Stage I damage is the development of surface relief due to microcracking at the grain boundaries. This roughening results in a "doughnut" shaped ring around the contact region (Figure 8). The doughnut shape is due to the fact that the innermost region never leaves contact, whereas the region outside the center deflects due to the high load application and experiences traction stresses. Stage II is the production of a Hertzian cone crack. Stage III is the removal of material from the propagation of the cone crack to the surface. [Note that if the surface finish is rougher, then Hertzian cracks will appear first, i.e., Stages I and II reverse order.]

The objective of this part of the study was to determine the critical crack size that can be tolerated without failure during service of silicon nitride in a hybrid bearing application. The approach for the experimental portion of this program is to simulate failure from spherical contact damage during cyclic loading to determine the severity of the cracks that lead to failure. In order to assess this damage, we need to determine selected physical and mechanical properties of the material.

500,000 Cycles at 15 GPa (2175 Ksi)

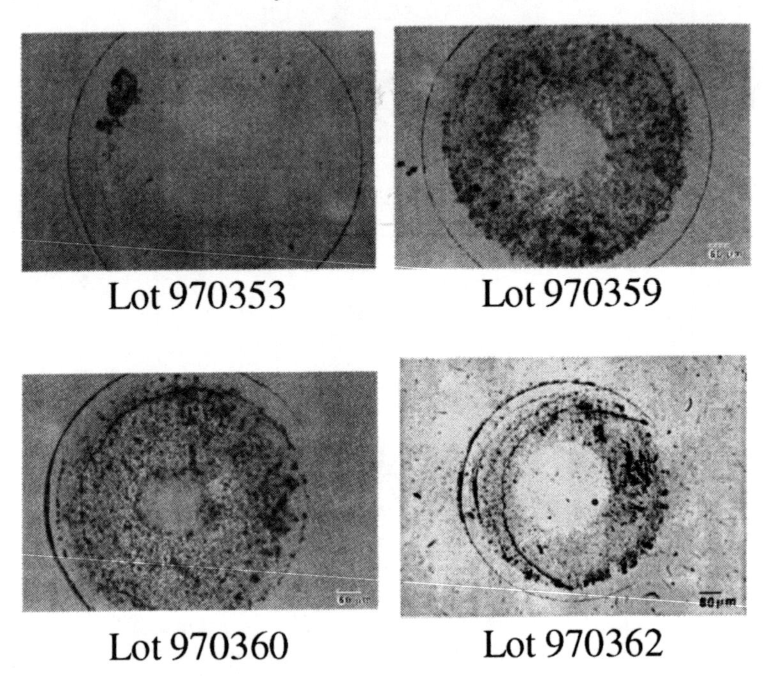

Figure 8. Photographs of wear damage from cyclic loading of a WC-Co ball on a silicon nitride bar. The area between the inner undamaged area and the outer un-damaged area is considered the amount of traction damage.

Figure 9 is a graph of the flexural failure stress versus the Hertzian contact stress in cyclic loading. Two distinct value regions appear on the graph. The first region incorporates the samples that failed with a failure stress of between 550 and 650 MPa. The second region consists of samples with failures below 450 MPa.

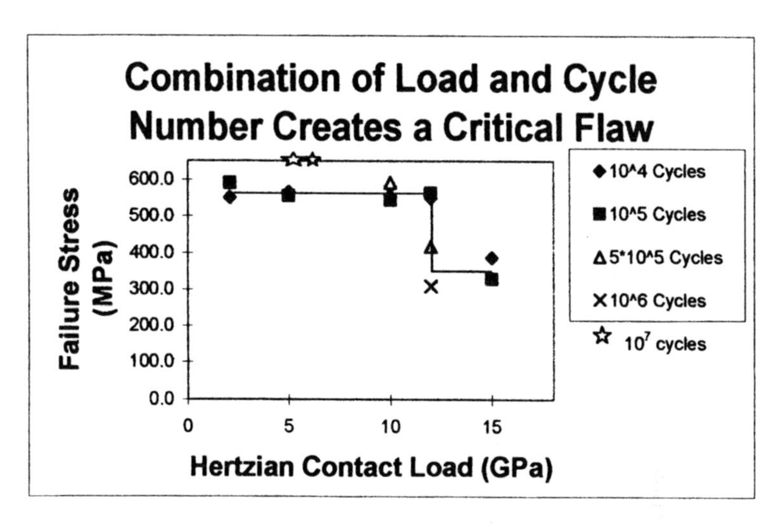

Figure 9. Flexural strength as a function of Hertzian contact stress.

Figure 10. Fracture Surface of Flexure Bar of Si_3N_4 subjected to Long- Term Spherical Indenter Cycling Test (10,000,000 cycles; 5 GPa). The vertical arrows indicate the width of the crack and the horizontal arrow indicates the depth of the crack from the surface.

The samples in the high failure stress range failed away from the Hertzian damaged area. The stress level and number of cycles were not large enough to create a critical flaw larger than the pre-existing cracks due to surface preparation at that location. Therefore, the samples simply failed from the flaws that were already present in the material. An example of the fracture surfaces showing the failure origins for bars that have been subjected to long term cyclic loading (10,000,000 cycles at 5 GPa) are shown in Figure 10. Note that the arrows provide a guide to the fracture initiating crack size. The vertical arrows are the width (2 b) of the crack and the horizontal arrow estimates the depth (a) of the crack. These fracture surfaces are representative of those bars that failed from "natural" and indentation cracks. However, the combination of contact stress above approximately 12 GPa and above approximately 10^5 cycles was such that a critical flaw was created larger than the pre-existing flaws. Thus, the cracks from the cyclic loading caused the low failure stress. Therefore, the Hertzian damaged area became the site of the critical flaw and was the source of failure. The size of the failure-inducing flaw depended upon both Hertzian contact stress as well as the number of cycles. There is a threshold value of contact stress of 10-12 GPa that exists for surfaces that are finished to bearing ball quality based on the present results as well as research results in the literature.
To determine a critical flaw size, the measured flaw size was compared with a calculated flaw size. The calculated flaw size was obtained using a fracture mechanics equation [10]:

$$c = K_C^2 / [\sigma_f Y]^2 \quad (5)$$

where Y is the crack and loading geometry constant, and c is the equivalent semi-elliptical flaw size measured on the fracture surface and is given by:

$$c = \sqrt{ab} \quad (6)$$

where a and b are the depth and half-width of the equivalent semi-elliptical flaw. The value of Y is the geometric and load factor, which has a value of 1.65 for cracks with local residual stress (from indentation) and 1.24 for cracks without local residual stress. The average toughness values of Si_3N_4 were estimated by measuring the crack sizes on the fracture surface of pre-indented bars and bars fractured in four-point flexure. The average value was determined to be about 7.2 $MPam^{1/2}$.

In order to try and establish a relationship between the ring crack diameter and the critical crack size, both were measured on the bars that failed after cyclic loading. An example of failure from a ring crack is shown in Figure 11. The crack that caused failure is similar to those shown in Figure 10. However, the origin of the crack is at the edge of the ring crack.

The above equations are based on fracture mechanics and have been adapted to predict the size of the critical crack for damage in cyclic loading. By combining experimental results of damage accumulation with finite element modeling of cracks in a bearing load simulation, we have been able to determine the size of the critical crack. Finite element analysis (FEA) allows us to change the angle and size of the crack in realistic geometric loading conditions. Combining the experimental and analytical approaches provide more reassurance that the size determined is reasonable for bearing applications.

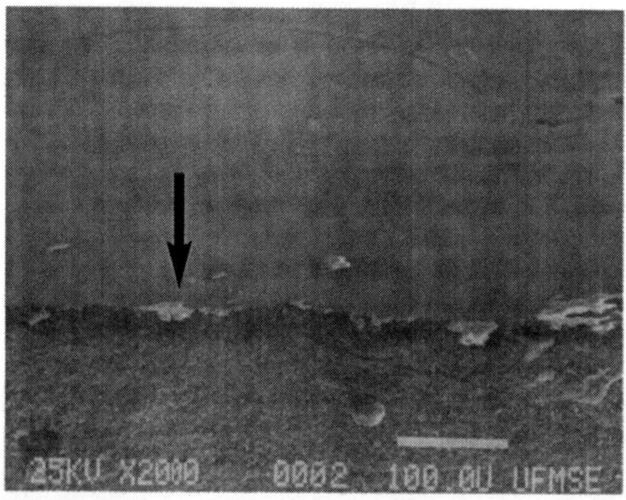

Figure 11. SEM Micrograph of the contact surface (upper micrograph) and fracture surface (lower portion of micrograph) of NBD 200. The arrow points to the fracture origin at the edge of the ring crack. Notice the Hertzian-type crack to the right on the fracture surface (black line from the surface at an angle of about 22 degrees from the horizontal.

The fracture surfaces of all of the bars fractured in the above cyclic loading test were examined and the crack sizes at the origin measured. The actual flaw sizes are graphed versus the calculated flaw sizes in Figure 13. The open circles in Figure 13 indicate the specimens that were subjected to Hertzian cyclic loading (as indicated in Figure 9). However, only the three with crack sizes larger than 150 microns failed from the Hertzian damage. The others failed from damage as a result of the surface finish. Note that bars with no contact damage (solid triangles) had the same flaw size as the open circles at ~ 80 microns. The solid diamonds show the crack sizes from the strength indentation tests for comparison, and the verification of the technique. Thus, the indentation technique results in cracks of similar magnitude as the Hertzian damage for the load levels used in the present study. The results of these tests imply that cracks smaller than about 150 microns will not result in damage due to contact fatigue.

Some explanation of what is meant by crack sizes is in order. The crack sizes above are the equivalent semi-circular cracks, which cause failure in flexure (Fig. 10). They are approximately perpendicular to the surface into the depth of the specimen, whether a plate or a ball. Note that for Hertzian type cracks, the cracks propagate into the depth at an angle of about 22° from the vertical, if they are loaded perpendicular to the surface. This angle change will most likely affect the Y parameter in Equation 5. The cracks, which can be inspected, are approximately in the plane of the surface extending down into the surface. Typically the Hertzian type cracks are circular or circular arcs on the surface. Thus, it is necessary to develop a relationship between the "C-crack" or circular crack that can easily be observed and the in-depth cracks that lead to damage.

The arrow indicates the location of the ring crack and failure origin on the fracture surface. The results for the comparison of ring crack size and the size of the crack on the fracture surface are presented in Figure 12. The measured crack sizes of the silicon nitride bars that were fractured in four-point flexure are shown in Figure 13 as a function of the calculated crack sizes. The cracks were a result of several treatments. The treatment designated as "Hertzian" means that the bars were subjected to cyclic loading at various stress levels before testing. They all failed from cracks associated with the surface finish of the bars. The bars designated with "Vickers" means that the bars were indented with a Vickers hardness diamond to produce controlled semi-elliptical cracks. The samples designated as "No Indent" are representative of the cracks produced by the surface treatment. All of the measured crack sizes below 150 microns failed from the surface finish and not from the other surface treatments. The cracks below 150 microns are all about the same size indicating the distribution of random flaws associated with the surface finish. Since the semi-circular crack is a radius, this means that the surface trace would be expected to be at least twice that size, i.e., 300 microns. An equation (Equation 5) was adapted to be used to determine the critical crack size for resistance to cyclic contact loading. In using this equation, it was determined that a crack size of about 300 microns on the surface, i.e., 150 microns in depth can be tolerated without failure from contact stresses for (pressure) loads less than 10 GPa. There is a threshold contact stress level below which failure from contact damage will not dominate the failure. The value of this threshold depends on the surface finish on the material. For surface finishes suitable for bearing applications the threshold contact pressure is about 10 GPa. The information obtained here along with the finite element analysis and rig tests should be sufficient to provide a critical crack size methodology for the non-destructive evaluation of silicon nitride balls.

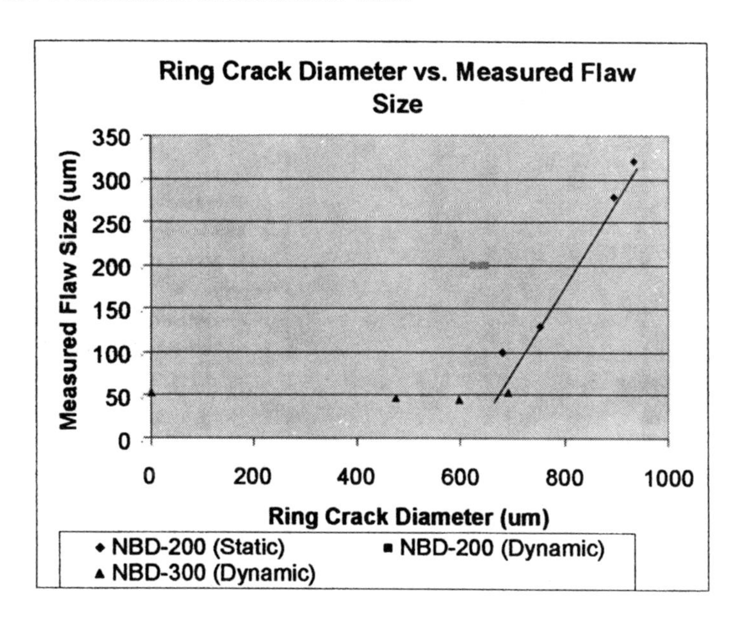

Figure 12. Crack size as a function of ring crack diameter.

This last example of the development of a flaw size methodology for the hybrid bearing demonstrates one of the many different processes that must be addressed before a material and structural design is implemented. After the size of the crack is determined, then the appropriate NDE techniques can be implemented. In addition, pilot scale testing, in bearing systems, must be performed. After these pilot scale tests are successful with their inherent *in situ* monitoring of the stresses and strains, then large-scale tests of bearing systems are implemented. This process of concept, testing, selection, development and insertion may take 10 years, if there are no major complications. In a for-profit company, this is a major investment. Thus, the input of the materials engineer at the onset is very important. In addition, it is the team of engineers, economists, business consultants that are necessary for the final decision to proceed with a new material and structural design.

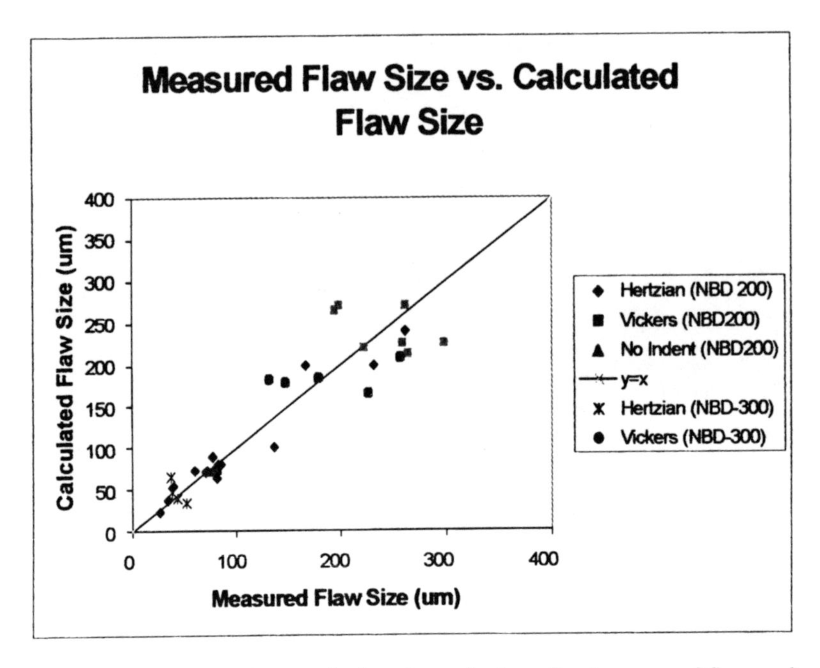

Figure 13. Measured vs. calculated crack sizes for fracture of flexure bars.

These two examples were selected to demonstrate that a combination of stress analysis, fracture mechanics and material selection principles to form a material design methodology. They also show that successful structural designs are possible using toughened ceramics. However, this success is dependent on a team effort. The materials engineer is just one, albeit a critical one, of the team members.

ACKNOWLEDGEMENTS

The author thanks the many colleagues, co-workers and students who have contributed to the work discussed in this paper. I especially want to acknowledge the former students who participated in the hybrid bearing program: Dr. Zheng Chen, Ms. Tammy Simpson and Gary Ross. The help of Dr. Ken Anusavice, Dr. Alvaro Della Bona and Dr. Thomas J. Hill with dental biomaterials information and slides is gratefully acknowledged. The support of DARPA and AFOSR on part of this work is also gratefully acknowledged. Dr. Herb Chin has been inspirational in his guidance in the design methodology for a high performance component. I also thank Pratt and Whitney and St. Gobain Industries for the use of testing facilities and materials, respectively.

REFERENCES

[1]M. F. Ashby, Engineering Materials Design, Cambridge University Press, Cambridge, UK 1997.

[2]Mecholsky, Jr., J. J. 1996 Fractography, Fracture mechanics and fractal geometry: An Integration; in Fractography of Glasses and Ceramics III, Varner, Frechette and Quinn, eds., American Ceramic Society, Westerville, OH pp. 385-93.

[3]Anusavice, K. J., (1996) Phillips' Science of Dental Biomaterials, 10th ed. Philadelphia: W.B. Sauders.

[4]Tsai, Y. L., Petsche, P.E., Anusavice, K. J., Yang, C. (1998), Influence of Glass Ceramic thickness on Hertzian and Bulk Fracture mechanisms, Int. J. Prosthodont., 11:27-32.

[5]Thompson J. Y., Anusavice, KJ, Naman, A, Morris, HF (1994). Fracture Surface Characterization of Clinically Failed All-Ceramic Crowns. J. Dental Res. 73:1824-32.

[6]Hucklenbroich I, Stein G, Chin H, Trojahn W, Streit E, (1999)High nitrogen martensitic steel for critical components in aviation , Materials Science Forum v. 318- pp. 161-166

[7]Z. Chen, J. Cuneo, J. J. Mecholsky, Jr. and Hu, S.F., "Damage Process of Si_3N_4 Bearing Material Under Contact Loading",Wear 198, 197-207 (1996).

[8]B. Lawn, "Fracture of Brittle Solids", Second Edition, University Press, Cambridge, (1995).

[9]H. Hertz, "Miscellaneous Papers by H. Hertz", London, Macmillan, (1986).

[10]K. Johnson, "Contact Mechanics", Cambridge University Press, Cambridge (1985).

[11]A. Higdon, E. Ohlsen, W. Stiles, J. Weese, W. Riley, Mechanics of Materials, Fourth Edition, John Wiley and Sons, New York (1985).

[12]Gundepudi, MK, Sankar, BV, Mecholsky, JJ and Clupper, DC (1997) Powder Technology 94(2) 153-161.

[13]Hu, SF, Chen, Z, Mecholsky, JJ (1996), On the Hertzian Fatigue Cone Crack Propagation in Ceramics, International J. of Fracture 79 (3) 295-307.

[14]D. Broek, "Elementary Engineering Fracture Mechanics", Sijthoff and Noordhohh, Alphen aan den Rijin, Netherlands (1978).

PROBLEMS

1. Calculate the expected crack size for a material to be used for dental crown veneer if the fracture toughness, K_C, is $0.7 MPam^{1/2}$ and the expected applied stress was determined from finite element analysis to be 21 MPa.

2. Determine the maximum contact stress in an elastic plate of silicon nitride, which is in contact with a 2.5 cm diameter spherical indentation of WC-Co loaded at 100N. [Si_3N_4 Properties: $K_C = 6\ MPam^{1/2}$, $E = 600$ GPa, $\nu = 0.24$; WC-Co Properties: $K_C = 12\ MPam^{1/2}$, $E = 700$ GPa, $\nu = 0.3$.]

3. There are generally considered to be three stages of the wear process in well-polished materials. Stage I is the development of surface relief, Stage II is the production of the Hertzian cone crack and Stage III is the removal of material. Why would a rough surface make a difference in the order of the stages, i.e., Stage II would occur before Stage I?

4. (a) Explain the shape of the graph in Figure 9 in terms of damage to the surface of the material experiencing the contact.
(b) How would the quality of the surface finish on the contacted surface affect the graph?

5. The graph shown in Figure 13 has three different types of cracks. Explain the source of these cracks, i.e., what caused the cracks? If the surface finish was improved (less rough) how would this change affect the graph shown in Figure 13?

6. How does the mesh size in a finite element analysis affect the calculation of stresses?

7. In a finite element analysis grid, should there be more or fewer mesh units near the corners and surfaces? Why?

8. There are two critical regions associated with a crown contacted by hard material as shown in Figure 3. What material properties and geometric changes will affect the value of stress in these regions? Will the stresses increase or decrease in each region with the changes? Why?

7

Role of Thermal Expansion and Conductivity in Design

D.P.H. Hasselman

K.Y. Donaldson

ROLE OF THERMAL EXPANSION AND CONDUCTIVITY IN DESIGN

D. P. H. Hasselman and K. Y. Donaldson
Department of Materials Science and Engineering
Virginia Polytechnic Institute & State University
Blacksburg, Virginia 24061 USA

ABSTRACT

A discussion, illustrated by many numerical examples, is presented on the role the thermal properties of ceramics play in a variety of engineering applications, including the accommodation of thermal expansion, thermal insulation, heat storage, and thermal management.

INTRODUCTION

Many applications of ceramic materials involve high temperatures and/or significant levels of steady-state or transient heat flow, which can present significant challenges in selecting appropriate materials for a given design. For example, a material may be required to optimize heat-flow, a criterion critical for heat exchangers. Other applications may require that heat losses be minimized, as in the design of buildings, furnaces, ovens and kilns. Some applications may require that prescribed temperatures not be exceeded, such as in the design of solid-state electronics and propulsion systems such as internal combustion or turbine engines. Thermal expansions may need to be limited or at least matched for multi-component systems made up of a number of different materials with differing thermal expansion behavior.

It is the purpose of this paper to present a number of specific examples of materials selection and design for a range of thermal environments and performance criteria. At the risk of over-simplification, each example will be discussed in the briefest manner. The sum total of these examples, however, will serve to demonstrate the complexity of the selection of ceramic materials for thermal applications. No attempt is made to present an exhaustive review of the literature on heat transfer. A listing of appropriate text books for further information is included at the end of this chapter.

ACCOMMODATION OF THERMAL EXPANSION

Structures designed for operation at elevated temperatures, such as kilns, furnaces and ovens, generally consist of individual components with different compositions and associated property values. Since such structures are usually assembled at room temperature, bringing them to the required operating temperatures will involve dimensional changes due to thermal expansion, which will differ from component to component. In order to avoid structural failure due to imposed tensile or compressive stresses, thermal buckling or other causes, the thermal expansions of the individual components need to be accommodated freely, without being constrained by neighboring components.

A well-known example of this problem is the use of silicon carbide heating elements in the construction of electrically heated laboratory furnaces. Generally, such heating elements are quite long and extend through the furnace walls to allow attachment to an electric power supply outside of the furnace. The strength of such elements is not particularly high, and furthermore silicon carbide is quite brittle. Accordingly, the strains-at-fracture of these elements is quite low, implying that even minor constraints to their free thermal expansion may well lead to premature failure. Also, their resistance to creep is quite high so that any stresses induced by constraints to thermal expansion are not easily accommodated by creep. Accordingly, these furnaces are generally constructed to allow adequate free space between the heating elements and the associated openings through the furnace walls so the heating elements can slide freely through the openings over the total range of operating temperatures. Any heat losses from the furnace through the gaps between the heating elements and the furnace wall can be reduced by packing the gaps loosely with a thermally insulating refractory wool.

The design of a fuel rod in an uranium oxide-based nuclear power reactor is a unique example illustrating the need to accommodate unequal thermal expansions. Uranium oxide in the form of right circular cylinders with a diameter of some ½ inch by some ½ inch high, referred to as pellets, is the usual fuel source. These pellets are inserted into a vertical zircalloy metal tube of ten or more feet in length. During operation the interior core of the fuel pellet may reach temperatures as high as 1600 °C or greater while the surface temperature reaches about 800 °C. For simplicity let's assume the mean temperature of the pellet is about 1000 °C. The zircalloy tube is surrounded by circulating pressurized cooling water which removes the heat generated within the nuclear pellet. During full-power operation the zircalloy tubing operates at a temperature near 600 °C. Not only do the pellets and surrounding tubing operate at a different mean temperature, but they also exhibit different thermal expansion characteristics. Insertion of the fuel pellets into the tubing takes place at the same (i. e., room) temperature. The differences in thermal expansions between the fuel pellet and surrounding zircalloy tubing upon reaching the operating temperature need to be accommodated by careful control of the outer dimension of the fuel pellet and the inside diameter of the zircalloy tubing.

As a simple numerical example, let us assume that at full-power operation of the reactor the fuel pellet and the zircalloy tubing are in direct contact. The diameter of the pellet in this case is identical to the value of the inner diameter of the zircalloy tube. We wish to calculate the maximum diameter of the pellet and minimum inside diameter of the zircalloy tube at room temperature prior to assembly so that they are just in contact at operating temperature.

The coefficient of linear thermal expansion, α, of a solid is defined by:

$$\alpha = \frac{\frac{(\Delta L)}{L_o}}{(\Delta T)} \tag{1}$$

where (ΔL) is the change in length, L_o is the original length and (ΔT) is the change in temperature. In effect, equation 1 defines the coefficient of linear thermal expansion as the thermal strain per unit temperature difference.

Equation 1 can be rewritten in terms of the change in length:

$$(\Delta L) = \alpha L_o (\Delta T) \tag{2}$$

Writing $L = L_o + (\Delta L)$, yields:

$$L = L_o(1 + \alpha(\Delta T)) \tag{3}$$

For the current example, $L = ½$ inch. Let us assume that the insertion of the pellet into the tube occurs at an ambient temperature of 25 °C.

With an assumed value for the mean operating temperature of 1000 °C, $(\Delta T) =$ 975 °C. The mean coefficient of linear thermal expansion for uranium oxide over the range of temperature from 25 to 1000 °C is approximately 10 x 10^{-6} (°C)$^{-1}$. Inserting these values into equation 3 yields $L_o = 0.4952$ inch. Similarly, for the zircalloy tubing, with $\alpha \approx 5.8 \times 10^{-6}$ °C^{-1} and $(\Delta T) = 575$ °C, the minimum inside diameter L_o

≈ 0.4983 inch.

The difference in the values for $L_o \approx 0.0031$ inch, which is the required gap between the fuel pellet and the inside surface of the tubing at room temperature. Of course, these values were calculated assuming an ideally constructed reactor exposed to ideal conditions during initial reactor heat-up. In reality, however, uncertainties in the dimensions, i. e., the tolerances, of the pellets and tubing need to be considered. Furthermore, uneven temperatures and cooling conditions within the reactor will introduce further uncertainties. For these reasons, the nominal diameter of the fuel pellet should be somewhat smaller and the corresponding inside diameter of the tubing should be somewhat larger than the values calculated above. It should also be noted that the length of time over which a reactor has been operated can add several more complicating factors. The fuel pellets tend to swell over time. Due to the pressure exerted by the surrounding pressured water, the zircalloy tubing will creep down onto the fuel pellet. As the reactor is repeatedly cycled in temperature, the fuel pellets tend to fracture and settle to the bottom of the vertical zircalloy tubing, an effect referred to as "ratcheting," which exerts an internal pressure on the tubing. To avoid this effect, nuclear fuel reactors generally are operated continuously to take care of the "base-load" of the demand for electrical power. In general, nuclear fuel element design requirements will depend on the design of the specific nuclear reactor and its intended operating schedule.

THERMAL INSULATION

Fiberglass wool for building insulation

Fiberglass wool is an excellent thermal insulator. Because of its structure, heat transfer by direct conduction through the fibers is reduced significantly when compared to the value for solid glass. Because a blanket of fiberglass wool interferes with air flow, convective heat transfer through a wall or ceiling is suppressed. This leaves conductive heat flow through the air contained within the fiberglass as the primary mode of heat transfer. As the ambient temperature rises, infrared radiation can make an additional contribution to the heat transfer. In building insulation this can be suppressed by the presence of an aluminum foil backing to the fiberglass wool blanket.

Numerical examples can demonstrate the effectiveness of fiberglass as a thermal insulator. A building wall or ceiling generally is constructed of a number of layers of different materials. These act in series to prevent heat flow. For practical purposes the effectiveness of a building material of given thickness as a thermal insulator is expressed in terms of its "R-value." Simply adding the R values for all layers yields the total R value for the wall or ceiling. The building industry typically uses British engineering units, so the R-value of 3½ inch thick glass wool is 10.9 $ft^2 \cdot hr \cdot °F/Btu$. Translating to metric units, this becomes 6.43 $°C/W/m^2$. The physical significance of this latter value is that if a temperature difference of 6.43 °C existed across the thickness of the insulation, 1 W would be conducted per second per square meter. Six inch thick glass wool has an R-value of 10.8 $°C/W/m^2$. Doubling the thickness to 12 inches yields an R-value of 21.6 $°C/W/m^2$.

Let us estimate the heat losses through the ceiling of a house of average size and the cost advantage realized by adding further insulation. Let us concentrate solely on the area of the ceiling consisting of sheetrock and fiberglass.

The total amount of energy, Q, lost can be calculated from:

$$Q = \frac{A(\Delta T)t}{R} \tag{4}$$

where A is the total surface area, (ΔT) is the total temperature difference, t is the time period of interest and R is the total R-value of the sheetrock and fiberglass wool.

Let us first consider 3½ inch fiberglass wool supported by ½" thick sheetrock with an R-value of approx. 0.26 $°C/W/m^2$. Adding the R-values yields a total resistance of $6.43 + 0.26 = 6.69$ $°C/W/m^2$. Let us consider a cool day in winter with an outside temperature of -10 °C. Assume the inside temperature is 20 °C. This yields a total temperature difference, (ΔT), of 30 °C. Further assume the total ceiling surface area, A, is 150 m^2 and a time period t, of one day (24 × 60 × 60 seconds).

Substituting the above values into equation 4 yields:

$$Q = \frac{150 \times 30 \times 24 \times 60 \times 60}{6.69} = 58 \times 10^6 \text{ J} \tag{5}$$

or since there are 3.6 x 10^6 J/kilowatthour (kWh):

$$Q = 16.1 \text{ kWh} \tag{6}$$

At \$ 0.10 per kWh, the cost of the heat loss through the ceiling is \$ 1.61/day, say about \$ 48.00 per month. Repeating the calculation for the 6 inch thick insulation yields a heat loss of about 9.7 kWh per day at a monthly cost of about \$ 29.00, while for the 12 inch thick insulation the heat loss is 4.9 kWh per day at a monthly cost of about \$ 15.00.

Comparison of the above values indicates that adding insulation to the ceiling of an existing home can result in considerable energy savings. The reader may wish to

obtain cost figures for such additional insulation and calculate the pay-back period. It will be found to be quite short indeed and probably no more than one heating season. Note that during the summer the heat flow goes the other way, so that additional insulation also can lead to large savings in air-conditioning costs.

Thermal protection for the space-shuttle

On re-entry into Earth's atmosphere, the surface temperatures of the space shuttle may reach as high as 1250 °C. The interior structure is primarily made of aluminum, which loses its load-bearing integrity when heated. Furthermore, the insulation chosen must survive multiple flights. Clearly, a ceramic material is in order. Silica fiber fills the bill in view of its very low thermal conductivity and low density. The silica fiber insulation is in the form of individual tiles bonded to the shuttle's outer surface and wings. A special coating on the front surface of the tiles promotes radiation of most of the heat generated by aerodynamic heating from the surface into space, instead of all of it being conducted towards the interior structure.

It should be pointed out that the silica tile is designed to be an effective thermal insulator only over the time duration of the re-entry of the shuttle. It is not designed for long-term flight through the atmosphere at low altitude, where significant heating of the internal structure can occur. In this case, active cooling of the surface and/or internal structure of the wing will be required. Such active cooling, accomplished for instance by pumping fuel through the wing, is an essential feature of the high-speed aircraft designs currently on the drawing boards.

It should also be noted that the silica tiles used for the space shuttle are inadequate for thermal protection of space modules re-entering the atmosphere during return flights from the moon, such as the recent Apollo flights. On re-entry from such flights the energy to be dissipated is far larger than for the relatively low altitude flights of the space-shuttle. Instead, on re-entry of the Apollo flights, thermal protection relied on a "heat-shield" composed of a low thermal conductivity fiber-reinforced polymer matrix composite, which allowed much of the heat generated by the aerodynamic friction to be dissipated through the processes of charring and ablation, i. e., the evaporation, of the polymer matrix. Clearly, such a heat-shield can be used only once. From a design perspective, it is critical to note that although destroyed in the process, such a heat-shield fulfills its design objectives over the duration of the period of re-entry.

HEAT STORAGE

Increasing the temperature of a material requires energy. This amount of energy for any given material is given by its "specific heat," defined as the amount of energy required to heat a unit mass of the material by a unit temperature difference. In the International system of units specific heat is defined in units of joules per kilogram per degree K. The value of specific heat differs from material to material. For any given material specific heat is also a function of temperature. The heat capacity is the total amount of heat required to increase the temperature of a structure of given mass over a given range of temperature.

Energy savings with fibrous insulation for industrial batch furnaces by minimizing heat capacity

Ceramic ware can be fired, i. e., sintered, in batches, by placing the "green" ware in a so-called "batch furnace" at ambient temperature. The temperature is then raised to the desired level, held there for a specified period of time, followed by cooling to ambient temperature. The fully sintered ware is then removed. This is in contrast to a "continuous kiln" which is kept at the desired temperature level for long periods, while the ceramic ware to be fired is placed on carts which are moved slowly through the kiln.

The energy required to bring a batch furnace up to temperature generally is not recycled but lost through the ventilation system. At the desired temperature the energy in the furnace is stored primarily within its walls, constructed from a thermally insulating refractory. Material selection and design for the walls of a batch furnace should minimize the energy required to bring them up to operating temperature. In practice, this is referred to as designing to reduce "thermal mass."

Let's compare two materials for use as insulation in a batch furnace. The first, a porous silica refractory, is a popular insulating refractory because of its low thermal conductivity value. However, its relative density may be of the order of some fifty percent. The second material, a high silica refractory fiber blanket, has a relative density of only about five percent.

Assume that the desired temperature of the hot-zone of our batch furnace is 1000 °C. Further assume the inner surface area of the furnace, i. e., the sides, bottom and ceiling, is about 6 m^2. Let us estimate the energy savings realized in bringing a batch furnace to a desired operating temperature after replacing the inner insulating layer of silica brick with a thickness of 0.15 meter with a silica blanket of the same thickness. The total volume of insulating material taken up by a thickness of 0.15 m is of the order of 0.9 m^3. We wish to concentrate on the decrease in total heat capacity on replacing the porous silica refractory with the silica fiber blanket. For this reason, let us assume that their thermal insulating abilities are the same. The theoretical density of silica is of the order of 2500 kg/m^3. The total mass of the 50% dense silica refractory then becomes $0.5 \times 0.9 \times 2500$ kg $\approx$ 1125 kg. The corresponding value for the silica fiber blanket is $\approx 0.05 \times 0.9 \times 2500 \approx 112.5$ kg. Let us set the mean temperature of the silica refractory and fiber blanket at 800 °C. For an ambient temperature of 25 °C the mean temperature rise becomes 800 - 25 = 775 °C. The mean value of specific heat of silica over the temperature range of 25 to 800 °C is approximately 950 J/kg·K. The total amount of energy, Q, required to bring the silica refractory up to temperature can be written in general as:

$$Q = mc_p(\Delta T) \quad (7a)$$

where m is the mass, c_p is the specific heat at constant pressure and (ΔT) is the change in temperature.

Substitution of the above values yields:

$$Q = 828 \times 10^6 \text{ J} \quad (7b)$$

The corresponding value for the silica fiber blanket is 10% of the above or 82.8×10^6 J.

The substitution of the silica fiber blanket for the silica refractory amounts to an energy savings of some 745×10^6 J. At 3.6×10^6 J/kWh, this energy savings comes to about 207 kWh per heat-up and cool-down cycle or, at $ 0.10 per kWh, $ 20.70 for each heating cycle of this furnace. Of course, other batch furnace configurations and fuel costs will yield different values for the cost savings that can be achieved. Regardless of the details, these savings should be significant. It should be noted that the reduced heat content resulting from the above substitution also reduces the time required to bring the furnace up to temperature, with associated additional savings in time and production costs.

Heat recuperator and air-preheater for truck turbine engines

The exhaust heat recuperator of a truck turbine engine is a prime example of the recycling of waste heat. Such a recuperator is essential to assure the economic advantage of this type of engine over the conventional internal combustion engine. The recuperator consists of a disk about two feet in diameter with a thickness of some four inches made of a ceramic material in a honeycomb or similar type of configuration with straight open channels. This disk is contained in a housing within which it rotates slowly. The hot exhaust gases pass through one side of the housing adjacent to the center of rotation. The heat from the exhaust gases is absorbed and stored in the ceramic honeycomb. The intake air passes through the disk at the other side of the housing and picks up the heat as the disk rotates by one half-turn.

The recuperator described above has a number of design and material requirements. A thin-walled honeycomb or similar structure with straight channels is chosen to provide minimum resistance to the flow of exhaust gases and intake air. The relatively short flow path of some four inches is chosen to further minimize resistance to the gas flow. The disk's large size assures plenty of material for temporary heat storage. In order to effectively store heat and to avoid excessive temperatures, the material of construction should have a specific heat as high as possible along with a melting point sufficiently high so that softening or actual melting at the temperatures of the exhaust gases is avoided. A low coefficient of thermal expansion is required to keep dimensional changes to a minimum, assuring a proper fit within the surrounding chamber and minimizing the magnitude of thermal stresses which will arise from the inevitable non-uniform temperature distributions within the disk. The choice of the recuperator material also depends critically on its resistance to corrosive constituents within the exhaust gases. Finally weight savings need to be considered as well. The material which most closely meets the above requirements is a lithium-alumino-silicate glass-ceramic, with very low thermal expansion characteristics, low density, high specific heat and reasonable corrosion resistance.

Heat recuperator and air-preheater for MHD-power generating system

Another example of the recovery of waste heat in order to improve energy generating efficiency is an experimental magnetohydrodynamic (MHD) system. In this type of power system, ionized gas generated by a coal- or natural gas-fired combustor is passed at near sonic velocity through a square channel with stationary electrodes on opposite sides of the channel. A magnetic field is applied oriented both perpendicular to the direction of travel of the ionized gas and parallel to the electrodes. The positive and negative ions within the gas are deflected toward the electrodes in opposite directions, thereby creating a potential difference. By connecting the electrodes to the electrical power distribution network via an appropriate DC-to-AC converter, the MHD channel, in effect, becomes an electric power generating system.

The MHD channel, however, will operate only at temperature levels sufficiently high that the gases coming from the combustor are ionized. This requires gas temperatures near 2000 K or higher. Even then, ionization needs to be promoted by seeding the gases with cesium, a material with low ionization energy. As energy is removed from the ionized gases, the gas temperature is decreased and ionization is suppressed. For this reason, only a fraction of the energy in the combustion gases can be extracted by the MHD channel. The remainder of the energy needs to be extracted by other means, such as a conventional power station, downstream from the MHD channel. Using an MHD channel in this fashion is referred to as a "topping cycle". If successfully operated the combined efficiency of the power plant could reach as high as some fifty to sixty percent, which compares favorably with the some thirty-five percent currently achieved with natural gas-, coal- or oil-fired generating systems. In order for gases from the combustor to reach temperature levels high enough for ionization to occur, the air drawn into the combustor needs to be preheated. For this purpose, the exhaust downstream from the MHD channel is circulated through recuperators, in the form of tall towers some 80 feet high and 12 feet in diameter. These towers have an inner core of ceramic material for storing the recovered heat and an outer layer of a thermally insulating refractory, the whole contained within a steel shell. The hot exhaust gases are blown through the tower from the top and heat is transferred from the gases into the refractory core. Once the tower becomes too hot to effectively recapture the heat, the exhaust stream is directed to a neighboring tower. The intake air is then drawn into the first tower and preheated before entering the combustion chamber.

The inner core of the recuperator in which the heat is stored and recovered has a number of material and geometric requirements. In order to store as much heat as possible the material used should have a high value of specific heat. For the same reason it should be fully dense. It must also contain channels for the effective transfer of energy from the MHD exhaust stream to the core and from the core to the incoming air for the combustor. One geometry developed for this purpose consists of closely packed hexagonal blocks, each block measuring some six inches across by some four inches high with holes some ¼ inch in diameter. The melting point of the material in the core should be very high to accommodate the high temperatures of the entering MHD exhaust stream. Its chemical stability should be very high to minimize possible corrosion by impurities within the MHD exhaust stream such as the cesium used for ionization of the gases from the combustor and/or corrosion by the slag within the combustion gases for coal-fired MHD generators. Clearly, its mechanical properties such as strength, creep and thermal shock resistance should be such that long-term trouble-free operation is assured. Its coefficient of thermal expansion is less critical as thermal displacements can be accommodated by a loose stacking of the individual components.

The only material which satisfies these requirements is stabilized zirconia. Magnesia, in view of its high melting point, may appear to be an alternative candidate material. However, when used in an air-preheater for a blow-down high-speed air-tunnel it was discovered that, due to its high vapor pressure, the magnesia evaporated in the hottest part of the recuperator and condensed in the lower part thereby plugging the air channels in the colder sections.

It should be noted here that such recuperators make up the major component for an MHD-power generation system, since the channel itself is only some 12 feet long with a tapered crossection some 2 to 3 square feet at the end where the gases enter and some 4 square feet at the exit end.

Because of some familiarity on behalf of one of these authors (DPHH) with the development of an MHD-electric power generator a general remark is in order. The development of this system was promoted because of a potential energy conversion efficiency much higher than for conventional electrical power generating systems. However, the improved efficiency of an MHD-system is rooted primarily in its much higher operating temperatures. If existing energy-generating systems could be modified to operate at similar high temperatures, corresponding increased efficiencies could be attained as well. Modifications of existing technologies is more likely to be successful than attempts to develop new technologies.

Energy storage for home electrical load-leveling purposes

The demand for electrical energy varies throughout the day, being highest in the daytime and evening and lowest at night. Installing generating capacity to meet peak demand is expensive as much of it would not be used during periods of low demand. Ideally, electrical utilities would like their customers to consume power at a constant rate throughout the day. Attempts to shift electrical power consumption from times of high consumption to times of low consumption is referred to as "load-levelling."

During times of peak power demand, the load for electrically-heated homes makes up a significant percentage of the total demand. One example of load-levelling currently being promoted is therefore aimed at electrically heated homes. The owners of such homes are encouraged to use a heat storage system and consume power for heating during periods of low demand. Such a system consists of a refractory brick with a checker-board geometry heated at night by electrical heating elements. During the day, heat is withdrawn as needed by blowing air through the pre-heated refractory brick. From the perspective of the electrical utility demand is reduced during the daytime, very effectively reducing the required generating capacity required to meet peak-power demand. Furthermore, by installing an appropriate time-of-day energy monitoring system, the cost savings to the home owner can be appreciable as electrical power costs are much lower during periods of low demand than during periods of high demand.

Such a heat storage system has a number of general requirements for the materials of construction, heat storage ability, size and weight. The system should be large enough to store sufficient heat to meet daytime heating requirements. This will be governed by the size of the home, local weather conditions, the insulating properties of the home's doors and windows, and all variables which affect home heat losses. At the same time it should be sufficiently small and light weight to allow convenient installation within the average home. These simultaneous requirements can be met by selecting a construction material exhibiting a high value of specific heat coupled with a low value of density. The total amount of material required for heat storage can be minimized by storing the heat at as high a temperature as possible, requiring a material with a high melting point, as well as a correspondingly high mechanical stability so that the material will not deform under its own weight at the temperature involved. This temperature will most likely be governed by the properties of the materials used in the heating elements, such as NiChrome or a similar high-temperature alloy, limiting the peak temperature to a value not much above 1000 °C. If these systems are to receive wide-spread use, the materials chosen must also be low in cost. An appropriate fully dense fire-clay refractory is expected to meet all of the above requirements. The required geometry for such a refractory can be quite similar to that for the MHD-recuperator, described earlier. Loosely stacked refractory bricks should serve as well.

If these systems are to be effective, heat losses during periods when no heat is withdrawn must be kept to a minimum. Furthermore, the external temperatures of the heat storage units must be sufficiently low as to present no safety hazards. The active part of the heat storage system should be surrounded by a material which is a very effective thermal insulator, such as porous silica refractory, or, in locations where the heat storage refractory does not need to be supported, i. e., at the top, a high silica refractory wool, of the type discussed earlier, could be used. The use of such wool would also help to minimize the total weight.

THERMAL MANAGEMENT

Subjecting a material to excessive temperatures generally has a negative effect on its chemical stability, creep and fatigue resistance and other performance criteria. As a result, long-term performance and durability can be seriously affected.

Avoiding excessive temperatures by appropriate design and materials selection is referred to as "thermal management." This is not a recently developed concept. For example, the cylinder walls of an internal combustion engine depend on thermal management for their durability. High engine efficiency depends on minimizing heat losses through the cylinder walls. This can be achieved by operating the engine with cylinder wall temperatures as high as possible. Unfortunately, increased temperatures also lead to increased wear rates and a corresponding decrease in engine life-time. Clearly, a trade-off needs to be made. Therefore, engine blocks and cooling systems have been designed so that cylinder wall temperatures are well below those of the combustion gases, so long as the coolant level is maintained. This enhances long-term durability at the expense of engine efficiency. One attempt to solve this dilemma some years ago was based on replacing the cylinder walls with sleeves made of stabilized zirconia, resulting in the so-called "adiabatic" (uncooled) internal combustion engine. Unfortunately, this solution, although possibly feasible in a technical sense, was regarded as impractical in view of the high cost of the zirconia.

The use of internally cooled blades in aircraft turbine engines represents another example of the use of thermal management. It is well known from thermodynamic principles that the efficiency of energy conversion systems, say from chemical to mechanical energy, increases with temperature. For this reason, the efficiency of a turbine engine can be increased significantly simply by increasing the turbine inlet temperature. Unfortunately, this would also increase the temperature of the blades in the turbine hot zone. In turn, the blades would exhibit increased creep deformation, the primary design variable in turbine engine design. To circumvent this problem the blades are hollow and are internally cooled by passing air through them. This permits an increase in the inlet temperature without a corresponding increase in blade temperature. Clearly, some energy is lost as the result of the blade cooling. This loss, however, is more than compensated for by the overall increase in engine efficiency due to the increase in turbine inlet temperature.

Note that material substitution offers an alternative approach to increased efficiency of aircraft, truck and automotive engines. This idea was the driving force behind the use of structural ceramics such as silicon carbide or nitride in these engines. These materials have a much higher melting point and greater creep resistance compared to the values for aerospace alloys, thereby allowing an increase of turbine inlet temperature of at least a few hundred degrees, resulting in a significant corresponding increase in efficiency. Unfortunately, in spite of much effort and expense, the effective use of structural ceramics in turbine engines has yet to be demonstrated, primarily because of their high degree of brittleness and associated low impact resistance in combination with low thermal shock and thermal fatigue resistance. Furthermore, the cost of structural ceramics still exceeds that of aerospace alloys by an order of magnitude. Unless their cost can come down and a solution is found for their brittleness, the use of structural ceramics will continue to be limited to low load-bearing applications.

We will present two specific examples of thermal management.

Basic-Oxygen-Furnace

Thermal management also finds applications in the control of corrosion rates and the durability of chemical process vessels. A specific example of this is the basic-oxygen furnace (BOF) used for the purification of molten iron from a blast furnace. The BOF operates in a batch mode, i. e, it is charged with molten iron, scrap steel, etc., which is purified and discharged, after which a new cycle begins. The life-time of a BOF is measured in terms of the number of charging and discharging cycles per refractory lining. The circular BOF lining generally consists of a tar-impregnated magnesia refractory. The cost of relining a BOF, in terms of both time and expense, is not insignificant. Accordingly, the cost of producing steel can be minimized by extending the life of the refractory lining. Chemical corrosion of the liner by the molten steel is the major cause of BOF liner degradation. Because chemical corrosion is a thermally activated process, its rate increases rapidly with increasing temperature. Accordingly, the life-time of the BOF-lining can be extended by minimizing the temperature at its hot face. This temperature is controlled by the thermal conductivity value and thickness of the magnesia refractory.

Let us make an estimate of the maximum thickness of magnesia lining to assure that the hot-face temperature does not exceed a specified value. For a first approximation, we can ignore the slight curvature of the wall and treat it as a flat plate. We'll assume that the liner at its cold side is in perfect thermal contact with a water-cooled steel shell at a temperature of 25 °C. The hot-face temperature is taken as 1625 °C. The magnesia refractory is assumed to exhibit a temperature independent thermal conductivity of 5 W/m·K.

From the basic definition of steady-state heat flow:

$$q = \frac{K(\Delta T)}{(\Delta x)} \tag{8}$$

where q is the heat flux, i. e., the heat flow per unit area per unit time, K is the thermal conductivity, and (ΔT) is the temperature difference across the thickness, (Δx), of the lining. The maximum thickness, $(\Delta x)_{max}$ of the liner can be derived to be:

$$(\Delta x)_{max} = \frac{K(\Delta T)}{q} \tag{9}$$

Assume the heat flux to be conducted through the magnesia liner is taken as 1.6×10^4 W/m^2. Substitution of the above values yields:

$$(\Delta x)_{max} = 0.5 \text{ m} \tag{10}$$

If the thickness were to exceed this value, say $(\Delta x) = 0.6$ m, (ΔT) would rise to 1920 °C with a corresponding hot-face temperature of 1945 °C, i. e., a rise in temperature of some 300 °C. Even for low values of activation energy for the corrosion process, such a temperature rise would increase the rate of corrosion appreciably, with a corresponding significant decrease in the life of the lining.

It should be noted that, in general, refractory linings for high-temperature processing equipment, such as ovens, furnaces and kilns, serve as thermal insulators in order to reduce the energy requirements. This is not the case with the BOF. In fact, increasing the thermal conductivity of the magnesia liner at the expense of increased heat flow would permit increasing its thickness, thereby extending liner life-time even further.

Solid-state electronics

The microminiaturization of solid-state electronics has resulted in an ever increasing density of active components and a corresponding increase in the rate of heat generation per unit volume, associated with increases in operating temperature. Because the electrical characteristics of semi-conductors are thermally activated, excessive temperatures could result in circuit malfunction. Consequently, solid-state electronic design is now placing increased emphasis on the efficient removal of heat from solid-state devices. Many methods of heat removal are being pursued, either singly or in combination, such as using support structures with increased thermal conductivity, high thermal conductivity CVD-diamond coatings placed directly on the circuitry, forced convective cooling or surrounding the circuitry with a high thermal conductivity gas, such as helium, or an appropriate fluid with a low dielectric constant, such as Fluorinert, made by the 3M Company.

As a simple illustrative example, let us look at a substrate used to support a silicon chip. Consider a unit area (1 cm^2) of a silicon chip operating at a uniform temperature not to exceed 25 °C. This chip is attached to a dielectric ceramic substrate with a thickness of 0.2 cm. The substrate is in perfect thermal contact with a coolant medium at a temperature of 20 °C. The silicon chip is attached to the ceramic substrate with a thermal adhesive having a thermal resistance (the temperature difference across an interface for a heat flux of 1 W/cm^2) of 0.2 $K/W/cm^2$. Assuming that the silicon chip generates heat at a rate of 10 W/cm^2, we wish to calculate the minimum value of thermal conductivity of the substrate so that the chip temperature will not exceed its design temperature of 25 °C, i. e., 5 °C above the temperature of the cooling medium.

First, we need to calculate the temperature differences across the thermal adhesive. For a heat flux of 10 W/cm^2 and a thermal resistance value of 0.2 $K/W/cm^2$, this temperature difference will be 10 $W/cm^2 \times 0.2$ $K/W/cm^2$ = 2 K (or 2 °C as temperature differences in K or °C are numerically the same). Accordingly, the substrate will be subjected to a temperature difference, 5 - 2 = 3 °C.

In general, the heat flux or flow per unit area per unit time, q, through the substrate is given by:

$$q = \frac{K(\Delta T)}{(\Delta x)} \tag{11}$$

where K is the thermal conductivity and $(\Delta T)/(\Delta x)$ is the temperature gradient. Substitution of $q = 10$ W/cm^2, $(\Delta T) = 3$ °C and $(\Delta x) = 0.2$ cm yields the minimum required thermal conductivity value of about 0.67 W/cm·K or 67 W/m·K.

Among various candidate materials for a dielectric ceramic substrate are cordierite, alumina, aluminum nitride, silicon carbide and beryllium oxide, with corresponding thermal conductivity values of about 5 and 30 W/m·K for the cordierite and alumina, respectively, and near 200 W/m·K for the aluminum nitride, silicon carbide and beryllium oxide. Clearly, for the above example the cordierite and alumina exhibit thermal conductivity values which will not meet the design requirements, leaving the AlN, SiC and BeO as possible choices. BeO is known to be very toxic and may be rejected for that reason. It should be noted that cordierite still is a material of choice for a number of electronic substrate purposes, because its coefficient of thermal expansion matches more closely with that of silicon. The example above, however, indicates that the effective use of cordierite is limited to those applications involving relatively low rates of heat generation.

FINAL REMARKS

It should be evident to the reader that even this rather brief discussion of the selection and design of ceramic materials can be quite complex. For materials intended for thermal insulation this process is probably the simplest. The primary function for thermal insulation is to minimize the rate of heat flow, without additional functional requirements. Material selection and design for heat storage and thermal management purposes is more complicated, as the material of choice not only must fulfill the requirements of high thermal conductivity, but must also fulfill load-bearing, electrical, thermal expansion and other simultaneous requirements. A specific material which meets all requirements may not always be found. If this is the case, appropriate trade-offs will then be in order.

RECOMMENDED LITERATURE

[1]R. A. Hinrichs, *Energy*, Saunders College Publishing, A Harcourt Brace Jovanivich College Publisher, Orlando, Florida, 1992.

[2] H. S. Carslaw and J. C. Jaeger, *Conduction of Heat in Solids*, 2nd ed. Oxford at the Clarendon Press, 1959.

[3]D. W. Richardson, *Modern Ceramic Engineering*, 2nd. ed. Marcel Dekker, New York, 1992.

[4]D. W. Richardson, *The Magic of Ceramics*, The American Ceramic Society, Westerville, Ohio, 2000.

PROBLEMS

1. This problem is a first (back-of-the-envelope, open-ended) estimate, with an answer whose accuracy depends on the assumptions made and the accuracy of the data acquired:

   a. Estimate the number of power plants that could be shut down in the state of Florida if each household replaced their tungsten filament light bulbs with fluorescent bulbs with the same light output. Present your answer in terms of the reduction in rates of peak-power consumption and annual energy savings.

   b. What is the total annual reduction in CO_2 emissions if all the power plants are assumed to be coal-fired?

   c. How many tons of coal are saved per year?

   d. Estimate the resulting reduction in air-conditioning needs and the average per household savings per year.

2. Estimate the total energy required to bring the lining of a basic-oxygen furnace up to temperature. State all the assumptions and simplifications you made in obtaining your answer.

3. During normal use, the electrical conductivity of a silicon semi-conductor should not vary by a factor of two. Using an introductory textbook on materials and the appropriate data for silicon, estimate the range of temperatures at which the silicon must operate in order to meet the above performance criterion. Note that your answer will depend on your assumptions.

4. Determine the thickness of insulation in the ceiling of an existing home (i. e., your parent's house, a friend's house, etc.) and estimate the annual energy savings realized if the insulation thickness were doubled. Is the required cost of the additional insulation a good investment? If so, estimate the rate of return of this investment and compare it with the interest rates earned on savings accounts,

certificates of deposit and other investments.

5. Make a survey of the houses on the street where you live and establish how many occupants are aware of their total heating and lighting costs. What is the potential savings for the street as a whole when the ceiling insulation in these dwellings is doubled in thickness? How does your answer compare to the cost of tuition for going to college?

6. A fireplace is constructed with a three foot long cast iron frame with damper held within a chimney with a firebrick lining. Make an estimate of the minimum spacing between the cast iron and firebrick so that the cast iron is free to expand without cracking the chimney. For a given spacing between the cast iron and firebrick estimate the maximum temperature rise of the fire such that fracture of the chimney is avoided.

8

Designing for Severe Thermal Stresses

D.P.H. Hasselman

K.Y. Donaldson

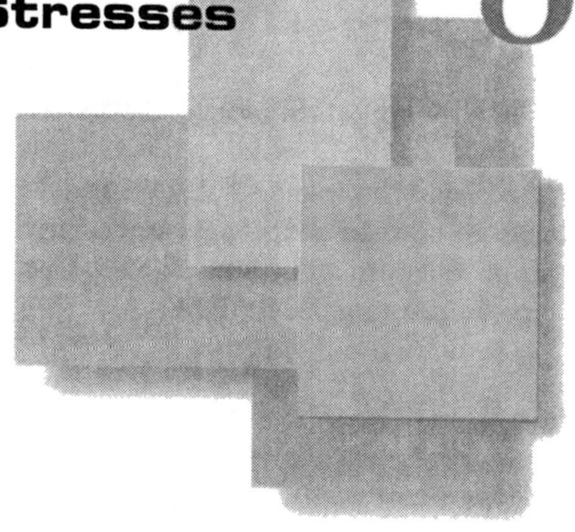

DESIGNING FOR SEVERE THERMAL STRESSES

D. P. H. Hasselman and K. Y. Donaldson
Department of Materials Science and Engineering
Virginia Polytechnic Institute & State University
Blacksburg, Virginia 24061 USA

ABSTRACT

This chapter serves to present an overview of the role thermal and mechanical properties play in the thermal stress resistance of brittle structural ceramics. Expressions are presented for thermal stresses for a wide range of thermal boundary conditions, including constrained thermal expansion under isothermal conditions and steady-state and transient heat flow. For a known criterion of failure, expressions for the thermal stresses are rearranged to obtain expressions for the maximum thermal conditions to which a structural component can be subjected without initiation of failure. In turn, a number of appropriate figures-of-merit, referred to as "thermal stress resistance parameters", are derived for the selection of the ceramic material with the optimum thermal stress resistance. For severe thermal shock, where the onset of failure cannot be avoided, material selection must be based on minimizing the resulting damage, such as loss of load-bearing ability. An appropriate "thermal shock damage parameter" is presented.

INTRODUCTION

Due to their extreme brittleness, structural ceramic materials are particularly susceptible to fracture by thermal stresses. Materials selection for designs subjected to high magnitudes of thermal stresses is quite complex and depends on the nature of heat transfer and failure mode. Design specifications, such as a maximum temperature difference or rate of heat flow under steady-state conditions, maximum ambient temperature changes on transfer from one medium to another, maximum rates of heating and thermal fatigue life need to be considered as well. Under conditions of thermal stress where the onset of thermal stress failure cannot be avoided, the main criterion of thermal stress resistance becomes the retention of load-bearing ability.

The purpose of this paper is to present a discussion of the relevant material properties which affect thermal fracture for a range of thermal boundary conditions. The appropriate material selection rules will be formulated in terms of so-called "thermal stress resistance parameters."

THERMAL STRESS DUE TO EXTERNAL CONSTRAINTS UNDER ISOTHERMAL CONDITIONS

Consider a beam with its ends fixed between rigid constraints and subjected to a uniform temperature change, (ΔT). Due to thermal expansion, the beam undergoes a thermal strain of $\alpha(\Delta T)$, where α is the linear coefficient of thermal expansion. Because the ends are fixed (i. e., are unmovable) the external constraints must apply a stress, σ, to compensate, resulting in a corresponding elastic strain, $\epsilon = \sigma/E$, where E is Young's modulus. The final length of the beam must be unchanged, so the sum of the thermal and elastic strain equals zero, such that,

$$\alpha(\Delta T) + \frac{\sigma}{E} = 0 \tag{1}$$

Rearranging yields:

$$\sigma = -\alpha E(\Delta T) \tag{2}$$

Note that when (ΔT) is positive, i. e., the beam is heated, the stress in the beam is negative (compressive). Conversely, when (ΔT) is negative, i. e., the beam is cooled, the stress is positive (tensile).

In order to avoid failure the stress should not exceed the strength. As thermal stress failure of brittle structural ceramics generally occurs in tension, the failure stress, σ_f, is the tensile strength. Substitution of the tensile strength, obtained experimentally or estimated from fracture-mechanical principles, for the stress in equation 2 results in the maximum temperature difference by which the beam can be cooled or heated:

$$(\Delta T)_{\max} = \frac{\sigma_f}{\alpha E} \tag{3}$$

Note that the result of equation 3 is based on the implicit assumption of a safety factor equal to unity. If a more conservative approach is taken, the result for $(\Delta T)_{max}$ is reduced accordingly.

A similar derivation for a biaxially constrained flat plate undergoing a uniform temperature change yields:

$$(\Delta T)_{max} = \frac{\sigma_f (1 - \nu)}{\alpha E} \tag{4}$$

where ν is Poisson's ratio.

Because the above examples involve thermal stresses resulting from uniform changes in temperature, without explicitly involving heat flow, $(\Delta T)_{max}$ is not affected by the thermal conductivity or other heat transfer variables. We can make the general observation that, regardless of the configuration and thermal conditions being encountered, the magnitude of thermal stress is always proportional to the product αE.

From the perspective of selecting the material with the highest resistance to failure by thermal stress, the relevant material properties should be such that the ratio on the right-hand side of equation 4 is a maximum. This ratio is referred to as the "thermal stress resistance parameter", denoted in the ceramic literature as "R", defined by:

$$R = \frac{\sigma_f (1 - \nu)}{\alpha E} \tag{5}$$

Note that high values of R require high values of tensile strength in combination with low values of coefficient of thermal expansion, Poisson's ratio and Young's modulus.

We will encounter the parameter R for other heat transfer conditions along with other parameters as well in subsequent sections of this paper. In general, these "thermal stress resistance parameters" serve as aids in materials selection for a component of given size and geometry subjected to known heat transfer conditions.

The following numerical examples will help illustrate the results of equations 3 and 4. First consider a beam made of a structural alumina, of the type used as an electronic substrate, with typical values of $\alpha = 7 \times 10^{-6}\ {}^{\circ}C^{-1}$, $E = 400$ GPa and $\sigma_f =$ 400 MPa. Substitution of these values into equation 3 yields the maximum temperature difference by which the beam can be cooled:

$$(\Delta T)_{max} = -143^{\circ}C \tag{6a}$$

For a biaxially constrained plate made of the same material with $\nu \approx 0.26$, equation 4 yields:

$$(\Delta T)_{max} = -106°C \qquad (6b)$$

If we had been more conservative and had selected a safety factor of three, the values calculated for $(\Delta T)_{max}$ in equations 6a and 6b would have been reduced to about 48 and 35 °C, respectively. These values are quite low indeed, and illustrate that alumina exhibits a rather low thermal stress resistance. High-strength structural ceramics such as silicon nitride and silicon carbide have values of the coefficient of thermal expansion about one half to two thirds of that of alumina, so the values calculated for $(\Delta T)_{max}$ will be correspondingly higher. This is the prime reason why silicon nitride and silicon carbide were considered as candidate materials for the proposed ceramic turbine engine, rather than structural alumina.

Let us pursue this somewhat further. If we were to consider a refractory material, a material generally used in high temperature furnace linings, with orders of magnitude lower strength values than those of the high-strength alumina in the previous examples, the corresponding values for $(\Delta T)_{max}$ on cooling may well be less than 10 °C. Accordingly, the resistance of refractories to thermal stress fracture is very low indeed. The operating environment of refractories, however, is such that thermal stresses of high magnitudes are expected. For this reason, an alternative approach to the materials selection and design of refractories had to be adopted. This subject will be discussed shortly.

Another example of constrained thermal expansion deals with thermal buckling, a failure mode most commonly associated with instability in structural members subjected to compressive loads. In thermal buckling compressive loads result from constrained thermal expansions.

Consider a beam with pinned ends which is uniaxially constrained from thermal expansion, but whose ends are free to rotate around frictionless pins. The length of the beam is L, its width is b, and its thickness is d. The beam has a corresponding "slenderness ratio", an indicator of the tendency for failure by buckling, of L/d. The beam is uniformly heated over the temperature range (ΔT). We wish to estimate the maximum allowable increase in temperature $(\Delta T)_{cr}$ of the beam before the onset of structural instability.

The thermal strain in the beam equals $\alpha(\Delta T)$, with a corresponding stress of $\alpha E(\Delta T)$. The total load, P, on the beam is the product of the stress and its crossectional area, which in this case is given by:

$$P = \alpha E(\Delta T) b d \qquad (7)$$

For a slender beam with boundary conditions corresponding to a column with pins at both ends as described above, the critical buckling load, P_{cr}, is given by the well-known Euler-formula as:

$$P_{cr} = \frac{\pi^2 E I}{L^2} \tag{8}$$

where I is the minimum crossectional moment of inertia of the beam, in this case the minimum of $bd^3/12$ or $db^3/12$ for a rectangular crossection. Substitution of P_{cr} from equation 8 into equation 7 yields:

$$(\Delta T)_{cr} = \frac{\pi^2 I}{\alpha L^2 b d} \tag{9}$$

The result of equation 9 is rather interesting and possibly counterintuitive. Although the critical load for instability, given by equation 8, increases with Young's modulus, the critical temperature difference for thermal instability, given by equation 9, is independent of Young's modulus. This occurs because the thermally induced load itself is directly proportional to Young's modulus. From the perspective of materials selection, thermal instability is a function only of the coefficient of thermal expansion, so the appropriate thermal stress resistance parameter, R_{tb}, for thermal buckling, is:

$$R_{tb} = \alpha^{-1} \tag{10}$$

As another example of constrained thermal expansion, let us examine the stresses resulting from a mismatch of thermal expansion between two materials bonded together. Consider the limiting case of a very thin film firmly bonded to a thick substrate. The film's coefficient of thermal expansion is α_f and the substrate's coefficient of thermal expansion is α_s. The difference in the coefficients of thermal expansion is $(\Delta\alpha) = \alpha_s - \alpha_f$. The film is placed on the substrate at a given temperature at which it is stress-free. The substrate plus film are then subjected to a uniform temperature change, (ΔT). The stress, σ, in the film as the result of the thermal expansion mismatch is:

$$\sigma = \frac{(\Delta\alpha) E (\Delta T)}{(1 - \nu)} \tag{11}$$

where E and ν are Young's modulus and Poisson's ratio, respectively, of the film.

When $(\Delta\alpha)$ is negative, i. e., $\alpha_f > \alpha_s$, and (ΔT) is negative, the stress in the film is positive, i. e., tensile. In this case the film will fail in tension. Similarly, if $\alpha_f < \alpha_s$ and (ΔT) is negative, the film is in a state of compression. Depending on the degree of adhesion between the film and substrate, the film may fail by compression for the case of strong adhesion (for which equation 11 is valid) or fail by forming blisters when the adhesion is weak. In general, a tensile stress state in the film is to be avoided, whereas low values of compression in the film are regarded as beneficial.

We are interested in obtaining an expression for the maximum temperature difference, $(\Delta T)_{max}$, by which we can cool or heat the substrate and film from the temperature at which the film is stress-free, i. e., the temperature of deposition of the film. Let us assume the film fails in tension. Setting the stress, σ, in equation 11 equal to the tensile failure stress, σ_f, yields:

$$(\Delta T)_{max} = \frac{\sigma_f (1 - \nu)}{(\Delta\alpha) E} \tag{12}$$

Equation 12 indicates that in order to avoid film failure, its tensile strength should be as high as possible, its Poisson's ratio, Young's modulus and coefficient of thermal expansion should be low, and the mismatch between the coefficients of thermal expansion of the substrate and film should be kept to a minimum. This conclusion is particularly relevant to solid-state electronic devices composed of a multitude of different materials, which frequently operate at temperature levels far different from those at which they were made. For such devices the prevention of premature failure due to the stresses which arise from thermal expansion mismatches is a key design criterion.

A general point needs to be made regarding the above examples. Thermal stresses which arise from uniform temperature changes under conditions where free thermal expansion is constrained are not affected by any of the material properties which affect heat transfer, such as thermal conductivity or diffusivity, emissivity, reflectivity or transparency, or by the external heat transfer conditions, such as convective or radiative heat transfer. Clearly, these latter variables come into play for thermal boundary conditions different from those for uniform temperature changes.

STEADY-STATE HEAT FLOW

Zero thermal stress state in an unconstrained solid undergoing linear heat flow

Heat flow through a structure inevitably involves non-isothermal conditions.

However, the resulting temperature non-uniformity does not automatically result in thermal stresses. Any temperature distribution in an externally unconstrained solid which is linear with respect to a rectangular coordinate system will result in a zero thermal stress state, since any thermal strains and displacements which result from such a temperature distribution can be accommodated by the appropriate elastic deformations of the solid. For example, an unconstrained beam or plate undergoing uniform linear heat flow through its thickness will undergo simple bending.

The importance of this concept cannot be overstated. Thermal stresses will arise in unconstrained solids only when temperature distributions are non-linear with respect to a rectangular coordinate system, such as the thermal stresses that arise from spatially non-uniform heat flow or temperature-dependent thermal conductivity. There seems to be a general misconception in the ceramics literature that thermal stresses arise from the existence of "temperature gradients." This cannot be supported on theoretical grounds. For more information regarding this issue, the reader is urged to consult the textbooks and reference 6 listed at the end of this chapter.

Thermal stresses in a constrained solid undergoing linear heat flow

Let us consider a simple example of spatially uniform linear heat flow through the thickness, d, of a beam which is free to expand laterally, but is constrained from bending. The temperature varies from $(\Delta T)/2$ at the hotter surface of the beam to $-(\Delta T)/2$ at the colder surface, where (ΔT) is the temperature difference through the thickness of the beam.

The thermal stress varies linearly along the thickness of the beam and exhibits a maximum (σ_{max}) at the surface, given by:

$$\sigma_{max} = \alpha E \frac{(\Delta T)}{2} \tag{13}$$

with α and E as defined previously. The stress given by equation 13 is compressive and tensile at the hot and cold surfaces of the beam, respectively. Note that for this particular example and thermal boundary condition, the magnitude of thermal stress is independent of the thickness of the plate.

As an illustration of the result given by equation 13, we are interested in selecting the material which will permit the highest value of $(\Delta T) = (\Delta T)_{max}$ without failure occurring. As failure is more likely to occur in tension than in compression we are looking for a material with optimum tensile strength, σ_f. Setting σ_{max} equal to σ_f yields:

$$(\Delta T)_{max} = \frac{2\sigma_f}{\alpha E} \tag{14}$$

As a numerical example, let us again take a structural aluminum oxide used for electronic substrate purposes with $\sigma_f = 400$ MPa, $E = 400$ GPa, $\nu = 0.26$ and $\alpha = 7 \times 10^{-6}$ °C^{-1}. Substitution of these values into equation 14 yields:

$$(\Delta T)_{max} = 286°\text{C} \tag{15}$$

In order to obtain a value for $(\Delta T)_{max}$ for a thin flat plate constrained from bending by an appropriate bending moment applied around its edges, equation 14 needs to multiplied by the factor $(1 - \nu)$. Using the values for aluminum oxide, $(\Delta T)_{max} \approx 212$ °C. Again note that these values for $(\Delta T)_{max}$ are based on the assumption of a safety factor of unity. Applying a common safety factor of three or higher reduces the values for $(\Delta T)_{max}$ by the same factor. The resulting values are less than 100 °C and far less than those frequently encountered in high-temperature technology. Again, this clearly indicates that optimum design should allow free thermal deformations to the fullest extent possible. A total absence of external constraints, resulting in a zero thermal stress state, is always preferred.

In general, as indicated by equation 14 multiplied by the term $(1 - \nu)$ to take into account multiaxial effects, material selection for maximum thermal stress resistance for steady-state heat flow based on the criterion of maximizing $(\Delta T)_{max}$ should be based on the thermal stress resistance parameter $R = \sigma_f(1 - \nu)/\alpha E$, as presented earlier.

For structures such as heat exchangers, the maximum allowable heat flux , i. e., heat flow per unit area per unit time, is expected to be the more critical design parameter rather than the maximum temperature difference across the plate thickness. The heat flux, q, can be related to the plate thickness, d, by:

$$q = \frac{K(\Delta T)}{d} \tag{16}$$

where K is thermal conductivity. Substitution of dq/K for (ΔT) in equation 14 yields the maximum allowable heat flux:

$$q_{\max} = \frac{2\,\sigma_f K}{\alpha\, E\, d} \tag{17}$$

Multiplication of the right term in equation 17 with (1 - ν) yields q_{max} for a plate which is biaxially constrained from bending. Note now that q_{max} is an inverse function of the plate thickness, as expected, as a thin wall will conduct more heat for a given value of (ΔT) than a thicker wall.

Material selection from the perspective of maximizing heat flow for a given value of plate thickness and temperature difference should be based on the thermal stress resistance parameter $\sigma_f K/\alpha E$, which when multiplied by the factor (1 - ν) yields the well-known thermal stress resistance parameter, R', given by:

$$R' = \frac{\sigma_f (1 - \nu) K}{\alpha\, E} \tag{18}$$

Note that in contrast to the form of the thermal stress resistance parameter, R, presented earlier and used for maximizing temperature differences, R' reflects the important role of thermal conductivity for maximizing heat flux.

A comparison of the thermal stress resistance parameters R and R' permits making a general point. Optimum material selection for reducing the probability of thermal stress fracture depends on the performance criterion of the component or structure under consideration. For this reason, the ranking of materials with optimum thermal stress resistance is governed by the performance criteria of the specific design under consideration. No general ranking of thermal stress resistance of materials can be given which would be valid for all geometries, designs and thermal boundary conditions. Furthermore, even the best material selected on the basis of the thermal stress resistance parameters discussed above may prove inadequate.

Thermal stresses for steady-state heat flow with internal constraints

A hollow cylinder, such as a tube in a heat exchanger, exposed to radially inward or outward heat flow represents a geometry which will exhibit thermal stresses as the result of internal constraints. A temperature distribution in a hollow cylinder is not linear with respect to a rectangular coordinate system. Any change in curvature in the cylinder wall due to the presence of the radial temperature distribution cannot be accommodated due to the constraints provided by the cylinder wall itself. Radial displacements cannot be accommodated freely due to the radial temperature nonuniformity. Strictly from a design perspective, the magnitude of thermal stresses can be minimized by reducing the ratio of wall thickness to the internal or external radius.

The resistance to thermal stress failure of a hollow cylinder undergoing radial heat flow can be based on the maximum temperature difference across the wall or the maximum heat flow per unit length, as we have seen before for the constrained flat plate. The respective thermal stress resistance parameters appropriate for a hollow cylinder or any other internally constrained geometry undergoing steady state heat flow are given by the parameters R and R', presented earlier.

TRANSIENT THERMAL STRESSES

Convective heat transfer

Quite frequently in practice a component is subjected to an instantaneous change in ambient temperature, colloquially referred to as "thermal shock." The "flame-out" of a turbine engine at normal operating temperature due to loss of ignition represents an extreme example of thermal shock. In this case, the turbine blades are suddenly subjected to the much colder non-combusted fuel mixture rather than the usual hot gases from the combustor. A much more mundane example is the immersion of porcelain or glass dinnerware or other household items in cold and then hot water or vice-versa during normal dishwashing procedures. Generally under these conditions heat transfer occurs by convection, described by the "convective heat transfer coefficient."

Let us consider a flat plate, initially at thermal equilibrium at a higher temperature. It is then instantaneously surrounded on both sides by a gaseous or fluid medium at lower temperature. A time-dependent spatially non-uniform temperature distribution and associated thermal stresses will arise. For this particular example the surface will be in a state of tensile stress, whereas the stresses at the center of the plate will be compressive. Failure is anticipated to occur in the surface.

For this plate, we are interested in the maximum temperature difference $(\Delta T)_{max}$ between the new medium and the plate's original temperature to which the plate can be subjected without failure due to the resulting thermal stresses. Solutions for the transient temperature and thermal stresses are available from the literature. These are mathematically rather complex and will not be repeated here. However, from the known values of maximum stress in terms of the relevant heat transfer variables and material properties an approximate expression for $(\Delta T)_{max}$ is:

$$(\Delta T)_{max} = \frac{\sigma_f(1 - \nu)\left(1 + \frac{4K}{ah}\right)}{\alpha E} \tag{19}$$

The ratio *K/ah* is the inverse of the Biot number, $\beta = ah/K$, where h is the heat transfer coefficient, a is the half-thickness of the plate and K is the thermal conductivity, all other symbols in equation 19 having been defined before.

Equation 19 can be written for two limiting regimes:

$$\text{For } \frac{4K}{ah} \ll 1: \quad (\Delta T)_{max} = \frac{\sigma_f(1 - \nu)}{\alpha E} \tag{20}$$

Equation 20 describes $(\Delta T)_{max}$ for the limiting case of a very thick plate with low thermal conductivity subjected to thermal shock in a medium with a high value of heat transfer coefficient. In general, this describes the situation in which the surface of the plate has attained the temperature of the medium well before the center of the plate has started to cool. In other words, the temperature difference between the surface and center of the plate, which governs the magnitude of the thermal stress, equals the difference between the plate's initial temperature and the fluid medium.

$$\text{For } \frac{4K}{ah} \gg 1: \quad (\Delta T)_{max} = \frac{4\sigma_f(1 - \nu)K}{\alpha E a h} \tag{21}$$

Equation 21 describes $(\Delta T)_{max}$ for a relatively thin plate with high thermal conductivity subjected to thermal shock in a convective medium with a low value of heat transfer coefficient. In this case, the temperature at the center of the plate is already decreasing well before the surface of the plate reaches the temperature of the fluid medium. It is the temperature difference within the plate which determines the magnitude of the thermal stress.

It is critical to note that the value of $(\Delta T)_{max}$ given by equation 21 can be much greater than that given by equation 20 because of its dependence on dimension, thermal conductivity and heat transfer coefficient. In practice, this is the general case, and describes the example discussed previously in which turbine blades are subjected to thermal shock by engine flame-out. $(\Delta T)_{max}$ of equation 20 applies to large bodies with low thermal conductivity subjected to thermal shock in a medium with a high value of heat transfer coefficient, such as water. The reader may have experienced this effect while washing large family-heirloom glass flower vases, fine china, or glassware using overly hot dish-water.

The general form of equation 19 for $(\Delta T)_{max}$ will also be valid for other geometries or thermal shock by heating, referred to as thermal "up-shock", by placing an appropriate constant in front and replacing the factor of four inside the brackets.

For any geometry, or whether thermal shock occurs by heating or cooling, material selection for optimum thermal shock resistance under convective heat transfer conditions again is based on the thermal shock resistance parameters, R and R', presented earlier.

It is critical to note that the relative thermal shock resistance of candidate materials subjected to thermal shock by convective heating or cooling is a function of the magnitude of the Biot number, $\beta = ah/K$. Experimental tests usually involve small laboratory samples, with test conditions best described by equation 21. These tests may indicate that one material is far superior to another. In the scaling up of these materials for actual designs, however, the accompanying significant increase in dimensions could create a thermal shock situation more appropriately described by equation 20 or by some expression intermediate to equations 20 and 21. If so, the relative thermal stress resistance of the candidate materials may well be interchanged.

It should also be noted that for thermal shock involving an instantaneous change in ambient temperature with convective heat transfer, the thermal stress resistance as measured by $(\Delta T)_{max}$ is not governed by the magnitude of the thermal diffusivity. The latter quantity enters the equations for the transient temperature and stresses in the general form of $A\kappa t/L^2$, where A is a constant, κ is the thermal diffusivity, t is the time and L is a dimensional measure of the component undergoing thermal shock. For this reason, the thermal diffusivity affects the time at which a certain temperature and stress event occurs, but not its magnitude. For materials with high thermal diffusivity values, the time duration of the thermal pulse is shorter than for those materials with low thermal diffusivity. It is critical to note, however, that under conditions of thermal fatigue, the increment of subcritical crack growth per thermal cycle is a function of the time duration of the thermal stress pulse. For this reason, keeping all other pertinent variables the same, improved thermal fatigue resistance measured in terms of number of cycles-to-failure is attained by selecting materials with high thermal diffusivity.

Radiative heat transfer

At the temperature levels encountered in many applications of structural ceramics, heat transfer is more likely to occur by radiation rather than by convection.

Of course, this will depend on many variables, such as the size of the component and whether the component is subjected to natural or forced convection. For large-sized bodies the temperature at which the radiative and natural convective components are equal can be close to room temperature. Near 1000 °C, however, natural convective heat transfer in air at atmospheric pressure represents a small fraction of the total heat transfer, radiative heat transfer being the primary component.

Thermal shock resistance under conditions of radiative heat transfer is expected to be affected not only by those material properties referred to above but by the relevant optical properties, such as transparency and absorptivity, as well. Furthermore, it should be noted that radiative heat transfer is a function of the value of absolute temperature rather than a temperature difference. For this reason, heat transfer by radiation is non-linear and, in general, transient temperature history cannot be described by general analytical equations.

Nevertheless, an approximate expression for the maximum temperature, T_{max}, can be derived for a spherical component at low initial temperature subjected to black-body radiation, with the condition that thermal fracture occurs when the component's surface temperature is still low enough that radiation from the component can be regarded as negligible. In effect, up to the time of fracture, the component is subjected to a constant heat flux. With these assumptions, T_{max} for a sphere can be derived to be:

$$T_{max} = \left[\frac{5\,\sigma_f (1 - \nu) K}{\alpha E \rho \varepsilon} \right]^{1/4} \tag{22}$$

where σ_f, ν, K, α, and E were defined previously, ρ is the Stefan-Boltzmann constant and ε is the absorptivity of the component surface or $\varepsilon = 1 - r$, where r is the reflectivity.

Examination of equation 22 reveals that a comparison of the relative thermal stress resistance of candidate materials subjected to thermal shock by radiation is governed by a total of six material properties. High radiative thermal shock resistance requires high values of tensile strength and thermal conductivity in combination with low values of Poisson's ratio, coefficient of thermal expansion, Young's modulus and absorptivity. Strictly on dimensional grounds, these conclusions regarding the role of material properties should also apply to other geometries and radiative cooling. Accordingly, an additional thermal stress resistance parameter can be defined:

$$R_{rad} = \left[\frac{\sigma_f (1 - \nu) K}{\alpha E \varepsilon} \right]^{1/4} \tag{23}$$

It should be noted that the absorptivity of structural ceramics, most of which are dielectric, is quite high compared to electrically conductive metals, with low absorptivity (high reflectivity). Metallizing a structural ceramic with a reflective coating offers one solution towards improving thermal shock resistance for radiative heat transfer for heating or cooling.

At least in principle, radiative thermal shock resistance can also be improved by enhancing the material's optical transparency. Any incident radiation transmitted through the surface and through the component's interior without internal absorption cannot contribute to the formation of thermal stresses. For instance, this approach should be applicable to windows and/or lenses for the purpose of concentrating solar energy.

TIME-DEPENDENT BOUNDARY CONDITIONS: INTERNAL CONSTRAINTS

Maximum rate of heating

The maximum rate at which one can heat or cool a ceramic component may be the critical design parameter. This represents a time-dependent boundary condition.

For a flat plate the maximum rate of increase of the surface temperature is given by:

$$\dot{T}_{max} = \frac{3\,\sigma_f (1 - \nu)\,\kappa}{\alpha\,E\,d^2} \tag{24}$$

where κ is the thermal diffusivity, d is the thickness of the plate, and all other symbols are as defined earlier. Note that for the criterion of maximum rate of increase in temperature, the thermal stress resistance is governed by the thermal diffusivity, rather than by the thermal conductivity. This would be expected strictly on dimensional grounds for any geometry and for cooling conditions as well. Also, note that the rate of change of surface temperature is prescribed, so the mode of heat transfer need not be specified.

For a numerical example let us consider a zirconia electrode for the MHD channel referred to in the earlier chapter by these authors. Assume the following typical values for zirconia: $\sigma_f = 100$ MPa, $\nu = 0.20$, $\kappa = 5 \times 10^{-7}$ m^2/s, $\alpha = 8 \times 10^{-6}$ °C^{-1}, $E = 300$ GPa and $d = 0.01$ m. Substitution of these values into equation 24 yields:

$$\dot{T}_{max} \approx 0.5\ °\mathrm{C/s} \tag{25}$$

Assume that the electrode surface temperature at normal operating conditions is near 1700 °C and is increased from ambient conditions, i. e., room temperature at 25 °C. At the maximum allowable rate of about 0.5 °C/s, it would take about one hour to reach the operating temperature if thermal fracture of the electrodes is to be avoided. This estimate is quite optimistic as it is based on a safety factor of unity. If the safety factor is taken as three, it would take some three hours to reach the operating temperature. The primary reason for these lengthy heat-up times is the rather low values for the thermal diffusivity of zirconia. Critical to note here is that zirconia represents about the only suitable candidate material for MHD channel operation, so substitution of a material with higher thermal diffusivity is not feasible. These long heat-up times preclude an MHD-channel from serving as an instant on-and-off power system to meet peaks in electrical demand. Steady-state operation to meet base-load requirements appears to be more suitable.

Maximum rate of internal heat generation

Thermal stresses due to internal heat generation are encountered in ceramic heating elements made from silicon carbide or molybdenum disilicide or in graphite electrodes for the metallurgical industry. Internal heat generation due to Ohmic heating coupled with heat losses which occur at the surface create temperatures which can be appreciably higher in the interior of the heating element than in the surface. Internal heat generation is also encountered in uranium oxide fuel pellets in atomic reactors, the heat being generated by the fission process. As all of these materials are quite brittle, the potential for thermal stress fracture resulting from excessive internal heat generation is a possibility which cannot be overlooked.

For a heating element or nuclear fuel pellet with a circular crossection, the maximum permissible rate of internal heat generation per unit volume, q_{max}, in order to avoid thermal stress fracture, can be expressed by:

$$q_{max} = \frac{8\,\sigma_f (1 - \nu) K}{\alpha\, E\, \Re^2} \tag{26}$$

where $\Re$ is the radius and all other symbols are as defined earlier.

Note that q_{max} is inversely proportional to the square of the cylinder radius. For this reason, graphite electrodes used in the steel making industry, because of their size, may be particularly prone to thermal stress fracture. Frequently the ratings of graphite electrodes with a known value of electrical resistivity are given in terms of maximum current density, amps/cm^2.

Materials selection for optimum thermal stress resistance, based on the criterion of maximum internal heat generation, is governed by the thermal stress resistance parameter, $R' = \sigma_f (1 - \nu) K / \alpha E$, which is identical to equation 18.

THERMAL SHOCK DAMAGE RESISTANCE

It is critical to note that for a given thermal shock environment, even when the optimum material is chosen using the selection rules outlined above, the avoidance of thermal fracture is not guaranteed. In fact, thermal fracture in ceramic components, especially those of large size, is relatively easy. In applications where thermal fracture cannot be avoided, an alternative solution had to be developed based on the concept that the damage, i. e., the extent of crack propagation, resulting from thermal fracture is minimized. This leaves the component, although partially cracked, still in a condition to render satisfactory service. Indeed, the development of refractories, a class of materials referred to earlier, is based on this concept. A survey of the properties of refractories used for blast-furnace linings and other applications involving severe thermal shock reveals that they exhibit values of tensile strength well below those considered acceptable for mechanical load-bearing requirements. However, these materials are used for their thermal insulation properties or resistance to corrosion and have no major load-bearing requirements other than supporting their own weight in compression. So even when fractured by some prior thermal shock event, they can continue to serve satisfactorily in their intended purpose.

The concept of "thermal shock damage resistance" requires its own selection rules. It appears evident that thermal shock damage resistance is based on the nature of crack propagation. High thermal shock damage resistance depends on a material's ability to arrest a propagating crack, so that it only partially traverses the component leaving the component essentially intact and still suitable for continued satisfactory service. The driving force for crack propagation derives from the elastic energy of the thermal stresses. The total extent of crack propagation is expected to be proportional to the elastic energy stored in the body undergoing thermal fracture. The arrest of the generated crack(s) is expected to be a function of the energy required to create new crack surfaces. Here surface roughness, plastic deformation and other energy dissipative processes play a role. Crack arrest is also expected to be a function of the number of cracks which are generated during fracture.

For a ceramic component of given size and geometry, subjected to a given thermal shock, the elastic energy (Q) should be proportional to the ratio:

$$Q \propto \frac{(\sigma_f)^2}{E} \tag{27}$$

where σ_f is the tensile failure stress and E is Young's modulus.

The total energy, W, required to propagate crack(s) over a given distance or area is proportional to:

$$W \propto N\gamma_f \tag{28}$$

where N is the number of cracks and γ_f is the total energy required to create a unit area of crack surface.

The total area, A_f, of newly created crack surface is expected to be proportional to the total elastic energy at fracture and inversely proportional to the energy required to create the new fracture surfaces or:

$$A_f \propto \frac{(\sigma_f)^2}{\gamma_f N E} \tag{29}$$

Note that the total extent of crack propagation is proportional to σ_f^2. Indeed, brittle structural ceramics with high values of tensile strength are known to fail in a spectacular and catastrophic manner.

Clearly, in order to minimize the extent of crack propagation the quantity A_f in equation 29 should be as small as possible. Accordingly, this requirement can be met by formulating the "thermal shock damage resistance parameter," defined in the literature as R'''' and expressed as:

$$R'''' = \frac{\gamma_f E}{(\sigma_f)^2} \tag{30}$$

Materials with high values of R'''', i. e., low values of tensile strength, high values of Young's modulus and high values of fracture energy, will exhibit superior thermal shock damage resistance compared to those materials with low values of this parameter. The requirement of low tensile strength and high Young's modulus for superior thermal shock damage resistance seems quite surprising as it is totally opposite to the requirement of high tensile strength and low Young' modulus for high thermal stress resistance from the perspective of avoiding the onset of fracture.

The above contradictory requirements present a major dilemma to the materials engineer. The avoidance of the onset of thermal stress fracture under any condition appears to be a desirable goal. For a given structural ceramic with given values of Young's modulus, thermal conductivity and coefficient of thermal expansion, thermal stress resistance can be attained only by increasing the tensile strength. If, however, for a severe thermal environment such increases in tensile strength are insufficient, the net effect is to create a condition under which thermal fracture occurs in an increasingly catastrophic and violent manner.

Note that decreasing tensile strength in order to increase thermal shock damage resistance can be achieved quite easily and at lower cost than increasing it to improve thermal stress resistance to avoid fracture. Indeed, examination of refractories reveals a microstructure of pores, large grains and the presence of cracks, features which should be avoided for high-strength ceramics used for structural purposes. Indeed, microcracked aluminum titanate is the material of choice for the ceramic lining of exhaust manifolds for high-performance internal combustion engines.

CONCLUDING REMARKS

The above discussion, even if brief, indicates that there is no general rule for the selection of materials with optimum thermal shock resistance which covers all possible situations. Each design needs to be examined on its own merits and the specific mode of heat transfer needs to be established (not always an easy task in itself). Furthermore, the performance criterion, such as rate of heating, etc., also needs to be formulated before the combination of material properties for optimum thermal shock resistance is established. The situation becomes even more complex when it is recognized that, with few exceptions, thermal shock encountered in practice is incidental to the primary function of the components or structure. Accordingly, high thermal shock resistance, from the perspective of avoiding the onset of fracture or minimizing thermal shock damage, may well be incompatible with the component's primary function. One specific example for such incompatibility is a turbine blade which is a load-bearing component and therefore its thermal fracture resistance cannot be based on minimizing thermal shock damage. In general, by promoting free thermal expansion and by keeping dimensions as small as possible, the probability of thermal stress failure can be minimized.

Materials selection and design for thermal shock is multidisciplinary in nature and should be conducted by a team made up of members with backgrounds in heat transfer, thermophysical properties, mechanical behavior including fracture, plus those responsible for the primary design.

RECOMMENDED LITERATURE

For more details the reader may wish to consult the following list of references:

[1]S. Timoshenko and J. N. Goodier, *Theory of Elasticity*, 2nd ed., McGraw-Hill, New York, 1951.

[2]B. A. Boley and J. H. Weiner, *Theory of Thermal Stresses,* John Wiley & Sons, New York, 1960.

[3]N. Noda, R. B. Hetnarski and Y. Tanigawa, *Thermal Stresses*, Lastran Corporation, Rochester, New York 2000.

[4]W. D. Kingery, H. K. Bowen and D. R. Uhlmann, *Introduction of Ceramics*, 2nd ed. John Wiley & Sons, New York, 1976.

[5]Z. Zudans, T. C Yen, W. H. Steigelmann, *Thermal Stress Techniques in the Nuclear Industry*, Elsevier Publishing, Amsterdam, the Netherlands, 1965.

[6]W. D. Kingery, "Factors Affecting Thermal Stress Resistance of Ceramic Materials," *Journal of the American Ceramic Society*, **38** [1] 3-15 (1955).

[7]W. R. Buessum, "Resistance of Ceramic Bodies to Temperature Fluctuations," *Sprechsaal,* **93**, 137-41 (1960).

[8]D. P. H. Hasselman, "Thermal Shock by Radiation Heating," *Journal of the American Ceramic Society,* **46** [5] 220-34 (1963).

[9]D. P. H. Hasselman, "Theory of Thermal Shock Resistance of Semitransparent Ceramics Under Radiation Heating," *Journal of the American Ceramic Society*, **49** [2] 103-04 (1966).

[10]J. P. Singh, J. R. Thomas., Jr., and D. P. H. Hasselman, "Thermal Stresses in Partially Absorbing Flat Plate Symmetrically Heated by Thermal Radiation and Cooled by Convection," *Journal of Thermal Stresses*, **3**, 341-49 (1980).

[11]D. P. H. Hasselman, "Elastic Energy at Fracture and Surface Energy as Design Criteria for Thermal Shock," *Journal of the American Ceramic Society,* **46** [11] 535-40 (1963).

[12]D. P. H. Hasselman, "Unified Theory of Thermal Shock Fracture Initiation and Crack Propagation in Brittle Ceramics," *Journal of the American Ceramic Society,* **52** [11] 600-04 (1969).

[13]D. P. H. Hasselman, "Thermal Stress Resistance Parameters for Brittle Refractory Ceramics, A Compendium," *American Ceramic Society Bulletin*, **49** [12] 1933-37 (1970).

PROBLEMS

The following problems are designed to be open-ended. You will need to consult a variety of literature sources for the required property data. Your answers will depend on the values found and the assumptions you make.

1. A fireplace is constructed consisting of a cast iron frame with damper held within

a chimney with a firebrick lining. The frame with damper is 3' long, 1' wide and ½' thick. Suppose that the mason who built the chimney did not provide any spacing between the cast iron frame and firebrick. You are asked to make an estimate of the total load between the frame and chimney for a blazing hot fire. Compare this load with the weight of an average automobile. Do you think that this load is sufficient to crack the chimney? Note that your answer will depend on the assumptions you make regarding the temperature of the fire.

2. Assume that equation 19 describes the stress in the blade of a turbine engine undergoing flame-out. Assume the blade is made from hot-pressed silicon nitride. From the literature or by other means find an approximate value for the convective heat transfer coefficient, as well as all other pertinent variables, for flame-out conditions. Calculate the thermal stresses for a range of blade thickness values. Make a graph of your results. For a range of safety factor values, estimate the maximum thickness value of the blade so that failure is avoided. Do you think it possible that the blades in a jet engine of the size used for a Boeing 777 can be made from silicon nitride? If so, would you go on the test flight? If no, why not?

3. Calculate the maximum rate of internal heat generation per unit length in a silicon carbide heating element as a function of its radius. Assume that such heating elements are used to heat a laboratory furnace measuring 3' × 3' × 3' with an inner hot zone of 2' × 2' × 2'. Assume that the walls of the furnace are insulated with a fire-clay refractory. The furnace needs to be operated at 1000 °C. How many of these silicon carbide heating elements of a given size are needed to maintain this temperature at steady-state operation? For a first estimate you can ignore any effects at the corners. Would you ever use only one or two elements? If not, why not? Why would a larger number of elements be preferred?

4. a. An improperly installed large glass pane in a picture window is prevented from free thermal expansion by its metal frame. The height of the pane is 1.9 m and its thickness is 0.005 m. Assuming the validity of equation 9, estimate the maximum allowable increase in temperature of the pane relative to the frame in order to avoid failure by thermal buckling. Note that this situation is a "worst-case scenario." Regardless of the value you have found in the literature for the coefficient of thermal expansion for the type of glass used in window panes, you will find your answer for $(\Delta T)_{cr}$ to be quite low. What is the compressive stress in the window pane at this value of $(\Delta T)_{cr}$? Will compressive failure occur before thermal buckling?

b. This problem illustrates the importance of installing window panes in a manner that permits free thermal expansion, since buckling failure can be a problem for improperly installed double-pane windows. The panes need to be carefully installed using soft rubber grommets. Installation over a heating vent is strongly discouraged. Thermal expansion of the gas phase within a double-pane window is a further contributing factor to its failure. Why would the use of fused quartz with a very low value of thermal expansion coefficient not be a feasible solution for the prevention of buckling failure in large window panes?

9

Thermal Protection Design Considerations for Human-Rated Reusable Space Vehicles

B.J. Dunbar

L. Korb

THERMAL PROTECTION DESIGN CONSIDERATIONS FOR HUMAN-RATED REUSABLE SPACE VEHICLES

B. J. Dunbar, Ph.D.
Assistant Director, NASA Johnson Space Center
Retired Astronaut
Lyndon B. Johnson Space Center
2101 NASA Road One
Houston, TX 77058

Larry Korb
Supervisor for Metallurgy and Ceramics, Retired
Rockwell International – Space Shuttle Program

ABSTRACT

The Space Shuttle Transportation System (STS) was developed to provide a reusable method for placing payloads and crew members into a low earth orbit in a single vehicle. Payloads consist of the following: experiments that take advantage of the microgravity environment in low earth orbit (e.g., Spacelab, Spacehab laboratories in the payload bay), instruments that study the earth (e.g., Side Imaging Radar), telescopes that study the university, and components of the International Space Station which are assembled in low earth orbit. One of the critical problems in developing this "revolutionary" concept in the 1060s was the design of a reusable thermal protection system (TPS) capable of surviving, not only the re-entry heating, but also the launch and landing mechanical loads. The major portion of the TPS consists of several thousands of pure silica fiber tiles covered with a ceramic coating and bonded onto the aluminum substructure. The design and manufacture of this system was not an easy process. Difficulties were encountered in achieving the aerodynamically smooth surface, structurally sound bonds and reduced mass. This chapter provides an overview of the rationale and thought processes that were instrumental in the design of the Shape Shuttle TPS from the late 1960s until its first launch on April 12, 1981. The authors have accessed a number of original studies and papers within NSASA and Rockwell International Space Division that described the design trades and rational involved in the development of the system. Additional improvements have been implemented in the last 20 years, but he design content and rational remains largely unchanged as one of the four shuttles is prepared for the 110th launch.

Fig. 1. Space Shuttle launch showing the Orbiter, External Tank and two Solid Rocket Boosters.

INTRODUCTION

The Space Shuttle System was designed to transport men and women with supplies into low earth orbit and to return all cargo to earth. The system (Fig. 1) consists of four major elements: the solid rocket boosters (SRB), the Space Shuttle main engines (SSME), the external tank (ET), and the Space Shuttle orbiter (SSO). The system was originally designed to be capable of launching up to 65,000 pounds of payload into orbit and returning 32,000 pounds of payload from orbit. There are four Space Shuttle Orbiters in the U.S. fleet: Atlantis, Discovery, Columbia, and Endeavor. The Space Shuttle is the only reusable launch vehicle operating in the world. The Thermal Protection System (TPS) was one of the major accomplishments of the National Aeronautics and Space Administration for enabling this reusability.

Nearly 80 percent of the thrust for launch is provided by the solid rocket boosters. The remaining 20 percent is supplied by the main engines, which burn hydrogen with oxygen supplied by the external tank. The orbiter, launched vertically from a piggyback position astride an external tank, contains the astronauts, the main engines, and the payloads. The orbiter must function as both a spacecraft and an aircraft. During entry from orbit, it must be protected from temperatures exceeding 2300°F on its lower fuselage and in excess of 2700°F along the leading edges and nose cone. At an altitude of approximately 150,000 feet, the orbiter will slow to about eight times the speed of sound and will pass its maximum heating. At 500,00 feet, the orbiter will enter into a level flight path and will be maneuvered aerodynamically to land as a glider. A typical mission profile is shown in Figs. 2a and 2b.

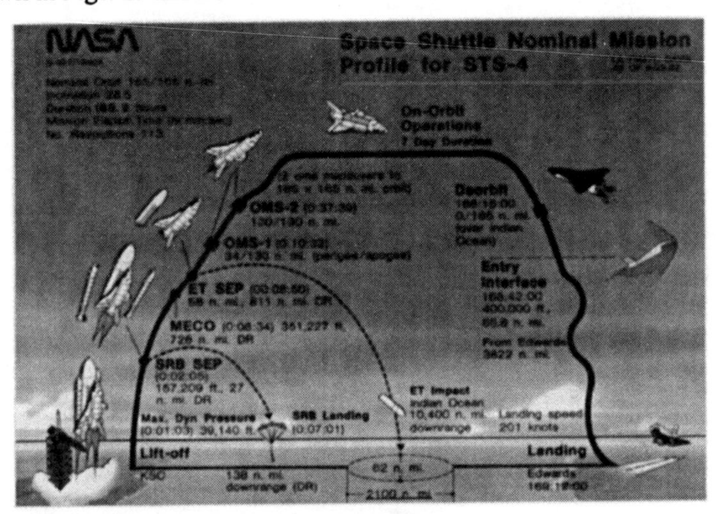

Fig. 2a. Typical Shuttle mission profile

Fig. 2b. Shuttle reentry attitude.

The requirement to achieve a minimum weight orbiter (165,000 pounds dry weight) necessitated use of the most efficient structural materials and processes. The requirement for 100-mission reuse has extended advancements in thermal protection materials well beyond the state of the art existing at the contract inception. Both weight and cost dictated that the basic orbiter structure be made from aluminum. In many areas, such as the cargo bay doors and orbital maneuvering subsystem (OMS) pods, graphite-epoxy was used to provide the minimum weight structure. Both aluminum and graphite are limited to a maximum of 350°F to avoid degradation. Thus, the orbiter thermal protection system (TPS) must function within the temperature regime dictated by the temperature limits of the TPS materials on its outer mold line and the temperature constraints of the vehicle structure at the inner mold line.

DESIGN REQUIREMENTS

The thermal protection system is designed to perform a variety of missions, each of which modifies the thermal environment of the TPS during its ascent, on-orbit, entry, maneuvering, and landing phases. The inclination of the orbit (e.g., polar vs. east-west), the total time in orbit, the requirements for vehicle altitude with regard to the earth and the sun, the payload requirements, and the range and cross range requirements are among some of the mission parameters that must be accommodated. Additional localized heating and impingement from plumes of the

solid rocket boosters, Space Shuttle main engines, orbital maneuvering subsystem, and reaction control subsystem engines (RCS) must also be accommodated. The TPS, under normal mission operations, must be capable of 100 missions with minimum refurbishment and must support a minimal turnaround for relaunch. The TPS must ensure survival in the event of a pad abort or an around-the-earth-once abort (AOA).

One of the keys to achieving a reusable thermal protection system is to minimize the temperatures. To do so, it is necessary to delay the transition from laminar to turbulent flow as long as possible into the entry cycle. This requires that very high levels of aerodynamic smoothness be built into the design.

The TPS not only must ensure that the substructure is kept below 350°F, but must be able to accommodate the stresses and strains resulting from both thermal, aerodynamic, and structural loads.

Finally, the TPS must be designed in such a manner so as to minimize fluid entrapment or absorption that could occur from a rainy or humid environment. The design must consider the potential for ice formation that could occur during cooling either from tanking (of the hydrogen and oxygen cryogens) or from evaporative cooling of entrapped moisture during launch.

DESIGN REQUIREMENTS SUMMARY

System Purpose

Protect the vehicle's structure and interior, particularly the human compartment, from the aerothermodynamic environment encountered during ascent and entry

Design Drivers

- Magnitude of Heating Rate: *Maximum Surface Temperature*
- Duration Of Heating Pulse: *Maximum Heat Load*
- Design Limitations: Thermal and Structural
- Mass
- Performance: Human Rated Reliability
- Loads/Structural Characteristics
- Cost
- Re-Use: Repair, Verification, Installation, Inspection

Design Constraints

- Limit Structure Temperature at 350°F
- 100 Mission Capability with Cost-Effective Unscheduled Maintenance/Replacement
- Withstand Surface Temperatures from –250°F to 2800°F
- Maintain the Moldlines for Aero and Aero-Thermo Requirements
- Attach to Aluminum Structure
- Economical Weight and Cost

Design Approach

Absorptive Systems: (absorb incoming heat by the surface material through heat sink, ablation, transpiration cooling, or convective cooling)
Radiative Systems: (Rejection of part of the incoming heat by thermal radiation from a suitable high temperature surface)
Combined Absorptive/Radiative Systems

SELECTION OF THE TPS CONCEPT

There have been a wide variety of thermal protection system concepts studied in the aerospace industry over the last 20 years. These concepts include ablative materials, hot radiative metallic structures, ceramic insulations, heat sinks, transpiration cooling systems, etc. Only the first three could be considered serious trade-off candidates for a system as large as that on the orbiter (e.g., a heat sink system would be too heavy; transpiration cooling would be too complex.)

Ablatives have been widely used for entry nose cones on military missiles, for linings of rocket engine nozzles, as well as for the primary heat shields for the Mercury, Gemini and Apollo spacecrafts (Fig. 3 and Table I). In these programs, the ablative material was encapsulated into an open-faced fiberglass honeycomb sandwich to ensure its structural integrity – that is, to prevent ablator loss from cracking at low temperatures and to retain the char surface at elevated temperatures. Although the extensive aerospace experience would support the choice of an ablator, the use of ablators for a major portion of the TPS presented two distinct shortcomings. First, ablators do not retain their aerodynamic surfaces through the entry trajectory. Use of ablators would result in costly refurbishment after each flight and would not readily support either the 100-flight reuse or the projected minimal turnaround requirements. Second, ablator materials such as those used on Apollo had nearly four times the density of the ceramic insulators being considered. In spite of the ability of ablators to operate at higher temperatures to permit faster rates of descent, it is doubtful that ablators were truly competitive from a weight-efficiency standpoint.

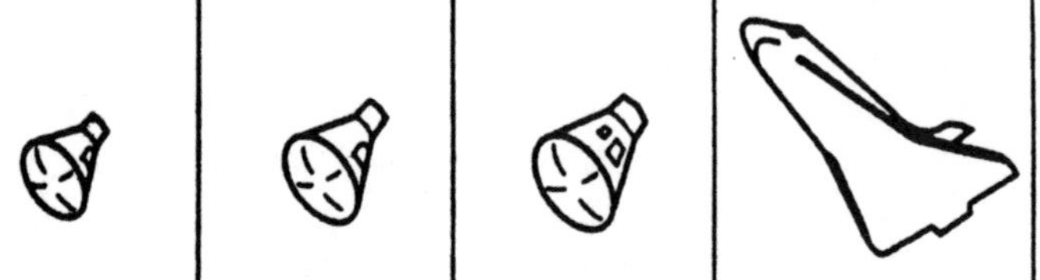

	Mercury	Gemini	Apollo	Shuttle
Area	32 ft^2	45 ft^2	365 ft^2	11 895 ft^2
Weight	315 lb	348 lb	1465 lb	18 904 lb
W/ft^2	10.2	7.5	3.9	1.7
Material	Ablator (Fiberglass-phenolic)	Ablator (Corning DC 235)	Ablator (AVCO 5026-39)	Rigidized silica fibers
Density	114 lb/ft^3	54 lb/ft^3	32 lb/ft^3	9-22 lb/ft^3
Usage	1 flight	1 flight	1 flight	100 flights

X101442M

Fig. 3. Comparison of spacecraft thermal protection systems.

Table I. Comparison of Single Use vs. Reusable Spacecraft

Orbital Return	Single/Re-usable	Veh/Entries
Mercury (1961-63)	Single	6/6
Gemini (1965-66)	Single	10/10
Apollo (1968-75)	Single	5/5
Shuttle (1981-2002)	*Re-usable*	*5/109*
Lunar Return		
Apollo (1969-72)	Single	10/10

Hot radiative structures have been used for many years in aerospace. The X-15 used Inconel alloy X-750 as both the aerodynamic and radiative surfaces to temperatures of approximately 1150°F (wing and nose areas included heat sinks; later models also employed a thin ablative layer). The Mercury and Gemini spacecraft used shingles made from Rene 41 on sidewalls to endure entry temperatures up to approximately 1700°F. Rocket engine nozzles, such as the columbium nozzles used on Apollo engines, were designed for service up to

2400°F. Yet, in spite of this experience base and the excellent development on a columbium metallic heat shield for the Shuttle (funded by NASA Langley Research Center), the use of a hot radiative metallic structure for a major portion of the TPS had several drawbacks. First, the concept was heavy. Second, it had limited over-temperature capability. Its fragile silicide coating is easily damaged. Localized loss of coating could result in rapid oxygen embrittlement at high temperatures and probably, catastrophic ignition above 2730°F, the melting temperature of its principle oxide, Nb_2O_5, formerly known as Cb_2O_5. Perhaps even more significant, however, was the design and manufacturing complexity. The problems of attaching panels of such a system to a typical aircraft substructure, yet maintaining the close step and gap tolerances necessary to avoid excessive heating from early boundary layer transition or plasma ingestion, are considered formidable. The design and thermal stress analyses of even the simplest structural panel is quite time-consuming. The analysis must ensure that panel buckling or excessive deflections would not occur on either initial or subsequent loading cycles. The installation is compounded by the large number of clips, brackets, stand-offs, frames, beams, and fasteners required. The complexities of packaging insulation, of controlling plasma ingestion while allowing for expansion and contraction, and of limiting fasteners to 1600°F (to permit reuse) are difficult challenges.

No single material approach could be used efficiently for the Shuttle orbiter thermal protection system. To a limited extent, both ablators and hot radiative metallic structures are used. Peak operating temperatures for various systems are compared in Fig. 4. The ablator used is the same material system as that used on the Apollo; however, its use is confined to the small area between the elevons where the combination of plasma flow and poor viewing factor (restricting reradiation) results in temperatures well above 3000°F. Hot radiative metallic panels of Inconel 625 are used in the engine-mounted heat shields (to approximately 1600°F) and for flipper doors and elevon seal panels (to approximately 1400°F).

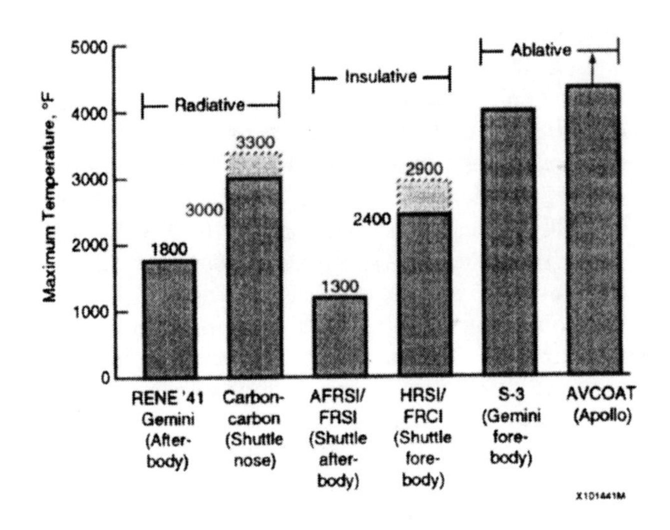

Fig. 4. Maximum Operating Temperatures for various Thermal Protection Systems.

Three material systems were originally used for the majority of the Shuttle orbiter thermal protection system (Figs. 5a and 5b). For temperatures in the 350-750°F range, felt blankets (FRSI) made from an aromatic nylon (Nomex*) and coated with a room-temperature vulcanizing elastomer are bonded directly to the aluminum substructure. These are located on the upper surfaces of the wings and along the fuselage side walls. (A typical isotherm map of the orbiter is presented in Figs. 6a and 6b.) Table II lists the operating temperatures for each component. For the temperature range of 750-2300°F, which represents by far the largest area of the orbiter, a coated tile system, made from amorphous silica, is employed. The wing leading edge and nose cone areas, where temperatures climb, in some locations, above 2700°F, use a coated reinforced carbon-carbon (RCC) composite material.

* Registered trademark, E. I.du Pont de Nemours and Company (Inc.).

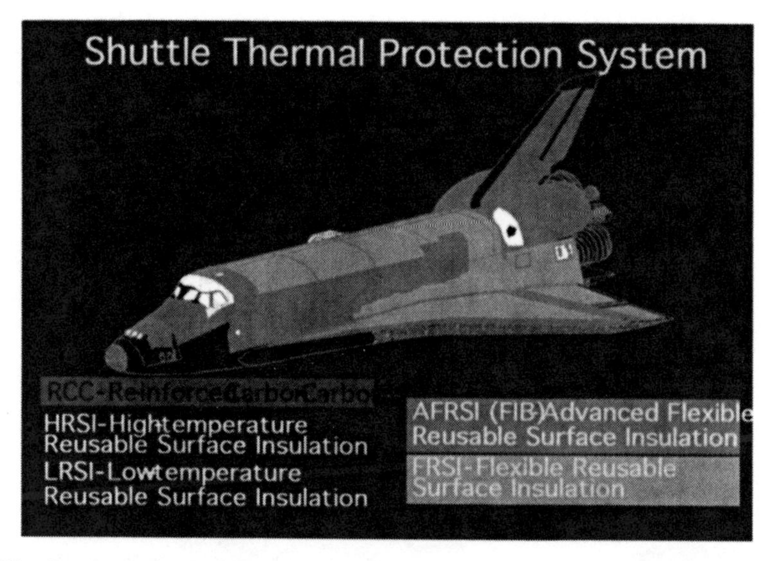

Fig. 5a. Overview of Shuttle Thermal Protection System "acreage"

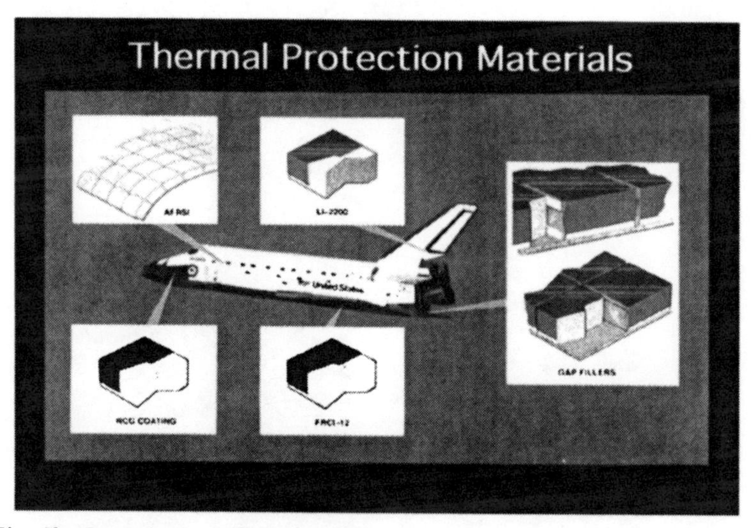

Fig. 5b. Overview of TPS materials

Each material system presents a variety of challenging and intriguing problems. Because the ceramic tile system constitutes the major portion of the Shuttle thermal protection and because this unique system offers such difficult challenges, it will be the primary focus of this discussion.

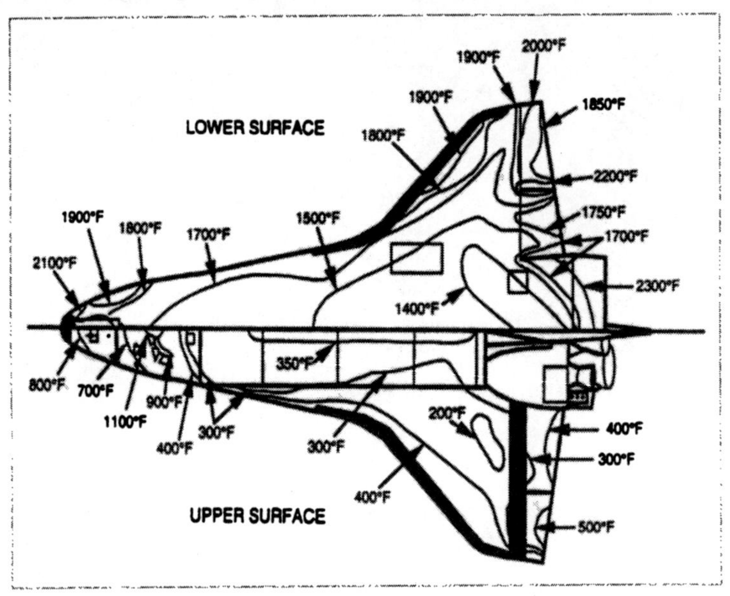

Fig. 6a. Lower surface isotherms.

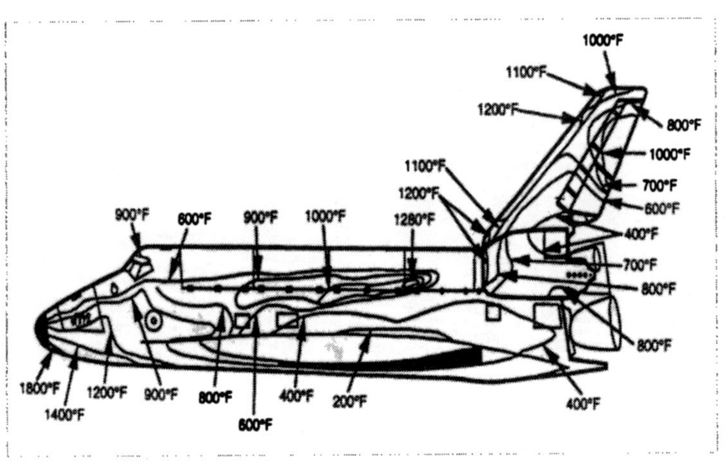

Fig. 6b. Port side isotherms.

Table II. TPS Design Temperatures

MATERIAL	Max Operating Temp (°F)		Min Oper Temp (°F)
	100 Mission Reusable	Single Mission Peak Use	
HRSI (high-temp reusable insulation)	2300	2600	-200
LRSI (low-temp reusable insulation)	1200	2100	-200
FRSI (felt or flexible reusable surface insulation)	700	1000	-200
RCC (reinforced carbon/carbon)	2700	3300	none
AFRSI (advanced flexible reusable surface insulation)	1200	1500	-200

THE TILE SYSTEM

The basic tile system is composed of three key elements: a ceramic tile, a nylon felt mounting pad, and a room-temperature vulcanizing elastomeric adhesive (RTV). The tile is coated with a high emittance layer of glass and functions as both an insulator and a radiator to limit heating of the structure. The felt mounting pad, called the strain isolator pad (SIP), provides for mechanical isolation of the tile from the vehicle deflections and strains. The RTV bonds the tile to the SIP and the SIP to the structure (see Fig. 7).

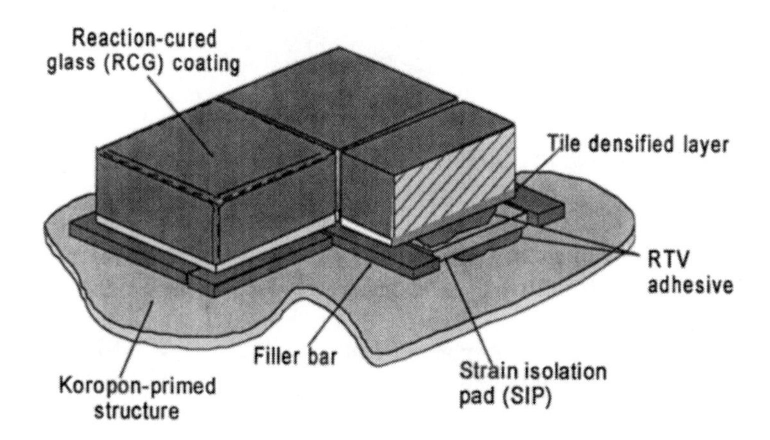

Fig. 7. Cross section of silica fiber tile system.

Silica Tiles

Ceramic tiles, made from pure amorphous silica or mullite, were under laboratory investigation for several years. Prior to the shuttle proposal, NASA had funded developmental studies on these materials. The system chosen for the Shuttle orbiter was that made from silica. From the beginning, it was clear that a silica tile system offered considerable advantages over the other systems; however, until designs matured in the late 1970's, some of its key design problems were not fully recognized.

The silica tiles are made from a very-high-purity amorphous silica fiber approximately 1.2 to 4 microns in diameter and up to 1/8 inch long (Fig 8).

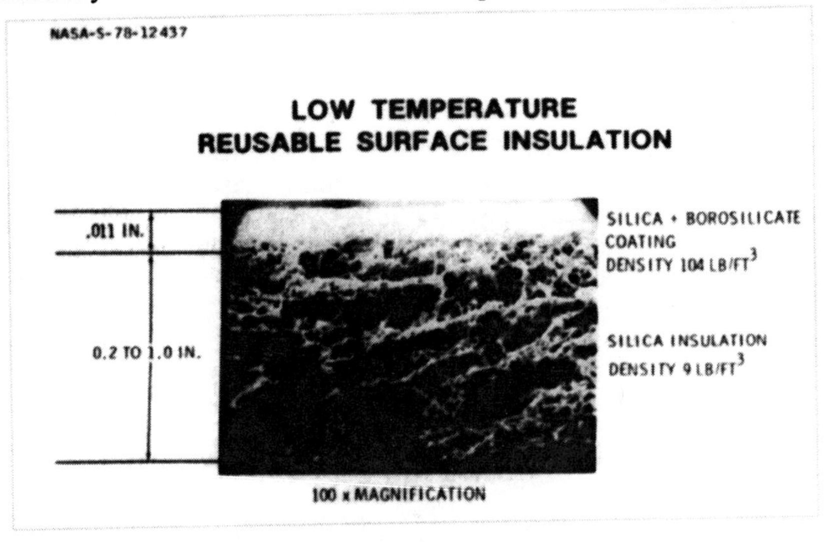

Fig. 8. Early micrograph (1970's) of silica fiber tile.

These are felted from a slurry, pressed, and sintered at approximately 2500°F into a tile production unit (PU) by Lockheed Missiles & Space Co., Sunnyvale, California.

The PU is the starting block of material from which tile banks and shapes are cut. Two densities of tiles are manufactured. Approximately 80 percent of these are made from a nine-pound-per-cubic-foot (PCF) density (LI-900) while areas requiring greater mechanical strength employ the 22 PCF density (LI-2200). The manufacture of these two products is similar, but there are major differences. In the fabrication of LI-900, it is necessary to add a silica binder to improve its strength and adjust its density. The LI-2200 requires impregnation with silicon carbide to reduce radiative heat transfer through the tile.

All tiles have a thin glass coating on five sides to provide the proper thermal properties (α, ε and the α/ε ratio, where α is the solar absorptance and ε is the total hemispherical emittance). In general, tiles on the lower surfaces of the orbiter are coated black for high emittance while those on the upper surfaces and sides have a white coating to limit solar heating. Typically, black-coated tiles are used in areas that exceed 1200°F, whereas white-coated tiles are used below this temperature. The tiles are sized in thickness to limit the temperature of the SIP, the adhesive bond line, and the aluminum structure. The platform size is limited to prevent tile cracking from the induced thermal and mechanical stresses of the vehicle structure and adhesive system. The majority of the tiles have a square platform; the black tiles are typically 6 inches by 6 inches and the white tiles are 8 inches by 8 inches. There are, however, many special shapes and sizes (as small as 1.75 inches square) where vehicle geometry dictates. There were as many as 30,759 tiles on the Shuttle orbiter. Over time, many of the upper surface tiles have been replaced by flexible glass fiber blankets. Fig. 9 illustrates a typical multi-tile installation.

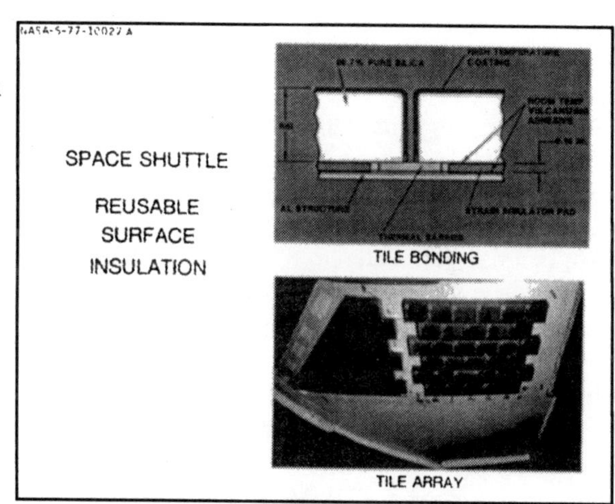

Fig. 9. Multi-tile installation on bottom surface of Space Shuttle Orbiter.

The silica tile system has many engineering advantages. The nine-pound-per-cubic-foot density makes the system extremely light in weight. Such a tile is approximately 93 percent void and therefore is an excellent insulator, having through-the-thickness conductivities of 0.01 to 0.03 BTU-ft/hr-ft^{2}°F. Although the low density is accompanied by low strength, the tiles are capable of tolerating the high g forces from the severe acoustic levels on the spacecraft, which in some areas approach 170 db and structural responses of 35 g^2/Hertz.

Amorphous silica is an ideal material for the tiles. Its coefficient of expansion is extremely low compared to other ceramics, resulting in reduced thermal stresses. This is particularly important because the tiles could have a ΔT within the tile of 2300-2400°F at the start of entry. The low density of the tile results in a low elastic modulus, also contributing to reduced thermal stresses. The silica chosen has a purity of greater that 99.62 percent to limit devitrification, for the formation of a significant crystalline phase, such as crystobalite, could result in tile cracking due to a large volume change in the 400-500°F temperature range.

Silica has high temperature resistance; it can survive excursions to 270°F. Since silica is already an oxide, it does not require further oxidation protection as do carbon-carbon and niobium (formerly known as columbium) materials. The tile coating is a borosilicate glass. The black coating has silicon tetraboride added for increased emittance. Subsequent oxidation of the silicon tetraboride in the glass yields boria and silica, the basic constituents of the glass itself.

The coating has an emittance well above 0.8. In addition to providing the desired thermal properties, it is a barrier to rain and atmospheric erosion. It tends to minimize tile-handling damage although it does not eliminate it. The coating thickness ranges from 0.009 to 0.015 inch. The coating terminates above the tile inner mold line (IML) on the sides to limit heat flow toward the IML and to permit the tiles to vent air pressure during both the ascent and descent phases of operation.

While the tiles must "breathe," they must not pick up water, for this could result in a vehicle overweight condition, in the loss of tiles in the vibroacoustic environment, or in coating damage under freezing conditions. Water repellency of the silica tiles is obtained from the vacuum deposition of a silane, Dow Corning Z6070, in a furnace heated to 350°F. The silane will burn out of that portion of a tile that exceeds 1050°F during launch or entry; therefore, the coating terminator is located below the 1050°F isotherm of each tile.

Strain Isolator Pad and Filler Bar

The strain isolator pad is made from Nomex*, an aromatic nylon material that neither supports combustion nor melts. The pad, manufactured by Albany International Research Company, a division of Albany International, is a felt product that is needled to provide strength in the thickness direction.

Two thicknesses of strain isolator pads are used for the majority of the vehicle, each with widely different properties. The 0.160-inch thickness is quite soft (a typical tensile modulus of 26** psi under a load of 5.3 psi), and is generally used under the LI-900 tiles. Where more mounting rigidity (or strength) is required, such as around doors or thermal barriers (discussed later), the 0.090-inch SIP is used. It is roughly ten times as stiff as the 0.160-inch SIP and is generally

* Registered trademark, E. I.du Pont de Nemours and Company (Inc.).

** Properties in the thickness dimensions

used under the higher-strength LI-2200 tiles. SIP will retain its room temperature properties, 35-60* psi typical tensile strength, without degradation with exposures to 550°F for the equivalent of 100 missions (25 hours). It is capable of supporting limited stresses up to 720°F for short times (i.e., on a single-mission basis).

The SIP, like the tile, must also "breathe" to permit equalizing pressures during ascent and descent yet must also avoid water pickup. The latter would result in ice formation during launch, excessive stiffness of the SIP, and overloading of the tile. Therefore, SIP is treated with a water-repellent agent, Zepel RN**.

The SIP is generally sized to a footprint an inch smaller than the tile; that is, a 6 by 6 inch tile has a 5 by 5 inch SIP under it, whereas an 8 by 8 inch tile uses a 7 by 7 inch SIP.

Underneath the tile edges, another Nomex** felt product, very similar to SIP, is used. It is called filler bar. The filler bar protects the surface of the aluminum substructure in the tile gap from radiation and plasma heating. The filler bar upper surface is coated with a thin layer of RTV-560. The filler bar is slightly thicker than the SIP and tends to form a mechanical moisture shield against the tile. The filler bar also becomes an attachment point for gap fillers (discussed later in the paper). The filler bar is given a final processing at temperatures of approximately 830°F by the manufacturer before the RTV coating is applied. This product is capable of surviving 100 missions of temperatures to 800°F (with some RTV damage) and a single mission to nearly 1000°F.

RTV Silicones

The heart of the TPS bonding system is RTV 560, a two-part condensation curing elastomeric sealant/adhesive with a unique combination of properties. Normally the first consideration for a room-temperature curing adhesive system would be an epoxy adhesive. However, the room-temperature curing epoxies are temperature limited. Since the tile bondline sees temperature extremes of –160°F to +550°F, temperature becomes the driver in selecting an adhesive. Fortunately, exceptional bond strength was not a requirement for the lightweight tile system, allowing consideration of room-temperature vulcanizers. The methyl phenyl RTV silicones have brittle points below –175°F, with a high-temperature capability to >600°F, depending upon exposure conditions. Therefore, the methyl phenyl RTV silicones as a class and specifically RTV-560, was selected early in the Shuttle program as the prime candidate for tile bonding. Early vehicle temperature predictions gave temperatures down to –200°F, which is below the brittle point of

* Properties in the thickness dimension

** Registered trademark, E. I. Du Pont de Nemours and Company (Inc.).

RTV-560. Therefore, early program tests were concerned with demonstrating that the low-temperature contraction and increase in modulus of the adhesive system would not crack the tile; the RTV-560 was satisfactory while other candidate adhesives were not.

Along with the use of RTV's came the need for a compatible primer, one that developed good adhesion over a variety of surfaces to which tile was to be bonded. The primer selected was SS-4155, a hydrolyzing titanate coupling agent. This primer was offered the challenge of providing good RTV adhesion on such surfaces as

- Chromated epoxy primed aluminum
- Bare aluminum
- Titanium
- Inconel
- Graphite epoxy
- Epoxy fiberglass
- Epoxy honeycomb edge filler

Any material that is to be used over the vehicle external surface must be weight-efficient and therefore perform in minimum thicknesses. RTV-560 is used in three bondlines beneath the tile:

- SIP transfer coating, 3 to 7 mils (1 mil = 0.001 inch)
- SIP to structure, 5 to 9 mils
- SIP to transfer coat-to-tile, 4 to 5 mils

The SIP transfer coat was developed to ensure a proper SIP-to-tile bond. Since both the SIP and tile are highly porous, it is not possible to apply a bond to either one of these and ensure consistent RTV penetration into both members at the same time. RTV penetration into the SIP is controlled during the application of the transfer coat in a subassembly operation. Since this penetration affects the SIP modulus, it must be controlled accurately. After the transfer coating has cured, it may then be bonded to the tile.

Other RTV's used in the TPS application are RTV-566 and RTV-577. RTV-566 is the low outgassing (but very expensive) version of RTV-560. It is used around windows to reduce contamination from outgassed products. Extensive vacuum tests on RTV-560 showed that the less expensive material could be used without jeopardizing payload experiments. RTV-560 and RTV-566 employ iron oxide as a filler material. RTV-577 has the same starting silicone polymer as RTV-560 and RTV-566, is filled with zinc oxide and calcium carbonate, and has an additive to provide a thixotropic "trowelable" consistency. RTV-577 is used as a "body putty" (screed) to fill in low areas of the vehicle structure and to achieve a contour compatible to the TPS installation. The RTV-577 material allows the use of structure with a ±0.062-inch as-manufactured contour to be faired to the ±0.032-inch waviness requirements for ceramic tile installation.

TPS STRUCTURAL INTEGRITY

It is essential to the function of the orbiter TPS that each tile remain attached to the vehicle through the critical entry heating phase, for the loss of a single tile in many areas could result in severe vehicle damage or even the loss of the orbiter. To ensure the integrity of this system, the loading spectrum to which each tile is subjected must be known, the tile and SIP properties must be adequately defined, and the method of analysis must be verified by laboratory testing. Further, it is important that no significant processing errors be made in the tile installation. To ensure the TPS structural integrity, nearly every tile on the orbiter has been subjected to a proof test.

Material Strength

There are several unique challenges in designing with tiles and SIP. The first shock to a materials engineer is the extremely low material properties. These are hard to put into perspective with one's aerospace experience base. The tile minimum tensile strength in one critical loading direction (through the thickness) is 13 psi. Compare this with typical airframe structural materials on the orbiter whose tensile strengths are in the range of 10,000 to 290,000 psi. It soon becomes obvious that even the most minor stresses introduced into the system or into laboratory test specimens must be accounted for.

A key property of the SIP is its elastic modulus, for vehicle deflections load the tiles in proportion to this value. One must learn to think in terms of modulus values of 20 psi where one's experience may be with aluminums or steels whose moduli are 500,000 to 1,500,000 times as great.

The second factor that meets the jaundiced eye of the materials engineer is the wide scatter in material properties. Tensile test values of coupons cut from LI-900 tiles typically range from 10 to 40 psi; moduli values of 0.160-inch SIP at identical loads often vary from 16 to 40 psi.

The combination of low properties and the wide property scatter impacts the reliability of the design. For example, the minimum tensile strength (99 percent probability, 95 percent confidence) of LI-900 in the thickness direction is 13 psi.

Using a design factor of safety of 1.4, the tile stress at limit load, 9.3 psi, lies less than one standard deviation below the minimum design tensile strength. In a aluminum design, the stress at limit load typically lies more than seven standard deviations below the design minimum tensile strength.

A third factor that soon becomes obvious is the difficulty of carrying structural loads with brittle ceramic materials. The tiles offer little or no forgiveness for stress concentrations, loading eccentricities, or minor defects in test specimens.

The 0.0160-inch SIP is a challenging material to characterize. Its stress-strain behavior is nonlinear, inelastic, and exhibits significant hysteresis changes. The modulus of SIP increases with load. As load is removed, the SIP does not return to its original dimensions but retains some permanent tension or compression set, depending upon which load was applied. Repeated cycling of SIP changes its stress-strain behavior, resulting in a soft zone ("dead band") in which little stress will result in rather large SIP extensions. .

As one would expect, the compressive modulus of SIP differs from the tensile modulus at the same stresses. The compressive behavior of SIP is typical of that of any batt; as compression increases, fibers are brought into more intimate contact, resulting in increasingly greater resistance to compression. Tensile behavior is controlled by the resistance to extension of bundles of threads needled in the thickness direction of the felt. As further extension occurs, fibers in more bundles are drawn taut, thus increasing its stiffness..

In effect, then, we have a tile system in which both major members have both low properties and wide property scatter. One member of the system, the tile, is brittle and incapable of deformation to reduce localized stresses. The other member, the SIP, is soft and compliant, up to a point, but its elastic properties depend on its load level and its previous loading history.

Design Loads and Stresses

The tiles experience stresses from a wide variety of loading conditions. For example, the tiles form the outer mold line of the vehicle and must react to the aerodynamic maneuvering pressures. They can experience positive or negative pressure loads from ascent or descent where the tile interior cannot equalize the external pressure rapidly enough. Similarly, aeroshock loads, resulting from impinging or dancing aeroshock waves, can add overturning moments and pressure loads to the tiles. Further, the tiles are subjected to high-level vibroacoustic loads from engines firing and air flow.

Pressure differences between the interior and exterior of the vehicle cause out-of-plane vehicle deflections which, in turn, load the tiles through the SIP. These deflections can be increased by additional in-plane stresses or thermal stresses in the vehicle substructure.

One potential source of large stresses is that resulting from mismatch during tile installation. Mismatch is the difference in local contour between the tile IML (Inner Mold Line) under the SIP and the structure OML (Outer Mold Line) to which it is attached. This difference must be accommodated by extension of the

SIP, resulting in tile loading. For example, a fit-up difference of 0.020 inch could result in residual stresses exceeding 3 psi in tile bonded to 0.160-inch SIP, whereas a mismatch as small as 0.005 inch could result in localized residual stresses exceeding 7 psi with 0.090-inch SIP.

Because the tile coating has a different coefficient of expansion than the tile, residual stresses are introduced into the tile upon cooling. The significance of these stresses is quite apparent as coated tiles are machined to a thickness of 0.5 inch or less; for as residual stresses are relieved, the tiles curl slightly, taking somewhat the shape of a potato chip. When such a tile is bonded to the SIP under pressure, the "reflatting" of the tile distributes some residual stress into the tile and introduces stresses into the SIP and its interface.

Finally, mechanical loads imposed on tiles by thermal barriers and gap fillers can result in overturning moments on the tiles, which must be reacted to by the SIP and tile. (Thermal barriers consist of ceramic cloth, Nextel 312*; knitted Inconel X750 springs and ceramic batt, Saffil**. They are used to ensure plasma-tight integrity around doors and penetrations. Gap fillers are made from ceramic cloth and batt and incorporate a piece of Inconel 601 foil to control shape. They are used to restrict plasma flow into tile gaps at pressure gradient areas.)

Tile System Strength

The strength of the tile system (tile-SIP-RTV) was originally thought to be equal to the strength of its weakest link, the tile. A typical tensile strength of an LI-900 tile averages 24 psi in the thickness direction, whereas the 0.160-inch SIP averages 40 psi and RTV is greater than 250 psi. When test failures occurred, they would occur in the tile just above the RTV layer that bonds the tile to the SIP transfer coat.

Structural element test specimens to confirm the analytical methods used a bending beam configuration. A tile and its SIP were installed on an aluminum plate that, in turn, was bent over various mandrels until failure occurred. The radius of the mandrel at tile failure was observed and stresses in the tile were computed based upon SIP deflections. Although these early tests tended to confirm the analytical methods, large errors could be introduced because the properties of the specific SIP or tile used in the test were unknown.

Subsequent testing later in the program indicated that a flatwise tension test was much more precise in defining the system properties since it eliminated the variability of the SIP stiffness. Flatwise tensile tests of various SIP and tile combinations revealed the average system strength was slightly less than 50 percent of the average tile strength.

---

* Registered trademark, 3M, St. Paul, Minnesota

** Registered trademark, ICI, United States, Inc.

Detailed studies showed that the bundles of fibers that gave SIP its short transverse strength were acting as localized stress concentrations just above the tile bondline, causing early failure. This revelation resulted in two activities: (1) a reappraisal of all tile safety margins on the vehicle; and (2) an effort to develop a higher strength at the tile-SIP bondline.

It was demonstrated that a thin layer of aluminum foil (0.010-inch thick) between the SIP and the tile would bring the tile system strength up to the tile strength. This approach would add approximately 1,000 pounds to the vehicle, yet it could not readily accommodate the vehicle contours machined into the tile IML's nor could it sustain the thermal cycling of the bondline without causing the thermal stress failure in the tile. A thin graphite composite shim was developed that prevented thermal stress failures, but its manufacture to tile contours was prohibitively difficult. The elimination of these detrimental stress concentrations was finally achieved by a process we called densification.

Densification

The tile is roughly 93 percent void. In the densification process, the voids in the IML surface of the tile are filled with a slurry containing a fine glass slip and a colloidal silica adhesive (Ludox*). This is analogous, in some respects, to putting a sealer coating on to concrete. It is important that the Ludox*-slip mixture be brushed into the tile and not be permitted to build up or the coating would flake off and have limited strength. Such a process increased the outer surface density to approximately 80 pounds per cubic foot and a density gradient penetrated inward approximately one-eighth of an inch. The glass slip is highly abrasive to the tile. Techniques were developed to limit tile IML contour changes to below 0.006 inch during densification. While the densification process was being developed, a second problem was encountered. The Ludox* inactivated the RTV catalyst, preventing its cure against the tile surface. Subsequent investigations revealed the Ludox* effect could be controlled by waterproofing the tile by vacuum vapor deposition of silane after densification in a furnace.

Once the densification process was fully developed, it was necessary to remove critical tiles on the Columbia OV-102, densify them and reinstall them in the same locations.

* Registered trademark, E. I. Du Pont de Nemours and Company (Inc.).

* Registered trademark, E. I. Du Pont de Nemours and Company (Inc.).

Proof Loading and Acoustical Emission Testing

While densification was required for a large number of tiles, the minimum system strength of the undensified LI-900 tile with 0.160 SIP, 6 psi, was considered more than adequate in some areas. To verify that these installations were strong enough, each tile was subjected to a proof load on the vehicle to a level 25 percent above the maximum load (limit load) it would see in service, wherever possible. Proof loading of tiles was accomplished by pulling the tile OML using a vacuum chuck. The load applied was limited by adherence of the vacuum chuck to the tile (theoretically 14.7 psi but approximately 10-12 psi from a practical standpoint) and the limit to which the substructure could be loaded. Acoustical monitoring transducers were in contact with the tiles during proof loading. Acoustical techniques were developed to ensure no detrimental bond or tile degradation occurred during proof loading. Nearly every tile was proof-loaded; however, acoustic techniques were not required for densified tiles.

Pulse Velocity Testing

The tile strength used in design and analysis is that minimum strength that 99 percent of the tiles will equal or exceed. With a vehicle covered by more than 30,000 tiles, one could expect more than 300 tiles could be below this minimum strength. Since the loss of a critical tile was so serious, a nondestructive method had to be developed to ferret out the weaker tiles. Such a process was developed based upon two known facts: (1) the speed of sound in a material is related to its elastic modulus; and (2) the strength of low-density ceramics is related to its modulus. All densified tiles were subjected to pulse velocity testing to screen out low-strength tiles.

Process Control

The use of RTV, like any other adhesive system, requires careful attention to process controls to ensure bonding strength requirements are met.

Surface Preparation. Most of the surface area of the Shuttle orbiter is coated with a chromated epoxy polyimide paint applied over the aluminum skin. It is necessary to solvent-wipe this surface with methyl ethyl ketone to remove organic contaminants. In addition, sanding of the surface with a fine mesh (400 grit) paper is required to remove the gloss and provide some mechanical attachment. Similarly, the glossy surface of screeded RTV-577 also requires roughening to achieve optimum bonding.

Epoxy surfaces were primed with two coats of a silicone primer, SS-4155, in addition to the sanding; RTV surfaces do not require an additional primer.

Pot Life. One of the most important controls applied to the RTV-560 adhesive is the control of pot life (working life). Pot life and curing times of RTV are decreased either by increases in temperature or relative humidity. In cool dry air, curing may take two to three days or more whereas in warm humid environments, bonding must be accomplished with 15 minutes of the mixing with the catalyst. Each RTV mix has its pot life recorded on the container when mixed, and inspection must verify the bonding is accomplished within the pot life. In addition, each batch is accompanied by a shore hardness coupon to verify proper pot life and cure.

Pot life and shelf life are susceptible to contaminates introduced by the manufacturer of the RTV. For this reason, both the manufacturer and Rockwell verify the pot life and shelf life on each order of RTV-560.

Penetrations

Penetrations pose a special challenge for maintaining the thermal integrity of the vehicle. Each penetration must be properly sealed prior to launch, and in the case of cycling doors (such as the Payload Bay Doors, Payload Bay vent doors, and Star Tracker Doors) must regain that integrity prior to reentering the earth's atmosphere. Several doors on the Shuttle, including the gear doors, are sealed with quartz fiber cloth covered inconel springs. (See Fig. 10)

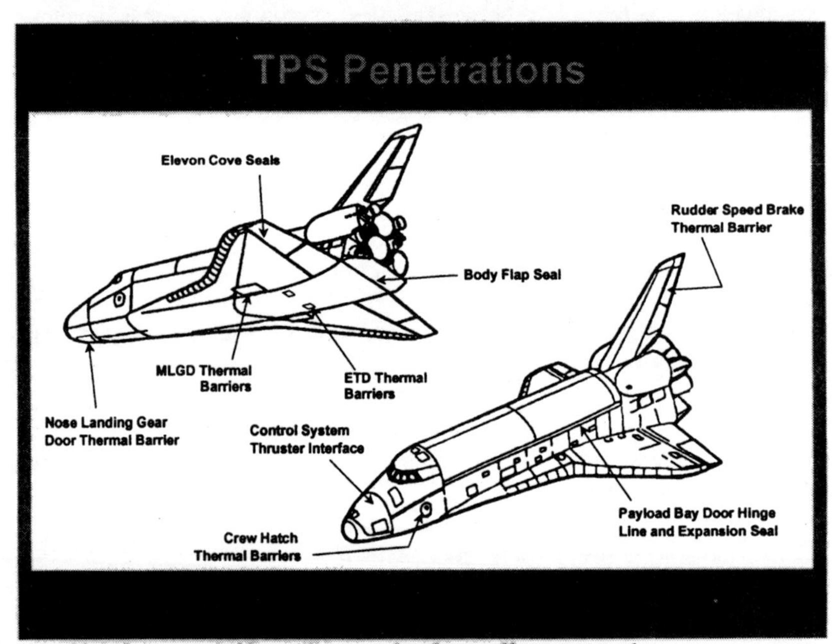

Fig. 10. Shuttle penetrations that must be thermally protected.

THE TPS TEAM---NASA, ROCKWELL INTERNATIONAL SPACE DIVISION, LOCKHEED SPACE MISSLES, AND UNIVERSITY OF WASHINGTON

The concept of a reusable ceramic thermal protection system was first briefed by Mr. Bob Beasely, Lockheed Space and Missiles, to Dr. Max Faget, of the NASA Johnson Space Center. Dr. Faget, who was later to become the principal designer of the Space Shuttle System, elected to incorporate the system into the world's first reusable space vehicle, the Space Shuttle, which launched in 1981. Dr. Faget later became NASA JSC Director of Engineering.

Four NASA Centers were primarily responsible for the TPS technology. The Lyndon B. Johnson Space Center (NASA-JSC) managed the overall program and made important contributions in materials and testing. The NASA JSC directed the development of the LI-900, an LMSC propriety process, developed and patented FRSI, surface densification and repair techniques for tiles, and several fastening techniques for the insulation. The NASA JSC also directed the development of the RCC (Reaction Controlled Coating) and provided significant progress in the development of Carbon-Carbon materials. Many materials and ceramic engineers contributed to this effort, including Mr. Robert Dotts and Mr. Calvin Shoberg. The NASA JSC and, most significantly, the NASA Ames Research Center (ARC), conducted testing of the TPS materials in large, plasma jet wind tunnels, which were NASA unique facilities. The NASA ARC developed AFRSI and many of the concepts that resulted in rigid Reusable Surface Insulation. The NASA ARC holds patents on the LI-2200 and FRCI tile materials, the RCG tile coating, and the layer type gap fillers. Dr. Howard Goldstein (Fig. 11) was the principal developer of the RCG along with Dr. Daniel Leiser.

Fig. 11. Dr. Howard Goldstein
NASA Ames Research Center

The Kennedy Space Center (NASA-KSC) has, over time, become responsible for the maintenance and repair of the TPS. Mr. Frank Jones has been a significant contributor to this effort. The Langley Research Center (LaRC) also conducted extensive environmental testing of the TPS and evaluated the structural integrity of the tile-SIP-RTV system. Dr. John Buckley (Fig. 12) supervised some of the external efforts along with Dr. James Gangler, NASA Headquarters.

Fig. 12. Dr. John Buckley with students at LaRC, circa 1970s

During the development of the Space Shuttle, the primary contractor to NASA, Rockwell International, Downey, CA, worked closely with NASA and its subcontractor, Lockheed Missiles and Space Company, to develop and manufacture the thermal protection system. The contract was awarded to Rockwell International with the Ceramic Tile system baselined for the Shuttle in 1972. Lockheed was selected to manufacture the ceramic tiles in 1973.

Additionally, NASA awarded several University grants to support the development of the TPS, specifically to the University of Washington, under the leadership of Dr. James I Mueller, Chairman of the Department of Ceramic Engineering. (Fig. 13)

Fig. 13. Dr. James I. Mueller,
University of Washington

The effort at the University of Washington, in Seattle, Washington, started in 1969 and continued well into the 1980's. Several undergraduate and graduate students were supported under this NASA grant, including the primary author of this chapter.

In summary, the development of a Ceramic Thermal Protection System for the United States Space Shuttle was what would be considered today as "revolutionary." It required a nation wide team effort between several NASA Centers, the contracting companies and universities. The system has proven to be more robust than ever envisioned and is serving as the basis for the design of other U.S. systems, such as the X-38 crew rescue vehicle for the International Space Station (Fig. 14), and international vehicles. Such international vehicles that have benefited, are benefiting, or will benefit from this technology are the Russian Buran (Fig. 15), the Japanese Hope (Fig. 16), and the French Hermes (Fig. 17).

Fig. 14. The X-38 Crew Rescue Vehicle (CRV) under development for the International Space Station.

Fig. 15. Russian *Buran*. Has flown one orbit unmanned.

Fig. 16. Japanese *Hope*. Reusable Launch Vehicle. Will launch and land from Japan. Also, will dock with the International Space Station.

Fig. 17. French Hermes (conceptual).

SUMMARY

The Space Shuttle System was developed to provide a reusable method for placing payloads and personnel into low earth orbits and returning them to earth. The Shuttle orbiter, the cargo-carrying element of this system, is launched into orbit and reenters as a spacecraft. One of the critical design problems has been the development of a lightweight, reusable thermal protection system capable of surviving not only the entry heating but also the launch and maneuvering loads.

The thermal protection system used on the Space Shuttle orbiter has been selected based upon the requirements for minimum weight, 100-mission reuse capability, and a minimal turnaround period for relaunch. The system chosen uses low-density, fired ceramic tiles made from fine filaments of pure silica. The tiles exhibit excellent characteristics: a low coefficient of expansion, low thermal conductivity, low density, excellent thermal stability, and compatibility with the environment. The design was developed to accommodate the shortcomings of this material: low strength, wide scatter in mechanical properties, and brittleness. The technique of bonding the tiles to a strain isolation pad and, subsequently, the pad to the structure, using a room-temperature curing elastomer, appears to provide a satisfactory approach. However, because the loss of a single tile could jeopardize the vehicle, extreme care is taken to fully understand specific tile loads and stresses, material strengths and behavior, and to verify analytical methods. Finally, close process control is maintained during tile installation. A final process evaluation of each tile bond must be made by proof-testing to ensure the structural integrity of the system.

At the end of 2001, the Space Shuttle Transportation System (STS) had flown 108 flights and will begin 2002 with a servicing mission to the Hubble Space Telescope. It is intended to continue flying the Shuttle through at least 2012 and possibly until 2020. Although some upgrades and modifications have been made to the TPS system, it is still largely the same system that was designed and installed on Columbia in the late 1970's prior to its first launch in 1981, more than 20 years before.

ACKNOWLEDGEMENTS

I am deeply grateful for the following persons who have contributed substantially to the development of this chapter.

L. J. Korb
Rockwell International
Retired

H. M. Clancy
Rockwell International Space Division
Retired

S. Rickman
Lyndon B. Johnson Space Ctr.
2101 NASA Road One
Houston, TX 77058

H. E. Goldstein, PhD
Senior Staff Scientist
NASA Ames Research CTR
Moffett Field, CA 94035

F. Jones
Kennedy Space Center, FL

D. Curry
Johnson Space Center
Retired

In Memoriam:

R. Dotts
Johnson Space Center

J. I. Mueller, PhD
University of Washington

QUESTIONS AND PROBLEMS

1. Compare the "colors" of the Shuttle, Buran, Hermes, Hope and X-38. What are the similarities and "why"?

2. Does the re-entry angle into the atmosphere make a difference? How might that relate to heating rate, peak temperature, and thermal loads on the TPS?

3. Why were the large area tiles on the bottom of the shuttle divided into approximately six-inch squares?

4. How could you modify the design of the existing tile system to improve its performance? (See Fig. 7.)

5. Assume that the ceramic tile system was replaced by a refractory metal system. How would the design change? Would active cooling be required? What might be some of the design challenges?

Space Shuttle docked to International Space Station (ISS). The X-38 (CRV) shown docked to the U.S. Module. For information on both systems, visit: http://www.spaceflight.gov

BIBLIOGRAPHY

Black, W. E., et al, "Evaluation of Coated Columbium Alloy Heat Shields for Space Shuttle Thermal Protection System Application," NASA CR 11219 (June 1972).

Dotts, H., "Materials and Fabrication Methods Used in the Gemini Spacecraft," Metal Progress, (March 1963).

Stockett, S.J., "Metals and Processing Used in the Mercury," Metal Progress, (June 1962).

Symposium on Reusable Surface Insulation for the Space Shuttle, NASA TMS 2720, 1972.

Greenshields, D.H., "Orbiter Thermal Protection System Development," NASA Johnson Space Center; AIAA Technical Information Service A77-35304, 1977.

Korb, L.J., Morant C.A., Calland R.M. and Thatcher C.S., "The Shuttle Orbiter Thermal Protection System," American Ceramic Society Bulletin, Vol. 60, No. 11, (1981) pp. 1188-1193, (Seven additional articles related to the TPS are also included in this Bulletin issue, pp. 1180-1217).

Banas, R.P., Gzowski E.R. and Larsen W.T., "Processing Aspects of the Space Shuttle Orbiter's Ceramic Reusable Surface Insulation," Proceedings of the 7th Conference on Composites and Advanced Materials, American Ceramic Society, Cocoa Beach, FL, (January 16-21, 1983).

Materials Properties Manual, Volume 3: Thermal Protection Systems Materials Data; Prepared by Laboratories and Test D/284, Rockwell International Space Transportation and Systems Group, Pub. 2543-W, Rev 5-79. (December 1982).

Arrington, J.P. and Jones J.J., compilers, Shuttle Performance: Lessons Learned, Conference Proceedings, NASA Langley Research Center, Report No. NASA CP-2283, (March 8-10, 1983), 759 pages.

Dotts, R.L, Curry, D.M. and Tillian, D.J., "The Shuttle Orbiter Thermal Protection System: Materials, Designs, and Flight Performance Overview," SAE 831118, 13th Intersociety Conference on Environmental Systems, San Francisco, CA, SSN 0148-7191, (July 11-13, 1983).

Goldstein, H.E., Smith M. and Leiser D., "Silica Reusable Surface Insulation," U.S. Patent 3,952,083, issued April 20, 1976.

Banas, R.P. and Cunnington G.G. Jr., "Determination of Effective Thermal Conductivity for the Space Shuttle Orbiter's Reusable Surface Insulation (RSI)," AIAA Paper No. 74-730; Presented at the Thermophysics and Heat Transfer Conference, Boston, MA, July 1974.

Banas, R.P. and Cordia E.R.. "Advanced High Temperature Insulation Material for Reentry Heat Shield Applications," Presented at 4th Annual Conference on Composites and Advanced Materials, American Ceramic Society, Cocoa Beach, FL, January 1980.

Beasley, R. M. and Izu Y.D., "Design and Construction Techniques for Radomes for Superorbital Missions," Presented at the OSU-RTS Symposium on Electromagnetic Windows, Columbus, OH, June 1964.

Beasely, R. M. and Clapper R.B., "Thermal Structural Composites for Aerospace Applications," Presented at the 67th Annual Meeting of the American Society for Testing and Materials, June 1964.

Buckley, J.D., Strouhal G. and Gangler J.J., "Early Development of Ceramic Fiber Insulation for the Space Shuttle," Am. Cer. Soc. Bull. 60:1196-1200, 1981.

Cooper, P.S. and Hollway P.F., "The Shuttle Tile Story: Astronautics and Aeronautics," 19:14-34, 36, 1981.

Designing Glass Fibers

W.W. Wolf

DESIGNING GLASS FIBERS

Warren W. Wolf, Ph.D.
Owens Corning
2790 Columbus Road
Granville, OH 43023

ABSTRACT

This is a review of the typical applications of glass fibers and the technologies used to produce them. The composition and properties of various types of glass fibers are discussed. Recent trends are emphasized with particular focus on the understanding of compositional effects on biopersistence in glass wool fibers.

INTRODUCTION

In the space of just over sixty years, glass fibers have grown from a novelty product with little commercial application into a material whose production is now measured in millions of tons annually. Optical fibers are not covered in this design review. An overview of compositions used in the fiberglass industry is given, along with the known relationships between product requirements such as strength, water durability, refractive index, dielectric content and biopersistence. Manufacturing requirements including temperature and crystallization characteristics are also included.

A brief overview of product applications and manufacturing techniques follow (see Table I).

Table I. Fiberglass Applications

	Important Product Parameters
Discontinuous fibers	
Insulation – thermal, acoustical	T, fib. dia., Kdis
Filters	T, chem., Kdis
Fabrics and mats	
Drapes, fabrics	fib. dia., mech.
Clothing	T, fib. dia., mech.
Cable and wire insulation, sleeves	elec.
Yarn and threads, screening	mech.
Battery-retainer mat	chem.
Reinforcement – continuous and chopped fibers	mech.
Plastics and polymers –	
boats, automotive, aircraft, consumer products	chem.
storage tanks	chem.
appliance housing	elec.
printed circuit boards, electronic	elec.
transparent panels	optic.
Teflon®* -- fabric structures	fib. dia.
rubber – tire cord, tires	
asphalt – shingles, road bond	
plaster, gypsum	
cement, mortar	chem.

Note: T = high temperature resistance; chem. = chemical durability; mech. = mechanical properties (modulus, strength, etc.); elec = electrical properties, (resistivity, dielectric, etc); optic. = optical properties (color, refractive index, dispersion); fib. dia. = controlled fiber diameter and/or fine fibers; and Kdis = biopersistence.

*TEFLON is a registered trademark of E.I. du Pont de Nemours & Co., Wilmington, DE.

PRODUCT APPLICATIONS

The first product application for fiberglass was a glass fiber mat for air filters in the 1930's. Table I gives a cursory list of applications for glass fibers. The applications are grouped into three categories:

(1) Discontinuous fibers packed as batts or boards
(2) Applications using fabrics and/or mats

(3) Fibers for reinforcement, either in a continuous form or as chopped fibers

Discontinuous fibers are mainly employed for thermal or acoustical applications. The use covers residential and commercial buildings as well as automotive and appliance applications.

It has been discovered that for glass fibers made either by a rotary or a wheel spinning process that the fibers must have adequate mechanical strength and chemical durability. Resistance to water attack is important in both cases since the insulation can encounter condensation and vibration conditions, such as in railroad refrigerator cars where if adequate water durability is not available, degradation could occur.

Fabric and mat uses involve glass fibers not placed in a composite matrix, but rather used by themselves, often with an organic coating. Glass mats are used in batteries to prevent electrical shorting during discharge. In this application, resistance to acid is paramount.

The third category, continuous or chopped fibers used as reinforcements, covers a diverse market where applications take advantage of the high strength to weight ratio of the glass fibers, and applications combine this property with other advantages of glass fiber reinforcement such as chemical inertness or low dielectric strength.

Table II lists the glass property needed for specific needs as well as glass compositional types developed to meet those needs. It should be emphasized that the glass types in Table II are mostly a set range of chemical compositions and not a single specific chemistry.

Table II. Glass Property Requirements and Glasses Developed to Meet These Needs

		Glass Types
Electrical properties	Conductivity	E
	Dielectric constant	
Mechanical properties	Strength	S, E, Boron Free E-CR ("BFE")*
	Modulus	(M)
	(Density)	
Chemical durability	Water	E, wool, BFE
	Acid	C, BFE
	Acid/leachable glass	
	Base/alkali	AR
Optical properties	Refractive index	(W1), W2
	Color	
Biosolubility	Dissolution in-vivo tests	Wool
Special properties	Radiation absorption	(L)

Note: Glasses enclosed in parentheses are not commercially available in large quantities.

*A boron free E-CR-glass was developed by Owens Corning and is sold by Owens Corning under its registered trademark Advantex®.

As can be seen from Table II, glasses have been developed to respond to many needs. Some of the needs were very specific such as S glass. Others were broader based, combining several properties such as in the case of the new Advantex® glass fibers and these glass fibers are used in a variety of applications.

MANUFACTURING

The steps involved in manufacturing glass fibers as well as the constraints imposed on glass composition by each step are listed in Table III. Batching and melting are fairly generally understood processes, common to the glass industry as a whole. Therefore, this discussion will focus on constraints imposed by the fiberizing process.

As indicated in the previous section, there are two types of glass fibers, discontinuous and continuous, and there are several discontinuous and continuous processes. These will be considered separately in the next section.

Table III. Manufacturing Constraints on Glass Composition

Step in Manufacturing Process	Constraints
Batching	Raw materials availability
Melting	Refractory life/refractory Corrosion Melting energy, conductivity
Fiberizing	Glass-forming viscosity compatible with available metals Glass-crystallization tendency Glass quality
Conversion into usable product	
Overall process	Economically acceptable cost

Continuous Processes

At one time two processes were used: marble melt and direct melt. Today most of the production of continuous fibers is by direct melt since it is more energy efficient. Marble melt is used in a few operations. It has some flexibility in that the marbles can be shipped from a central point to many fiber-producing locations.

Commercial glass fibers are made by extruding molten glass through an orifice, usually 0.8 to 3 mm in diameter. This extruded glass is still molten and it is rapidly attenuated into a finer diameter by use of mechanical winders running at speeds up to 200 km/hr (see Figure 1). The diameter range produced is typically between 4 to 20 microns.

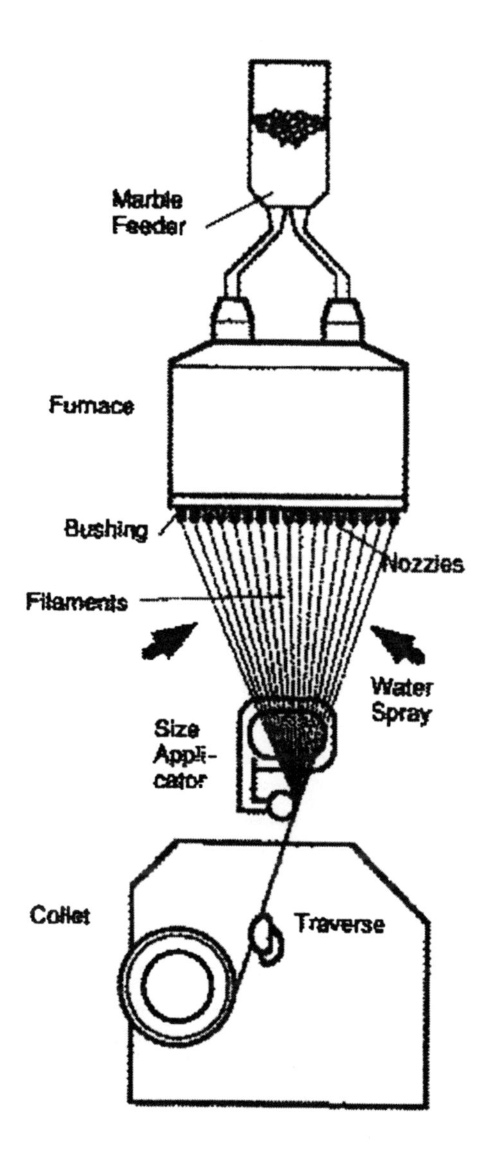

Figure 1. Continuous drawing process for manufacturing textile fibers.

The glass filaments are pulled through orifices in the base of a platinum-alloy "bushing." The number of orifices can range from a low of about 204 into the thousands. As the individual filaments are pulled through each orifice, they are combined or gathered into a strand, which is the basic building block to make glass-fiber products. During the process of pulling the filaments into a strand, sizes (coatings) are applied to the filaments to minimize abrasion and degradation of the glass fibers.

The most important properties to ensure efficient fiberization are:

(1) High Glass Quality
(2) Low Liquidus
(3) Sufficient Viscosity/Temperature Relationship

Excellent glass quality is critical since submicrometer inclusions can break glass fibers of dimensions 20 microns or less. Inclusions could be small refractory particles from melter side walls, a piece of devitrified glass or even metal dust. Regardless, the problem in breaking one filament leads to the tail breaking others and an interruption of the whole fiberizing process.

A low liquidus is important because temperature fluctuations inherent to the process must not lead to glass crystallization and the formation of small pieces of devitrified glass mentioned earlier. The viscosity/temperature relationship is a defining property since glass is fiberized between 500 to 1500 poises. For large commercial processes such as E or Advantex® glass, the bushings need to last a long enough time, like a year, if possible to ensure economic production. Hence the bushing temperatures are defined around 1300° C or lower. S glass is a special glass and is fiberized at much higher temperatures with a significant impact on the life of the platinum-alloy bushings.

Discontinuous Processes

Discontinuous processes are used not only to make glass wool insulation products, but also have been used to make a variety of products using "so called" rock or slag products. The latter are used not only in insulation applications but also find use because of their ability to crystallize at high temperature and provide a degree of protection in ceilings in the event of a fire. The phrase "so called" is used because these compositions take their names from the fact that conventional glass batching processes are not used, but often either a rock, such as basalt, or a slag from perhaps a steel mill is the principal component of the chemistry and the rocks or other large pieces such as coke or slag are layered together and then melted.

Flame Attenuation (see Figure 2)

This process is still in use in some glass fiber companies. Primary fibers about 1 mm in diameter are drawn through a large bushing in the bottom of the glass tank. The filaments are then aligned in a uniform array and exposed to a jet flame from an internal combustion burner. The flame breaks the larger filaments into fine fibers. Special glass fibers for filtration, etc. have been made in this manner.

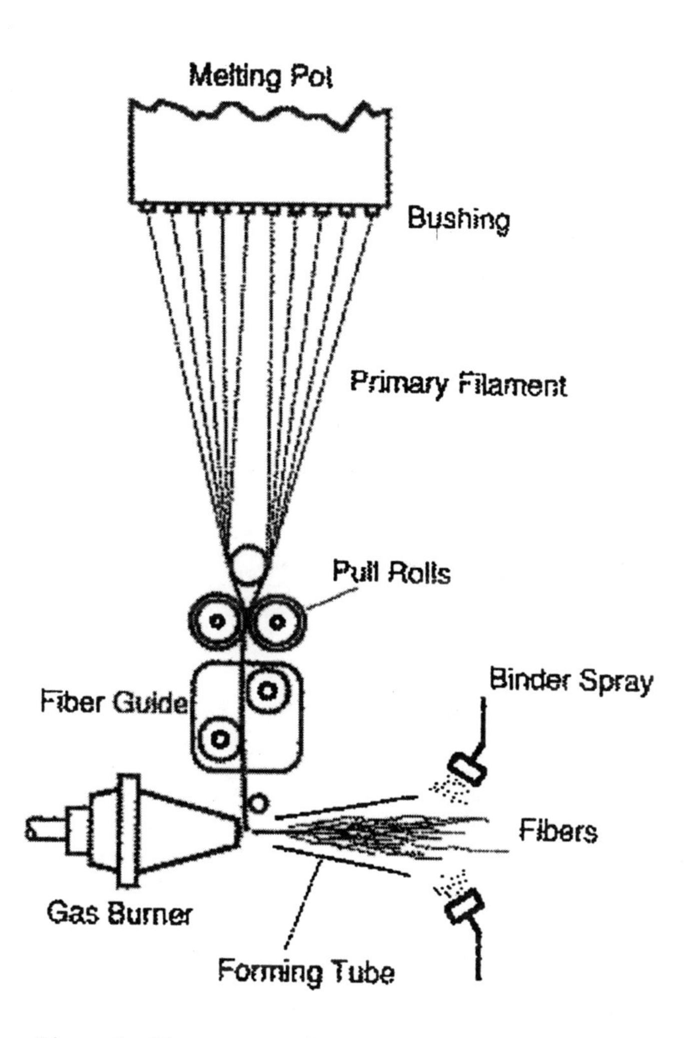

Figure 2. Flame attenuation process.

Wheel Spinning Process (see Figure 3)

There are several variations including a three and a four-wheel process in use. The molten glass source is often a cupola furnace where fist-sized rocks of basalt, other siliceous materials, dolomite and/or limestone can be interlayered and mixed with the fuel source coke. More modern operations have tended to better compositional control, and although non-fibrous "shot" formation is still a problem, the quantity of that has been considerably reduced in these operations.

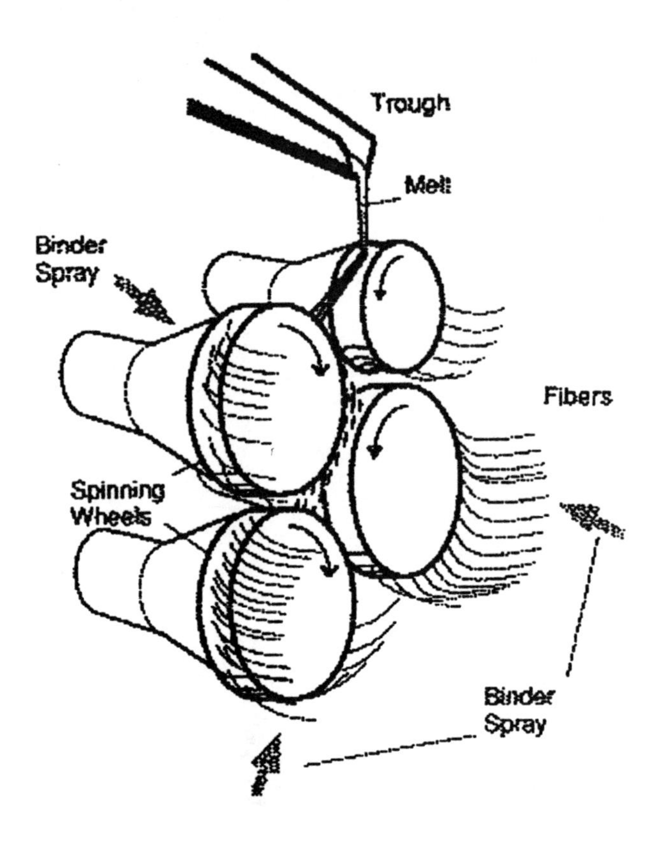

Figure 3. Wheel centrifuge or spinning process.

Rotary Process (see Figure 4)

The largest production of discontinuous fiber is made by the rotary process. The molten glass drops into a rotating cylindrical unit (the spinner). The spinner has numerous holes in its side walls. Centrifugal force extrudes the molten glass and it streams laterally into the path of a high velocity gas stream that attenuates and breaks it into discontinuous fibers. The advantage is a high throughput process that produces no shot. But the process is more demanding of the glass. The temperature at the forming viscosity of the glass must be low enough to meet the mechanical limitations on the base metal spinners. The liquidus must be significantly below the fiberizing temperature to prevent crystals growing in the spinner, which would plug the spinner holes and stop production.

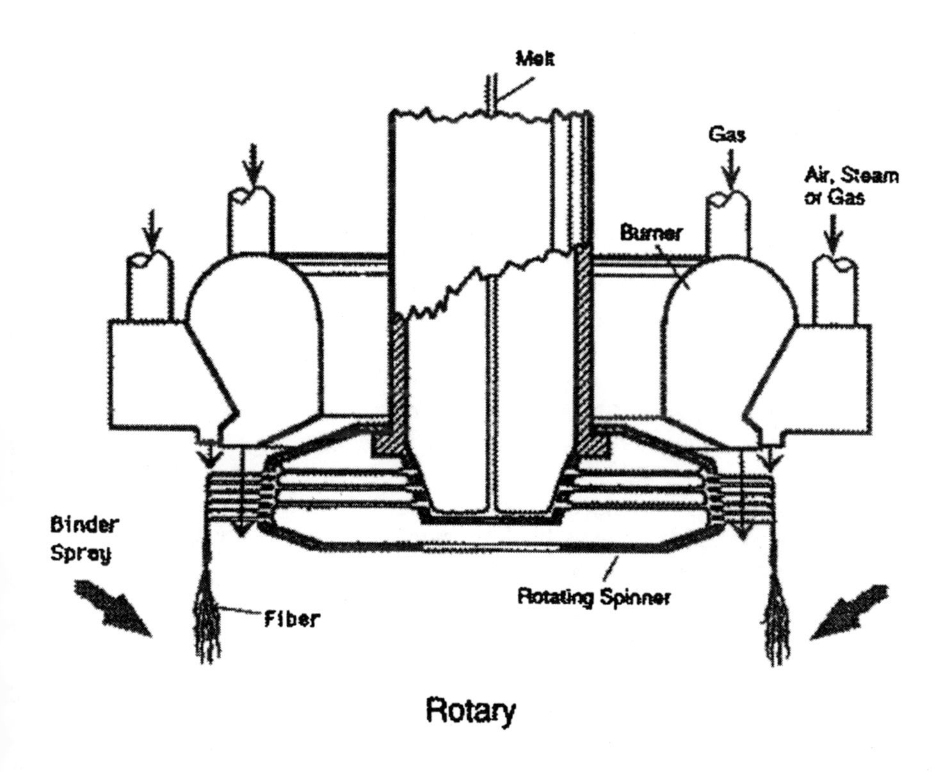

Figure 4. Rotary process for manufacturing glass wool fibers.

GLASS COMPOSITIONS

Two glass types provide the predominant materials for the production of continuous glass fiber: E glasses and S glass.

E Glass

E glass was first formulated in the 1930's where there was a need to use a continuous fiber to insulate fine wires at high temperature. E glass predominates in continuous glass fiber making, comprising into the high nineties, as a percent of total continuous glass fiber production.

E glasses are lime alumina silicate compositions. They were developed to have both high bulk electrical resistance and high surface resistivity as well as good fiber forming characteristics. E glass is based on the eutectic in the CaO-Al_2O_3-SiO_2 system which occurs at

62.2% SiO_2
14.5% Al_2O_3
23.3% CaO

The total alkali content ($Na_2O_2K_2O$) is held typically to less than 2 weight percent.

E glass need not be a simple definite composition, but may be many compositions within a single range as shown in Table IV. The composition may be tailored by each producer to the economics of raw material supplies and the details of the production. Originally until the 1990's, E glass always had a fair percentage of boric oxide (B_2O_3). B_2O_3 has had little influence on finished glass properties. Rather boric oxide was added to help reduce melting and fiber forming temperatures. With better refractories, better tank design, more knowledge on batch formulations and melting, and better bushing materials and design, it has been possible to actually eliminate boric oxide as illustrated by the Advantex® glass fibers made by Owens Corning.

Table IV. E Glass Composition, Wt%

	Compositional Range	Schoenlaub 1940		Tiede/Tooley 1948		Advantex®
SiO_2	52-56	52-56	54	52-56	54	59-62
Al_2O_3	12-16	12-16	14	12-16	14	12-15
Fe_2O_3	0- 0.8					
CaO	16-25	16-19	17.5	19-25	22	20-24
MgO	0- 6	3- 6	4.5			1- 4
B_2O_3	0-10	9-11*	10*	8-13*	10*	
$Na_2O + K_2O$	0- 2					0- 2
TiO_2	0- 1.5					
F_2	0- 1	0-1.25				
T_3 (°C)**		1200		1200		1200-1370
Liquidus (°C)		1120		1065		

*1 to 3% of fluorspar and 1 to 3% alkali may be substituted for part of the B_2O_3.

** Temperature at a viscosity of 1000 poises.

The original E glass patent from Dr. Schoenlaub was issued to Owens Corning in 1943.[1] The range of compositions and a preferred composition is given in Table IV. The patent revealed the use of fluorides and sulfates as forming agents. At the time of the Schoenlaub patent, it was felt B_2O_3 could not be reduced below 9% by weight in the finished glass.

In 1948, Ralph Tiede and Fay Tooley developed for Owens Corning an E glass composition in which most of the MgO was replaced by CaO.[2] The range and a specific composition are shown in Table IV. By eliminating MgO, dolomite did not have to be used in batch formulations. In particular, the new glass compositions were superior because of a significant reduction in liquidus.

It should be noted that CaO and MgO may be used interchangeably on a weight percent basis with regard to their effect on viscosity/temperature behavior.

Various modifications of E glass have been patented. Many have dealt with maintaining properties while reducing B_2O_3. Indeed it has been found that the Advantex® glasses can be fiberized without any B_2O_3 basing the composition on the E glass eutectic in the CaO-Al_2O_3-SiO_2 system.[3] A small added amount of MgO is advantageous.

Prior to Advantex® glass another approach for producing boron and fluorine free E glass was patented by Erikson and Wolf for Owens Corning.[4] It allowed forming temperature/viscosity properties similar to E glass using Li_2O and TiO_2 to reduce viscosity and ZnO, SrO and BaO to reduce the liquidus. It has been found that glasses such as these, as well as Advantex® glass, which does not have B_2O_3, have much better acid durability.

S Glass

S glass was developed to provide a commercially fiberizable glass with higher strength and elastic modulus than E glass.[5] S glass is compared to E glass in Table V. There is a modulus increase of 15% and a tensile property improvement of almost 20%; plus S glass has a slightly lower density. S glass has no fluxes and is fiberized between 1500 to 1600° C.

Table V. Composition of Glasses Used in Fiber Manufacture and Their Basic Properties

Constituent or Physical Property	E Glass (wt%)	S Glass (wt%)	S Glass (mol%)
SiO_2	52-56	65	69
Al_2O_3	12-16	25	15.5
B_2O_3	0-10		
MgO	0- 5	10	15.5
CaO	16-25		
Na_2O	0- 2		
K_2O	0- 2		
Fe_2O_3	0- 0.8		
F_2	0- 1		
Tensile strength of single fiber (MPa)			
At 22°C	3800	4500	
At 370°C	2600		
At 540°C	1700		
Tensile strength of strand (MPa)	2400	3100	
Young's modulus of fiber at 25°C (MPa)	75×10^3	85×10^3	
Density (g/cm^3)	2.52-2.62	2.47-2.49	
Refractive index	1.546-1.560	1.524-1.528	
Coefficient of linear thermal expansion per °C	5×10^{-6}	3×10^{-6}	

Extensive research has been conducted to increase the tensile and modulus of glass fibers. The addition of BeO was found to be useful to raise modulus, but the toxic properties of BeO prevented commercial use. Group II oxide as cerium oxide, vanadium oxide, or lanthanum oxide have been tried, but the increase in modulus is more than offset on a specific basis due to density increases.

Wool Glass

A typical early soda-lime-silicate composition is given in Table VI (Glass A). This glass was rapidly replaced by T, or thermal, glass with improved water and acid durability and lower viscosity. The glass has a comparatively high liquidus, which was acceptable for the Owens steam-blowing process. The composition range of this glass is disclosed in the Bowes patent.[6] Patented range and recommended composition are given in Table VI. The viscosity was reduced by adding CaO and B_2O_3 and reducing SiO_2; durability was increased by adding B_2O_3 and Al_2O_3 and reducing Na_2O; finally liquidus and crystallization tendencies were reduced by adding Al_2O_3 and B_2O_3 and keeping the ratio of MgO/CaO to around 0.25.

Over the years this glass was modified to reduce batch cost and viscosity. Boron was, as now, the most expensive raw material. The Dingledy patent describes a composition with lower viscosity obtained by increasing alkaline earth content at the expense of SiO_2.[7] Durability was maintained at an acceptable level by adding TiO_2. A low-cost modification of this glass was developed by Tiede.[8] To reduce cost, TiO_2 and B_2O_3 levels were kept very low; as a result glass durability was not as good. Product properties were maintained and, in fact, improved by using this glass in conjunction with a silicone binder. [9]

Viscosity and forming temperature of the initial T glasses were reduced by replacing part of the alkaline earths with iron oxide. This glass contained 5 to 12% iron oxide, and viscosity and liquidus were maintained at a low level by proper control of the FeO/Fe_2O_3 ratio. This glass was developed for the production of fine fibers using the Johns Manville spinning process. [10] The low forming temperature (925-1095° C) allows the use of base metal to form the spinning apparatus.

The introduction of the rotary process in the mid-1950s dictated drastic changes in glass composition to reduce viscosity and liquidus while maintaining good water durability. Such compositions are disclosed in the Welsch patent shown in Table VI.[11] One of the several typical compositions listed in the patent is also given. Viscosity was reduced by increasing the B_2O_3 and alkali levels. The increase in boron also contributed to improved durability. This patent also discloses a series of options: (1) addition of MnO, ZnO, and Fe_2O_3, to improve durability and reduce viscosity and (2) addition of TiO_2 and ZrO_2 to improve durability, liquidus, and viscosity. The high alkali content of these glasses also improves melting behavior.

Table VI. Wool-Glass Compositions, Wt.%

	A Glass	T Glass (Bowes, 1939)		(Welsch, 1955)	
SiO_2	72	60-65	53	50-65	58.6
Al_2O_3	2	} 2-6		0-8	3.2
Fe_2O_3			5	0-12	
CaO	5.5	} 15-20	14	3-14	8.0
MgO	3.5		3	0-10	4.2
BaO				0-8	
B_2O_3		2-7	5	5-15	10.1
$Na_2O + K_2O$	16	8-12	10	10-20	15.1
ZnO				0-2	
MnO				0-12	
TiO_2				0-8	
ZrO_2				0-8	
F_2					
T_3 (°C)*	≈ 1200		1050		1015
Liquidus (°C)	≈ 850		1160	870-980	925

*T3 = Temperature in °C at which the glass viscosity is 10^3 P.

The Welsch patent is very broad, and his work provided the foundation for most of the glasses developed for the rotary process. The main extensions of this work involved reducing B_2O_3 to below 5% to reduce cost, and adding fluorine to maintain or improve durability and liquidus, further reduce viscosity, and improve melting behavior[12], and/or TiO_2 and ZrO_2[13-15] to maintain or improve properties. Other modifications involved a large increase in ZnO and a reduction in alkali level to improve durability.[16] Fibers with improved durability and lower viscosity were obtained by keeping boron and alkali levels high and adding ZnO.[17]

Table VII presents a broad chemical range for commercial glass wool compositions that have been made throughout the world.

One significant change in wool glass processing and composition was the development and implementation by Owens Corning of a novel process of manufacturing bicomponent glass wool fibers. The resulting product is sold by Owens Corning under its registered trademark Miraflex®. Currently the Miraflex® product is made at only one manufacturing facility in the world that is located in Mount Vernon, Ohio.

Key glass composition and process/product patents are listed below and the reader is encouraged to go to these patents since space does not permit an extensive discussion in this paper.

1. R.M. Potter, "Glass Compositions for Producing Dual-Glass Fibers", U.S. Patent No. 6,017,835, January 25, 2000.
2. R.H. Houpt, et al, "Dual-Glass Fibers and Insulation Products Therefrom", U.S. Patent 5,431,992, July 11, 1995 and U.S. Patent 5,536,550, July 16, 1996.

Miraflex® glass fibers are unique in that the process requires melting two separate glasses of differing thermal expansion. A special delivery system and rotary fiberizer is designed so that the two glasses merge into a combined single fiber and the difference in thermal expansion creates enough change in fiber packing, etc. that a glass wool binder is not required in the processing step to form a glass batt or other insulation products.

The two glasses used in Miraflex® fibers are each designed to meet biopersistence criteria discussed in the next section, and do have SiO_2, Al_2O_3, B_2O_3, and Na_2O values that fall outside the ranges given in Table VII.

Table VII. Commercial Range for Wool Glasses Utilized Throughout the World

	Chemical composition of wool glass in weight percent
SiO_2	55 -70
Al_2O_3	0 - 7
B_2O_3	3 -16
K_2O	0 - 2.5
Na_2O	13 -18
MgO	0 - 5
CaO	5 -13
TiO_2	0 - 0.5
Fe_2O_3	0.1- 0.5
Li_2O	0 - 0.5
SO_3	0 - 0.5
F_2	0 - 1.5
BaO	0 - 3

Probably the most significant event since the present author did a review of Glass Fibers with P. F. Aubourg in 1984 for the volume, Advances in Ceramics, Volume 18, Commercial Glasses,[18] has been the consensus within the international scientific community that the biopersistence of fibers plays a significant role in their potential biological activity.

Of course it has been realized that fibers must also be of the right length and diameter as well as present in significant amounts in the air for there to be a concern. But it is realized that discontinuous glass forming processes, such as the rotary or any of the wheel processes, do produce fiber diameters of small enough diameter to bc respirable.

With this point made, it is now the consensus that a fiber must persist in biological tissue for a sufficient time to have the potential to produce biological effects. [19]

There is considerable literature on the chemical durability of silicate glasses in various solutions[20-21] and many of the reported findings may be applied to the case of silicate synthetic vitreous fiber (SVF) in biological fluids. [22]

In this document, chemical durability of fibers refers to the rate at which a fiber reacts with a solution and dissolves in it, with the result that the fibers finally disappear. The durability or dissolution rate is therefore different from the solubility, although they are often confused. Solubility has to do with the maximum amount of material that can dissolve in a given amount of fluid, before it reaches equilibrium with the solution. The dissolution rate of a fiber depends mainly on its composition and is generally proportional to the fiber surface area.[23-25]

There have been a number of studies of the dissolution rate of various SVF in simulated biological fluid. These studies have found a considerable variation in dissolution rate depending on the composition of the fiber, generally in agreement with trends of dissolution of silicate glasses in neutral solutions. Substitution of sodium, potassium, calcium, and magnesium into a silicate composition modifies the silicate network and leads to increased dissolution rates. Alumina tends to strengthen the network and to decrease the dissolution rate. The natural, crystalline fibers of various kinds of asbestos are found to be very durable, with dissolution rates much less than commercial SVF. Among SVF, those with high alumina and silica content, such as E glass and traditional refractory ceramic fibers, have lower dissolution rates than the rock and slag wool fibers and the glass wool insulation fibers.

The attached list of references[22-27] gives a number of citations from the literature linking glass composition and biosolubility for synthetic vitreous fibers. Among the important items to mention is the early work of Scholze in Germany,[24] the work by Leineweber,[23] and the work of Potter and Mattson.[26] Comparisons of in-vivo and in-vitro have been done by Eastes and Hadley,[27-29] as well as others [e.g., L.D. Maxim, et al. "Hazard Assessment and Risk Analysis of Two New Synthetic Vitreous Fibers", *Regulatory Toxicology and Pharmacology*, 30, pp. 54-74, 1999], and have shown the ability to predict glass fiber effects in test animals based on their compositions and predicted biosolubility from in-vitro studies. Of particular importance to the reader may be the articles by Eastes, Potter, and Hadley[30-32] which link composition and predictions of dissolution rates.

THE CERAMIC ENGINEER

Glass is an oxide material and is studied in the United States as a branch of ceramic material science and engineering. Most of the books published on silicate glasses are also pertinent to glass fibers, which are silicate glasses modified to meet commercial needs.

The individual who wants to design a specific glass fiber composition for a commercial use should approach the problem similar to any material problem. The main two criteria are going to be a combination of evaluating economic factors and technical factors.

Economic Factors

The problem is most likely going to be designing a glass formulation for an existing facility and an existing application. But the approach would not be that different if the individual has the assignment for a greenfield facility and a new application.

The first area to consider is the location of the facility, and what raw materials are most available and at the best cost that can be delivered to the manufacturing location.

Generally for a wool glass or a continuously formed fiber, the composition will be selected from within the ranges given in this article. Immediately though the economic and the technical factors begin to interact. Both factors are equally important and have effects on each other leading to best compromises.

Technical Factors

You have selected a particular type of glass that needs to be manufactured. You know if it is a wool glass, an E glass, or a boron free version of E glass, such as Advantex® glass.

The wool glass will be slightly easier, so let us select it first. You need to first look at the primary ingredients: SiO_2, Al_2O_3, B_2O_3, Na_2O, CaO, and perhaps MgO. You may be fortunate and have access to algorithms or other modeling techniques that may allow you to help zero in on final composition. But regardless, you are going to do some actual small melt measurements and final glass property measurements before you are going to attempt to use the glass full scale and begin to make tons of it.

The Interplay

If you are designing a wool glass, you are going to probably focus on the following:

Raw Materials Available

Do you use sand, feldspar or recycled cullet to supply SiO_2?; clay if you need Al_2O_3?; a calcium carbonate or a dolomite for the CaO and how much MgO will come from the dolomite?; you will probably need sodium carbonate and a boric oxide source.

Raw Material Quality

Are the sources reliable and continuous? Is the recycled cullet in particular clean and reliable? Are the trace oxides added such as iron oxide excessive for the type of melter being used? What is the particle size of raw materials available?

Meltability/Formability

What is the viscosity/temperature curve and the liquidus? Is the liquidus far enough away from the forming temperature? Are melting aids such as niter or sodium sulfate necessary? Are there environmental issues with niter or with volatiles that can be addressed by composition?

Physical Properties

Is the water durability of fibers made from the glass melt adequate?; have you ensured that the biosolubility calculated from known compositional factors is more than adequate for the standards of good stewardship?

A continuous fiber like E glass or Advantex® glass might be similar in development although the raw material sources would be different, or differently processed such as ground to finer size. You would not be concerned with the biopersistence calculations since continuous fibers are formed at diameters greater than 3 microns, the upper inhalable limit for humans. You may also need to approach the problem quite differently if a new organic coating or "size" is needed for a product application. But that is a different game, since now we are talking about designing composite materials.

SUMMARY

Glass fiber compositions have been in development for over 60 years. The practicing engineer or chemist needs to be familiar with the basic material science and engineering principles to design the best glass fiber products. Both technical and economic factors need to be understood to achieve optimum results. The practicing engineer or scientist should also acquaint themselves with the history of glass fiber composition development reflected in the patents and articles at the end of this paper.

QUESTIONS AND PROBLEMS

1.Outline an approach to design a new glass formulation for a new glass wool operation being installed in an existing facility?

Assume there is an existing furnace to melt the glass but the furnace has been shut down for three years and is being brought up again for this new operation.

Please answer the following questions as you complete the design for the glass formulation:

(i) What is the first questions you need to answer that are not specifically linked to the final formulation?

(ii) What questions do you need to ask to ensure that the final formulation ensures both quality and cost matters are addresses?

2.Repeat the above design of a glass formulation but now assume it is for a new glass fiber that will be used in a new reinforcement opportunity in the automotive field?

(i) In addition to answering the same questions as you did in 1., please contrast what questions are different in the two design problems?

REFERENCES

[1]R.A. Schoenlaub, "Glass Composition", U.S. Patent No. 2,334,961, December 5, 1940.

[2]R.L. Tiede and F.V. Tooley, "Glass Composition", U.S. Patent No. 2,571,074, November 2, 1948.

[3]W.L. Eastes, D.A. Hoffman, and J.W. Wingert, "Boron Free Glass Fibers", U.S. Patent No. 5,789,329, August 4, 1998.

[4]T.D. Erickson and W.W. Wolf, "Glass Composition Fibers and Methods of Making Same", U.S. Patent No. 4,026,715, August 22, 1984.

[5]R.S. Harris and G.R. Machlan, "Glass Composition", U.S. Patent No. 3,402,055, July 12, 1965.

[6]V.E. Bowes, "Sodium Calcium Borosilicate Glass", U.S. Patent No. 2,308,857, December 20, 1939.

[7]D.P. Dingledy, "Glass Composition", U.S. Patent No. 2,664,359, June 1, 1951.

[8]R.L. Tiede, "Glass Fiber Products", U.S. Patent No. 3,253,948, February 12, 1962.

[9]J. Stalego, U.S. Patent No. 2,990,307, June, 1961.

[10]W.P. Hahn and E.R. Powell, "Glass Composition", U.S. Patent No. 2,756,158, September 9, 1952.

[11]W.W. Welsch, "Glass Composition", U.S. Patent No. 2,877,124, September 25, 1955.

[12]W.W. Welsch, "Glass Composition", U.S. Patent No. 2,882,173, June 20, 1955.

[13]S. de Lajarte, "Glass Composition", U.S. Patent No. 3,013,888, September 6, 1959.

[14]B. Laurent and C. Haslay, "Fiberizable Glass Compositions", U.S. Patent 3,508,939, January 14, 1965.

[15]B. Laurent and C. Haslay, "Silicate Glass Compositions", U.S. Patent No. 3,853,569, July 1, 1970.

[16]C. Haslay, Agnety, and J. Paynal, "Manufacture of Borosilicate Fibers", U.S. Patent No. 3,523,803, August 11, 1970.

[17]L.V. Gagan, "Glass Composition for Fiberization", U.S. Patent No. 4,177,077, September 17, 1978.

[18]P.F. Aubourg and W.W. Wolf, "Commercial Glasses" pp. 51-63 in *Advances in Ceramics*, Vol. 18 edited by D. C. Boyd and J. F. MacDowell. The American Ceramic Society, 1986.

[19]J.M.G. Dans in *Non-Occupational Exposure to Mineral Fibers,* p. 33, edited by J. Bignin, J. Petro, and R. Saracci, WHO/IMC, Lyon, 1989.

[20]R.H. Doremus, *Glass Science*, p. 242, John Wiley, New York, 1973.

[21]D.E. Clark, C.G. Pantano, and L. Herch, *Corrosion of Glass*, Magazines for Industry, New York, 1979.

[22]H. Scholze, *Glastechnische Berichte*, 61, pp. 161-171, 1988.

[23]J.P. Leineweber, *Biological Effects of Man-Made Mineral Fibers*, p. 87, WHO/IARC Conference, Copenhagen, 1982, WHO, Copenhagen, 1984.

[24]H. Scholze, *Glass: Nature, Structure, and Eigenschoften*, second edition, p. 265, Springer, New York, 1977.

[25]Nomenclature of Man-Made Vitreous Fibers by the Nomenclature Committee of TIMA, Inc, April 15, 1991.

[26]R.M. Potter and S.M. Mattson, "Glass Fiber Dissolution in a Physiological Saline Solution", *Glass Technische Berichte,* 64, pp. 16-28, 1991.

[27]S.M. Mattson, "Glass Fibrosis Simulated Lung Fluid: Dissolution Behavior and Analytical Requirements", *Annals of Occupational Hygiene*, 38, pp. 857-877, 1994.

[28]W.L. Eastes and J.G. Hadley, "Role of Fiber Dissolution in Biological Activity in Rats", *Regulatory Toxicology and Pharmacology*, 20, pp. 5104-5112, 1994.

[29]W.L. Eastes and J.G. Hadley, "A Mathematical Model of Fiber Carcinogenicity and Fibrosis in Inhalation and Intraperitoneal Experiments in Rats", *Inhalation Toxicology*, 8, pp. 323-343, 1996.

[30]W.L. Eastes, R.M. Potter and J.G. Hadley, "Estimating In-Vitro Glass Fiber Dissolution Rate from Composition", *Inhalation Toxicology*, 12, pp. 269-280, 2000.

[31]W.L. Eastes, R.M. Potter, and J.G. Hadley, "Estimation of Dissolution Rates from In-Vivo Studies of Synthetic Vitreous Fibers", *Inhalation Toxicology*, 12, pp. 1037-1054, 2000.

[32]W.L. Eastes, R.M. Potter and J.G. Hadley, "Estimating Rock and Slag Wool Fiber Dissolution Rate from Composition", *Inhalation Toxicology*, 12, pp. 1127-1139, 2000.

11

Designing for Nuclear Waste Containment

G.G. Wicks

DESIGNING FOR NUCLEAR WASTE CONTAINMENT

G.G. Wicks
Waste and Environmental Remediation Programs
Westinghouse Savannah River Company
Aiken, SC 29808

INTRODUCTION

Ceramics have been an important part of the nuclear community for many years. On December 2, 1942, an historic event occurred under the West Stands of Stagg Field, at the University of Chicago. Man initiated his first self-sustaining nuclear chain reaction and controlled it. The impact of this event on civilization is considered by many as monumental and compared by some to other significant events in history, such as the invention of the steam engine and the manufacturing of the first automobile. Making this event possible and the successful operation of this first man-made nuclear reactor, was the use of forty tons of UO_2. The use of natural or enriched UO_2 is still used today as a nuclear fuel in many nuclear power plants operating world-wide. Other ceramic materials, such as ^{238}Pu, are used for other important purposes, such as ceramic fuels for space exploration to provide electrical power to operate instruments on board spacecrafts. Radioisotopic Thermolecric Generators (RTGs) are used to supply electrical power and consist of a nuclear heat source and converter to transform heat energy from radioactive decay into electrical power, thus providing reliable and relatively uniform power over the very long lifetime of a mission. These sources have been used in the Galileo spacecraft orbiting Jupiter and for scientific investigations of Saturn with the Cassini spacecraft. Still another very important series of applications using the unique properties of ceramics in the nuclear field, are as immobilization matrices for management of some of the most hazardous wastes known to man. For example, in long-term management of radioactive and hazardous wastes, glass matrices are currently in production immobilizing high-level radioactive materials, and cementious forms have also been produced to incorporate low level wastes. Also, as part of nuclear disarmament activities, assemblages of crystalline phases are being developed for immobilizing weapons grade plutonium, to not only produce environmentally friendly products, but also forms that are proliferation resistant. All of these waste forms as well as others are designed to take advantage of the unique properties of the ceramic systems.

In January 2001, an important milestone was reached in the U.S. program to safely and effectively store and dispose of high level radioactive waste (HLW). A unique glass making operation in an equally unique vitrification facility, the Defense Waste Processing Facility (DWPF) in Aiken, South Carolina, poured its 4 millionth pound of environmentally safe nuclear waste glass, converting a significant portion of a very large inventory of HLW stored in a liquid form, to solid and inert products. These products will ultimately be disposed of in carefully selected geologic repositories, where they will be buried as part of a multi-barrier isolation system, designed to not only be safe for our generation, but all generations yet to come. This paper will describe key features of this program and some of the important roles that ceramics play in this effort, including waste form development, processing and equipment design.

Also, it is necessary to quantify performance of nuclear waste glass systems in complex environments and for very long time periods. A strategy has been developed in the nuclear community to accomplish this important and challenging task. As a result of collaborations at last's years meeting in Cocoa Beach with colleagues working on biomaterials, we noted that important and challenging problems, such as being able to predict long term behavior of materials in aggressive environments, are common to many fields. We thought it might be worthwhile to see if the strategy, developed over many years for predicting long term performance of 40-component nuclear glass systems in a geologic environment for long time periods, could be useful in predicting long term performance of other ceramics and material systems in the body. Using an interdisciplinary approach, a master flowsheet is being developed. This will piggy back on the experience and strategy of the nuclear field and see if any of these elements can be applicable to those in the biomaterials area. This new initiative and collaboration will be introduced.

CASE STUDY- VITRIFICATION OF HLW

BACKGROUND

Every major country involved with long-term management of high-level radioactive [HLW] waste has either considered or selected glass as the matrix of choice for immobilizing and ultimately, disposing of potentially hazardous, high-level radioactive material. There are many reasons why glass is preferred. Among the most important considerations is the ability of glass to accommodate and immobilize the many different types of radionuclides present in waste, and make them into a highly durable, integrated, solid waste form. The waste glass product produced not only has outstanding technical properties but also possesses excellent processing features, which allow glasses to be produced with relative ease even under difficult remote-handling conditions, necessary for processing highly radioactive material. The single most important property of the waste glass is its outstanding chemical durability and its corrosion resistance to a wide range of environmental conditions. The following treatise will provide an introduction into the field of HLW management and emphasize the behavior and performance of nuclear waste glass systems studied under a wide variety of conditions.

High Level Waste Inventory

Located in the United States today, are approximately 100 million gallons of high-level radioactive waste [HLW] containing over 1 billion curies of radioactivity. Most of this waste has been generated from defense programs and results from the reprocessing of spent fuels used in the production of tritium and plutonium, both of which are used for military applications. Most of the defense waste is stored as a liquid or semi-liquid in large underground tanks at the Hanford Site in Richland, Washington and at the Savannah River Site [SRS] in Aiken, South Carolina. While the largest volume of this waste is located at Hanford, the largest amount of radioactivity within this waste is contained at Savannah River. In addition to these inventories, a smaller quantity of defense waste has also been generated from the reprocessing of fuels used in naval reactors, and is stored in stainless steel bins as calcine at the Idaho Chemical Processing Plant in Idaho Falls.

In addition to the large amounts of defense waste currently on hand, another potential source of HLW is from reprocessing associated with nuclear power plants used to produce electricity. While reprocessing is in progress in other countries, commercial reprocessing was discontinued in the United States in the early 1970's. Hence, only 612,000 gallons of HLW were generated, all of which is stored at West Valley, New York. Existing inventories of defense and commercial HLW stored in the United States and their locations are summarized schematically in Figure 1.

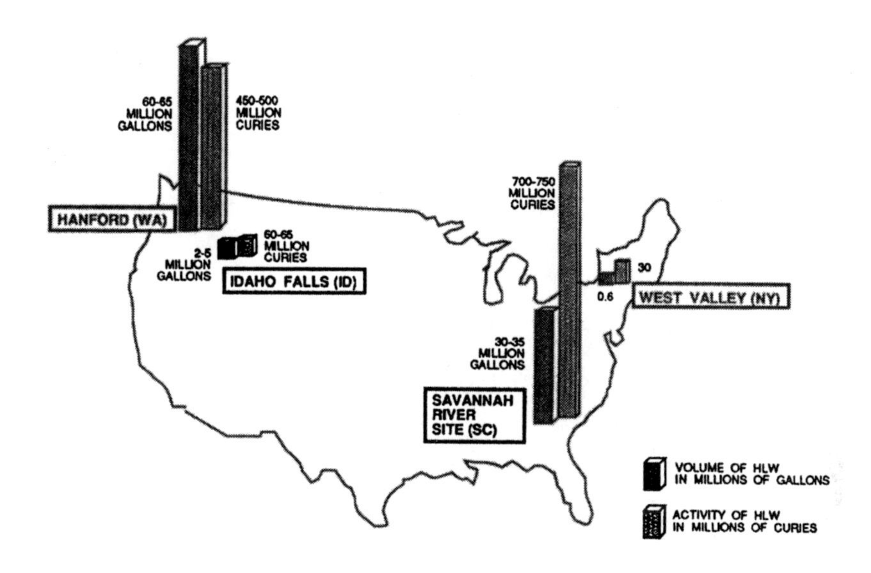

Figure 1. Volumes and Activities of High Level Radioactive Wastes in the U.S.

High Level Waste Characteristics

High-level waste [HLW] often consists of three major components; *sludge*, *supernate* and *saltcake*. The sludge is the most important component of the waste in that it contains most of the radioactivity, including fission products and long lived actinides. Sludge comprises about 10% by volume of the entire waste inventory and settles to the bottom of the underground storage tanks as a thick, gelatinous precipitate. It consists mainly of aluminum, iron and manganese oxides and hydroxides and contains actinides, fission products and Sr-90. The supernate makes up the remaining 90% of the waste and consists mostly of sodium nitrate and sodium nitrite along with Cs-137. In order to reduce the total amount of waste stored, the supernate is concentrated by evaporation and returned to the cooled tanks where saltcake crystallizes out of solution. Hence, in waste tanks today, these three phases can co-exist. It is these forms of the HLW that must be considered in immobilization tasks.

ANALYSES

HIGH LEVEL WASTE DISPOSAL STRATEGY

Since the Manhattan Project of World War II, HLW in the U.S. has been stored mainly in a liquid slurry form in underground storage tanks (Figure 2). Many of these original tanks are now reaching the ends of their projected lifetimes and as a result, some leaks have developed. Even though no injuries from radiation have occurred as a result of these practices, it should be clear, that a new means to more effectively and more permanently isolate the waste is needed. The strategy for long-term management of HLW is now changing from a program developed in the 1940's, of temporarily storing the waste in a relatively mobile liquid form, to a policy of the late 1990's, of immobilizing the waste into a solid glass product and ultimately, permanently disposing of the waste glass forms in deep geologic burial, as part of a multi-barrier waste isolation system.

Figure 2. HLW Storage Tanks at SRS

At the center of the proposed permanent disposal system is the durable, high integrity waste glass product. Adjacent to the waste glass, is the casting canister [304L stainless steel for the first HLW vitrified in the U.S.], followed by an engineered barrier tailored for the geologic conditions of the repository [may include an overpack and possible backfill], and finally, the geologic barrier itself. Geologic formations studied include salt, basalt, shale, clay, granite, and tuff, with the tuff site at Yucca Mountain in Nevada being selected as the reference repository setting. Each of the elements of the multi-barrier isolation system is designed to prevent the harmful release of radionuclides into the accessible environment and to allow the waste products to meet Federal release rate criteria. Among the most important regulations relating to waste form and waste package performance are 10 CFR 60 [Nuclear Regulatory Commission] and 40 CFR 191 [Environmental Protection Agency]. An attempt to define a time schedule for permanently disposing of HLW in the United States was provided by the Nuclear Waste Policy Act of 1982 and subsequent legislation.

Unlike other types of hazardous wastes, high level radioactive waste can actually become less hazardous with time, due to decay of radionuclides. For example, about 99% of the many radionuclides and daughter products in SRS HLW decay with half-lives ranging from microseconds to less than 30 years. After a time period of about one thousand years, the activity of the waste decreases approximately 4 to 5 orders of magnitude and for much longer time periods, the activity can actually approach and eventually becomes less, than the natural uranium from which the waste was originally derived.

ADVANTAGES OF GLASS

Many different disposal options and potential waste forms, including both vitreous and crystalline ceramics, were evaluated to determine the best system and strategy for safely and effectively disposing of HLW. Borosilicate glass was selected over other waste form alternatives which included calcines and supercalcines, cements, titanates, cermets, Synroc as well as other types of glass systems for immobilization of the first HLW to be treated in the United States, the waste at SRS. This decision was made based upon data and information supplied by more than three decades of national and international research and by reviews supplied by independent committees. Among the peer review committees used in this process, were groups sponsored by the National Academy of Sciences, the American Physical Society, and the U.S. Department of Energy, who all concluded that borosilicate glass would be a good material for immobilizing the SRS HLW. The selection of borosilicate glass was based on good characteristics in two categories; a) *processing considerations* and b) *technical performance* in five main areas. Following is a brief description of waste glass performance in each of these categories:

PROCESSING CONSIDERATIONS

Vitrification Facilities & Practical Operating Experience

The only major HLW immobilization facilities in the world are glass-making operations. These include successful production facilities at Marcoule and LaHague in France, as well as a plant located at Mol in Belgium. There are also additional plants either in production, under construction or being planned in other countries such as Japan, Germany, India, the United Kingdom and the United States. Following is a brief description of the vitrification process and facility at SRS, which is currently in production, immobilizing the first HLW in the United States.

SRS Vitrification Process

An important feature of the long-term management strategy of SRS HLW is to first separate the waste into radioactive and non-radioactive parts. Since most of the waste is non-radioactive, this allows the large bulk of the waste to be handled by less expensive and easier processing and disposal techniques. After a series of processing steps, the decontaminated portion of the waste is segregated and ultimately, mixed with a cement-based material to produce a waste product called saltstone, that is buried in engineered vaults on site, as shown in Figure 3. This disposal strategy is designed to produce a durable waste form and subsequent system, that even allows groundwater at the disposal location to meet EPA drinking water standards.

Figure 3. Saltstone Facility at SRS

Only the highly radioactive part of the waste is sent to a remote processing facility for conversion into borosilicate glass. After processing operations, which concentrate radionuclides and produce a waste stream amenable for vitrification, glass frit is then added to the HLW stream. The slurry is then liquid-fed into a joule-heated ceramic melter in the vitrification facility. Here the waste mixture is melted at a temperature of 1150°C, using a unique off-gas system that is able to remove 99.999999% of the cesium in the original waste. The molten glass is then discharged from the melter and cast into Type 304L stainless steel canisters, 2-ft. in diameter and approximately 10-ft. in height. The units are then decontaminated by a frit blasting technique, which then recycles the used frit back to the front end of the process to help produce subsequent waste glass products. The canisters are welded shut using a resistance upset welding technique. This process and products produced are described in more detail elsewhere.

SRS DWPF

The Defense Waste Processing Facility [DWPF] represents the first major vitrification facility for immobilizing HLW constructed in the U.S. The facility, which contains 5 million cubic feet of volume and represents more than a billion dollar investment, is currently in operation at the Savannah River Site in Aiken, South Carolina. The DWPF, shown in Figure 4, is the largest plant of its type in the world today, and now the most prolific, recently pouring its 4 millionth pound of environmentally acceptable waste glass product. The design of the facility is based on the 35 years of successful operating experience of reprocessing plants within the Department of Energy (DOE) complex. Additional facilities are also planned at other HLW sites in the U.S.

Figure 4. The Defense Waste Processing Facility (DWPF)

One of the most important pieces of equipment in the DWPF is a joule heated ceramic melter. Continuous melters using ceramic vessels have been used successful for many years in industry in production of a variety of different glasses. The joule heated ceramic melters operate by passing an electric current through the molten glass, causing heating. Continuous electric ceramic melters have resulted in higher production rates, more homogeneous products, and reduced volatilization for many applications. The DWPF melter is round in shape and consists of Monofrax K3 contact refractories and Inconel 690 electrodes.

Another important feature of the melter is that it is liquid fed, in which slurry of HLW and glass forming frit are injected directly onto the surface of molten glass at 1150°C, where the components are then melted. The Joule heated ceramic melter after installation in the DWPF is shown in Figure 5.

Figure 5. Joule Heated Ceramic Melter in DWPF Melt Cell

SRS DWPF STATISTICAL PROCESS CONTROL (SPC) SYSTEM

The waste glass forms produced in the DWPF must be of high quality and able to be manufactured in both a highly reliable and effective manner. Therefore, properties of the waste glass must satisfy both processing constraints, to assure that production is optimized, and product quality considerations, to assure that a durable and stable wasteform is produced for storage and ultimate disposal. However, these important properties cannot be measured directly in situ and are instead, predicted from models relating the properties to composition, which can be routinely measured.

To further complicate controlling the DWPF vitrification process, the repository requires that product quality constraints be satisfied to a very high degree of confidence, demanding a statistically-based control system for DWPF process and product control. The system developed for the DWPF is called the Product Composition Control System or PCCS. To provide the necessary confidence, all pertinent sources of variation are estimated and incorporated into PCCS, which simultaneously provides sufficient flexibility to process varying waste types and feed compositions from the SRS Tank Farm. Furthermore, the statistical process control algorithms at the foundation of PCCS are rigorous to assure that only acceptable waste glass will be produced. This is especially important since once waste glass products are made, the glass cannot be re-worked or re-melted. Finally, PCCS maximizes the efficiency of waste vitrified, so waste glass volumes and associated storage and disposal costs are minimized. No commercially-available alternative exists that can provide the rigor necessary for DWPF control while simultaneously furnishing the flexibility required to process all probable feeds. The PCCS is an important contribution to DWPF operations, because it represents a rigorous, statistically-defensible management strategy for a highly complex, multivariate system with multiple constraints imposed.

It is not possible to conduct routine measurements of key processing parameters, such as melt viscosity, electrical resistivity, and liquidus temperature, during radioactive operations. The key product parameter, glass quality, is able to be routinely measured using a seven-day crushed glass durability test on samples of the glass produced. Since the key process parameters are impossible to measure and glass quality requires more than a week to determine, models were developed from designed experiments that relate these key parameters to parameters that are easily measured in the process, i.e., composition. Consequently, these models help form the basis for the DWPF process and product control strategy.

TECHNICAL PEFORMANCE

As mentioned earlier, waste glass products that are fabricated must perform well in five main categories of technical performance. These include a) flexibility, b) thermal stability, c) mechanical integrity, d) radiation stability and e) chemical durability. These features will be discussed briefly below, with special reference to the designing of glass forms tailored for the vitrification process and for the ability of accommodating and immobilizing hazardous radionuclides.

Flexibility

Glass has demonstrated the ability to accommodate not only the 40 or more different elements that are found in individual waste tanks, but also the large variations in waste composition that occur. The reason for this feature is a result of the relatively open random network structure that characterizes glass systems and its ability to accommodate elements or radionuclides of different sizes, charge, and characteristics, as well as differing amounts of constituents.

The diversity of the feed is illustrated in Table I, which shows the many different elements found in SRS waste and frit formulations, which encompass many elements of the periodic table.

Table I. Waste Elements in SRS Waste Tanks Flexibility

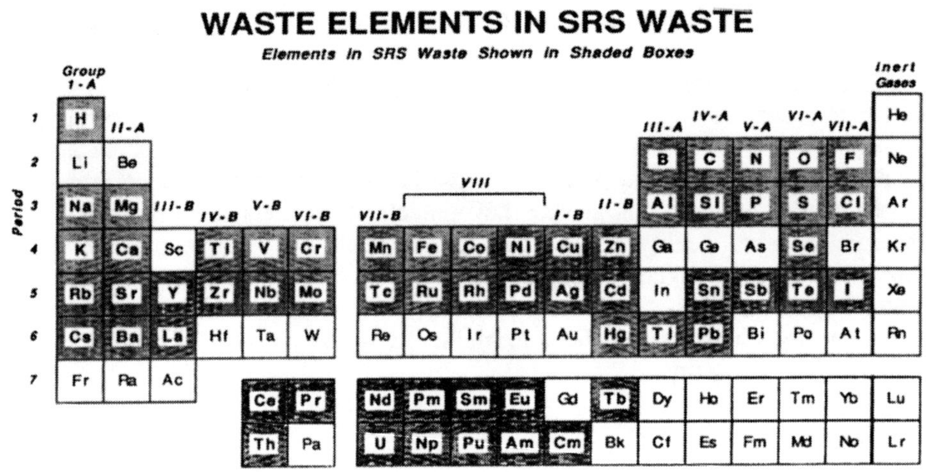

WASTE ELEMENTS IN SRS WASTE

Elements in SRS Waste Shown in Shaded Boxes (shaded elements shown in bold below)

Period	Group 1-A	II-A	III-B	IV-B	V-B	VI-B	VII-B	VIII	VIII	VIII	I-B	II-B	III-A	IV-A	V-A	VI-A	VII-A	Inert Gases
1	**H**																	He
2	Li	Be											**B**	**C**	**N**	**O**	**F**	Ne
3	**Na**	**Mg**											**Al**	**Si**	**P**	**S**	**Cl**	Ar
4	**K**	**Ca**	Sc	**Ti**	**V**	**Cr**	**Mn**	**Fe**	**Co**	**Ni**	**Cu**	**Zn**	Ga	Ge	As	**Se**	Br	Kr
5	**Rb**	**Sr**	**Y**	**Zr**	**Nb**	**Mo**	**Tc**	**Ru**	**Rh**	**Pd**	**Ag**	**Cd**	In	**Sn**	**Sb**	**Te**	**I**	Xe
6	**Cs**	**Ba**	**La**	Hf	Ta	W	Re	Os	Ir	Pt	Au	**Hg**	**Tl**	**Pb**	Bi	Po	At	Rn
7	Fr	Ra	Ac															
					Ce	**Pr**	**Nd**	**Pm**	**Sm**	**Eu**	Gd	**Tb**	Dy	Ho	Er	Tm	Yb	Lu
					Th	Pa	**U**	**Np**	**Pu**	**Am**	**Cm**	Bk	Cf	Es	Fm	Md	No	Lr

In order to develop glass matrices with the ability to immobilize this diversity in waste composition, various frit formulations have been developed world-wide. These are illustrated in Table II.

Table II. Selected International Glass Frit Compositions for Immobilization of HLW

Component	Compositional Range	French AVH [R7T7]	German PAMELA [SM 513]	US SRS [165 Defense Waste]
SiO_2	45-77%	54.9%	58.6	68.0
B_2O_3	5-20%	18.9%	14.7	10.0
Na_2O	1-20%	11.9%	6.5	13.0
Li_2O	0-7%	2.4%	4.7	7.0
MgO	0-4%	-	2.3	1.0
ZrO_2	0-1%	-	-	1.0
Al_2O_3	0-3%	5.9%	3.0	-
CaO	0-7%	4.9%	5.1	-
ZnO	0-10%	3.0%	-	-
TiO_2	0-5%	-	5.1	-

As indicated in the table, borosilicate glasses have been the most widely studied systems for incorporation of nuclear wastes. Each component of the glass frit plays an important role by complementing the roles of waste constituents and by either assisting glass melting or by improving durability of the solidified product. Some of these key effects are shown in Table III.

Thermal Stability, Mechanical Integrity and Radiation Stability

Waste glass products possess good *thermal stability*. Upon cooling from the melt or from self-heating due to radionuclide decay, waste glasses can phase separate or crystallize. Many different studies have been performed to assess the effects of these processes on performance of the glass. In one series of experiments, SRS waste glass was purposely devitrified, even though extensive devitrification is not expected for this system. The resulting time-temperature-transformation [TTT] curves identified phases formed and leaching tests showed that even in a 'worst-case' scenario, the effects of the crystalline phases on chemical durability of this system, were not significant.

Table III. Effects of Frit Components on Processing and Product Performance

Frit Components	Processing	Product Performance
SiO_2	Incr. viscosity signif., reduces solubility	Incr. durability
B_2O_3	Reduces viscosity, Incr. waste solubility	Incr. durability in low amts, reduces in large amts.
Na_2O	Reduces viscosity & resistivity, incr. waste solubility	Reduces durability
Li_2O	Same as Na_2O but larger, incr. devitrif.	Reduces durability, but less than Na_2O
CaO	Incr. then reduces viscosity & waste solub.	Incr. then decr. durability
MgO	Same as CaO, reduces tendency for devitif.	Same as CaO
ZrO_2, La_2O_3	Reduces waste solubility	Incr. durability signif.

Waste glass products also possess good *mechanical integrity*. Cracking can occur due to stresses induced during fabrication or from accident scenarios during handling, transportation or storage operations. A variety of different mechanical tests have been performed on waste glass systems. The testing has ranged form laboratory-scale studies to drop tests of full-size canisters. Based on these data,

several important observations can be made. First, glass forms fracture into relatively large chunks inside canisters and fracture is generally localized to the area of impact. Second, the amount of increased surface area produced is low along with the amount of resulting fines or small particles. Waste glass products possesses more than adequate mechanical stability under anticipated as well as accident scenarios.

As radionuclides are incorporated into glass structures, a significant radiation field can be produced, so waste glasses must possess good *radiation stability*. Many studies of effects of radiation have been performed on waste glass systems and properties studied during these investigations include chemical and mechanical integrity, stored energy, helium accumulation, density changes and radiolysis. Based on all existing data, waste glass forms perform very well under all of the radiation conditions expected during all stages of solidification and isolation of the HLW.

Chemical Durability

The most important and most studied property of solidified waste glass forms is its *chemical durability*. This provides a measure of how well the waste glass will retain radionuclides under anticipated as well as accident scenarios. Nowhere is it more important than in the final resting place of the waste, the geologic repository. The two most important observations made after evaluating the chemical durability of a variety of waste glass systems under many different repository conditions is that 1) *leaching of glass is very low when subjected to realistic scenarios and conditions* and 2) *not only is chemical durability of waste glass good, but durability actually improves with time.*

Improved analytical capabilities have allowed investigators to obtain more insight into the corrosion processes of glass. In the late 1960's, an integrated study approach was first applied to studies of nuclear waste glasses, which combined solution analyses with a variety of bulk and surface studies of leached glasses, to obtain a more complete picture of the leaching process. In subsequent years, new and more sophisticated surface analytical tools appeared along with more accurate solution analyses. The integrated study approach provides important information on the chemistry or structure of species of interest, on and within leached surface layers, as well as in solution. This, therefore, provides as complete a picture as possible of the leaching behavior and corrosion mechanisms of complex nuclear waste glass systems. Analytical tools used to study leached nuclear glasses are depicted schematically in Figure 6, along with their sampling depths.

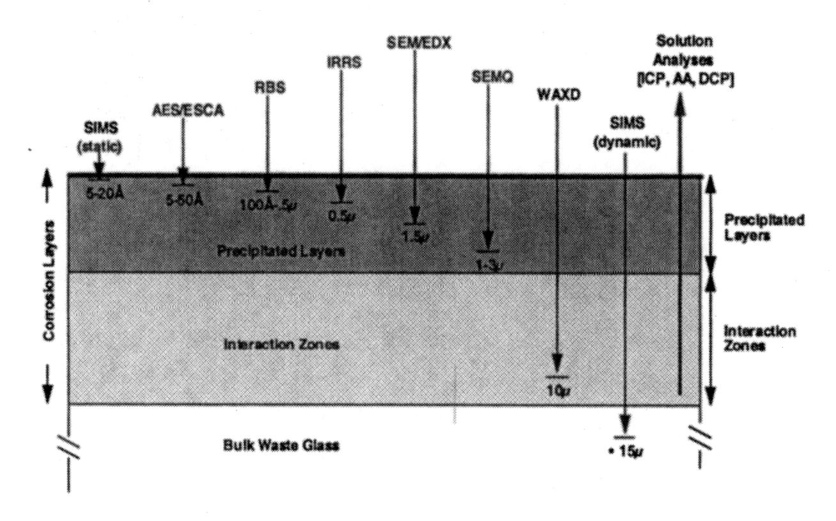

Figure 6. Integrated Study Approach

In order to assess, understand and ultimately, be able to predict the long term reliability of waste glass systems, chemical durability has been assessed as a function of important parameters that would be encountered during each stage of the solidification, transportation and interim and permanent disposal scenario. These important variables affecting the chemical durability of waste glass include time, temperature, solution pH, Eh, composition (waste, glass, leachate and homogeneity), devitrification, waste loading, surface area of sample to leachant ratio [SA/V], flowrate, pressure surface finish, glass cracking and fines, radiation effects, geology, hydrology, and package components [canister metal, possible overpack, potential backfills]. Based on all data currently available, the chemical durability of waste glass forms should be extremely good when subjected to realistic values of these parameters.

WASTE GLASS STRUCTURE

As discussed earlier, it is the random network structure of glass that helps to make this structure so accommodating for waste constituents (Figure 7). Waste glass forms are composed of approximately 70% glass forming chemicals or frit added to about 30% waste constituents. Although there are many individual elements that comprise nuclear waste glass compositions, these components can play only one of three basic roles in the glass structure; *network formers*, *intermediates* or *modifiers*.

Constituents such as silica and boric oxide, are generally added to the waste as major components of the frit. The silicon and boron atoms are *network formers* and become located in the center of oxygen polyhedra in the configuration of tetrahedra or triangles. These polyhedra are then tied together by sharing corners, generally in accordance with Zachariasen's rules, which then makes up the 'framework or skeleton' of the random network structure of the solidified waste glass form. Another structural role that both glass frit and waste elements can play is that of *intermediates*,

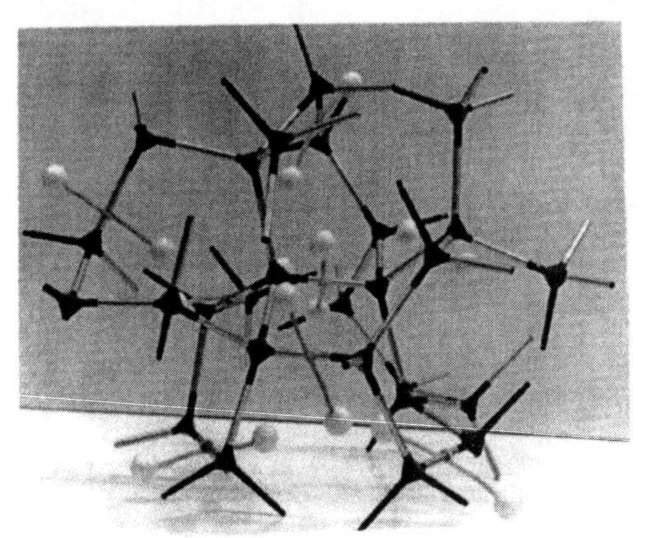

Figure 7. Random Network Structure of Glass

which is exemplified by major components found in the waste such as alumina. These components can replace the network formers and still retain the framework structure of the glass. Other cations can move to the singly bonded oxygen ions that are created, for charge neutrality. The final role that components can play is the most prevalent, that of *modifiers*. In this case, important waste components such as cesium and strontium, along with alkali and alkali earth constituents, are located within the holes of the random network structure, and can also be associated with nearby singly bonded oxygen ions.

An important point to note is that both glass and waste components become an integral part of the random network structure of the glass. Components are incorporated by primary and/or secondary bonding which helps explain why the glass is able to retain radionuclides so well during leaching and why different elements can leach at different rates. Based on this picture of waste immobilized in glass, a 3-stage corrosion process for nuclear waste glass was proposed, consisting of *Interdiffusion*, *Matrix Dissolution* and *Surface Layer Formation*. Both kinetic and thermodynamic modeling has been used successfully to describe and predict long-term behavior of these complex systems.

IN-SITU TESTING AND FIELD PERFORMANCE OF SRS WASTE GLASS

The ability of geologic formations to retain radionuclides is not without precedent. Two billion years ago, a chain reaction was started in a natural uranium ore deposit located at Oklo in Africa, and continued for hundreds of thousands of years before burning out. The waste that was generated by this natural event was trapped within this site. Hence, nature not only produced the first fission reactor, but also the first geologic repository to contain the waste generated.

The first field tests assessing the ability of waste forms to retain radionuclides in geologic formations was begun at Chalk River in the early 1960's by Canada. Following this pioneering effort, subsequent in-situ programs were characterized by cooperative, international undertakings, such as the burial of glasses, waste forms and package components in granite in Sweden, in clay in Mol, Belgium, in a limestone formation in the United Kingdom, and in salt, at the Waste Isolation Pilot Plant [WIPP] in Carlsbad, New Mexico. This later effort, the WIPP/SRS in-situ testing program is the first field tests involving burial of simulated nuclear waste forms to be conducted in the United States. It also represents the single largest, most cooperative venture of this type yet undertaken in the international waste management community.

Materials Interface Interactions Tests [MIIT]

The WIPP/SRS Materials Interface Interactions Tests [MIIT] represents a joint effort managed by Sandia National Laboratories in Albuquerque, New Mexico and the Savannah River Site in Aiken, South Carolina and sponsored by the U.S. Department of Energy. MIIT involves the field testing of simulated or non-radioactive waste forms and waste package components supplied by seven different countries. Included in MIIT are over 900 waste form samples comprising 15 different systems, almost 300 individual metal samples of 11 types of potential canister and overpack materials and over 500 geologic and backfill specimens. In total, there are 1,926 relevant interactions that characterize this effort.

The MIIT program is part of a larger effort at Savannah River aimed at understanding and being able to predict long term performance of DWPF waste glass in realistic repository environments. All samples, including waste forms, canister and overpack metals and geologic specimens, were fabricated in the shape of pineapple slices. They were then stacked onto heater rods in various stacking sequences in order to produce interfaces and interactions of interest. There are fifty assemblies in MIIT in seven different stacking sequences, which comprise a 7-part MIIT program, each part with its own set of specific objectives. The assemblies were inserted into brine-filled boreholes about 655 meters below the surface at WIPP in bedded salt deposits of the Salado Formation and heated to a temperature of 90°C [Figure 8] Samples and aliquots of brine were obtained and analyzed after 0.5, 1 and 2-years as part of this 5-year burial study.

Post-Test Analyses of SRS Burial Waste Glasses

Based on both solution analyses and surface studies, the performance of SRS waste glass was found to be very good during the 5 years of field testing. There were two distinct regions observed in surfaces of leached glasses; an outermost precipitated layer consisting of two sublayers, and an inner glass reaction zone containing 3 sublayers. Analyses of the leached layers correlated well with solution data in the field experiments and findings were consistent with a variety of earlier laboratory support studies. Hence, many different simulated waste glass systems were seen to perform well not only in controlled laboratory tests, but also in a variety of multi-year field experiments.

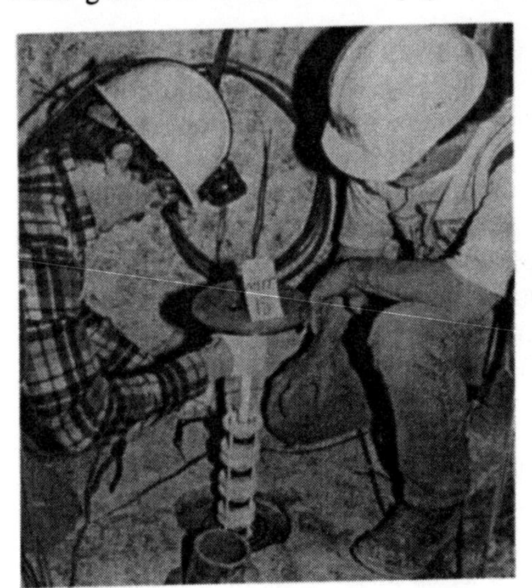

Figure 8. MIIT Field Tests at WIPP

PROPOSED NEW INTERDISCIPLINARY PROJECTS;
LONG-TERM PERFORMANCE OF NUCLEAR WASTE GLASSES AND BIOMATERIALS- A new interdisciplinary approach to improving and predicting performance

A common problem in many different fields and disciplines is to assess, understand and predict the long-term performance of materials of interest. This is necessary in order to determine the reliability of materials, often in very severe environments, and ultimately, to be able to extend their useful lifetimes.

During last year's ACerS Cocoa Beach meeting, a small group of interested individuals met to discuss a potential interdisciplinary strategy for extending lifetimes of materials of interest. One objective included examining the methodology developed after more than 30 years of R&D development for assessing and predicting long term behavior of complex nuclear waste glasses in a complex repository environment, and to see if it is also applicable to other fields.

Of special interest was to define ways of improving the performance of biomaterials and medical devices, within the challenging environment of the human body. For example, materials such as those used in hip replacements, work well for older, less active individuals, but for young adults and especially children with medical needs in this area, hip replacements may only last 8 to12 years. After that time period options for medical treatment are reduced and as these individuals approach middle age, their prognosis often becomes poor. If one can find a means to improve development of these and other important biomaterials and devices, and ultimately, improve their long-term performance and reliability, this could have a very important beneficial impact to the public.

After assessing the strategy for assessing and predicting long-term performance of nuclear waste glasses, it was found that this methodology may have relevance to other fields, including biomaterials and medical devices. A large flowsheet was developed examining potential similarities by a person who has worked for many years with nuclear waste glasses (George Wicks- Senior Advisory Scientist at DOE's Savannah River Site) in collaboration with someone who has much experience with biomaterials (Gary Fischman- Director of Biomaterials UIC, also previously with FDA). The team also received input from many other ACerS and NICE colleagues and may develop an interactive workshop at a future meeting on this subject, if there is enough interest.

The overall mission of this collaboration and interaction can be stated as follows:

MISSION

Examine a methodology to demonstrate long-term performance and reliability of important materials and systems in their final environment in an interdisciplinary manner, using experiences and expertise gained from the Nuclear Waste Management community.

A common, interdisciplinary approach for assessment and development of improved nuclear waste glasses and biomaterials/ medical devices was defined. Key common elements of this strategy and methodology include the following considerations:

INTERDISCIPLINARY TEAM

Nuclear Waste Glass Disposal vs. Human Implants, Prosthetics, Diagnostics, etc.

In the nuclear community, a strategy and experimental program was developed and executed over a time period of more than 30 years, and involved many national and international collaborations. Stakeholders involved in key contributions came from national and federal laboratories, academia, government labs, agencies and organizations, the public, various peer review groups, as well as the industrial sector. Each of these groups provided important contributions and perspectives of the overall program, and allowed the main objectives of the program to be met using all resources available, in a relatively cost effective, time efficient and productive manner.

INTERDISCIPLINARY OBJECTIVES

Nuclear Waste Glass Disposal

Predict the long-term performance of a 40-component waste glass system within a multi-barrier package in a complex geologic repository, in situ, out to time periods of 1,000 years (NRC regulations) and 10,000 years (EPA regulations).

Human Implants, Prosthetics, Diagnostics, etc.

Predict and improve the long-term performance and compatibility of materials and devices, in vitro, in a complex human system, out to time periods of 100 years.

INTERDISCIPLINARY R&D PROGRAM APPROACH

a) General 3-Stage Experimental Program

A 3-stage experimental program was used in the nuclear waste glass community that included thousands of individual and specific experimental studies. This overall program parallels that which is currently used to various degrees, in investigations of biomaterials and other related systems. This involves;

I. Laboratory Tests
II. Prototype Materials & Systems Tests
III. In-Situ Testing/ Clinical Trials

b) General Areas Performance

Important general areas of performance associated with R&D activities common to both fields include:

Nuclear Waste Glass Disposal	vs.	Human Implants, Prosthetics, Diagnostics, etc.
Compatibility	vs.	Bioactivity & Interactivity
Chemical Durability	vs.	Biocompatibility & Stability in Fluid Environments
Mechanical Integrity	vs.	Mechanical Integrity
Flexibility	vs.	Material & System Diversity
Thermal Stability	vs.	In-Vitro Stabilization
Radiational Stability	vs.	Photonic & Nuclear Effects
Recycle/ Reuse	vs.	Retrievability/ Reuse

c) General Experimental Features

There are general experimental features and considerations that were found to be especially important in nuclear waste glass studies. These considerations would also be anticipated to be relevant to other fields, including studies of biomaterials. These include:

- Standardized/ approved tests and procedures
- Accelerated tests
- Standards
- Natural systems
- M&I's
- Statistical significance of data
- QA
- Electronic databases
- Documentation and accessibility of data
- Interdisciplinary programs
- Collaborative efforts
- Regulatory compliance issues
- Modeling for long-term predictions
- Peer review processes

SUMMARY

Based on all data currently available, the performance of nuclear waste glass systems is excellent, when tested under realistic conditions, as determined by many studies performed by many different investigators. In addition to possessing excellent chemical durability, the durability also improves with increasing time. This behavior has been observed not only in laboratory tests, but in actual field experiments as well. The observed time-dependent leaching behavior has been described by a variety of existing kinectic models, geochemical models and by the use of thermodynamic analyses, which relates the behavior of waste glass systems to natural glasses such as basalt, which have been stable for millions of years. While glass systems have demonstrated the ability to immobilize and retain radionuclides very well, it must also be emphasized that waste glass forms are only one part of a multi-barrier isolation system. This disposal scenario provides an extremely safe and effective means for disposing and isolating potentially harmful radioactive elements for not only our generation, but for generations yet to come.

Finally, the strategy and methodology used in more than 30 years of experimental studies for assessing, understanding and predicting nuclear waste glass behavior, was examined for possible applicability to other fields. It is believed that using the experience and expertise gained in the national and international waste management field, may be very helpful for predicting and optimizing the long-term performance of other materials in other areas, such as in development of biomaterials and medical devices, in the medical community.

BIBLIOGRAPHY/ SUGGESTED READING

(1) I.J. Hastings, D.T. Rankin, and G.G. Wicks, in Ceramics Innovations in the 20th Century, J.B. Wachtman ed., pp. 226-231, The Am. Cer. Soc. (1999).

(2) U.S. Department of Energy, "Integrated Data Base for 1988: Spent Fuel and Radioactive Waste Inventories, Projections, and Characteristics," Report DOE/RW-0006, Rev. 4 (1988).

(3) J.L. Crandall, and H.J. Clark, "Integrated High-Level Waste Immobilization Plan," Savannah River Laboratory, Aiken, South Carolina (1977).

(4) G.G. Wicks, "Nuclear Waste Glasses", in Treatise on Materials Science and Technology - Glass IV, (M. Tomozawa and R.H. Doremus, eds.), 26, p. 57 (1985).

(5) U.S. Nuclear Regulatory Commission, "Final Rule for the Disposal of High-Level Waste in Geologic Repositories," Code of Federal Regulations, 10 CFR 60, June (1983).

(6) U.S. Environmental Protection Agency, "Environmental Standards for the Management and Disposal of Spent Nuclear Fuel, High-Level and Transuranic Radioactive Waste, " Code of Federal Regulations, 40 CFR 191, Sept. (1985).

(7) Federal Register, "Nuclear Waste Policy Act of 1982; Proposed General Guidelines for Recommendation of Sites for Nuclear Waste Repositories," 10 CFR Part 960, 48, No. 26, U.S. Department of Energy, Washington, D.C. (1983).

(8) G.G. Wicks and D.F. Bickford, "High Level Radioactive Waste- Doing Something About It," DP-1777, Savannah River Laboratory, Aiken, SC (1989).

(9) E.L. Wilhite, C.A. Langston, G.F. Sturm, R.L. Hooker, and E.S. Ochipinti, Spectrum '88- Nuclear and Hazardous Waste Management International Topical Meeting, American Nuclear Society, p. 99 (1986).

(10) P.D. d'Entremont and D.D. Walker, Waste Management '87, Univ. of Arizona, Vol. 2, p. 69 (1987).

(11) J.E. Haywood and T.H Killian, Waste Management '87, Univ. of Arizona, 2, p. 51 (1987).

(12) H.H. Elder, D.L. McIntosh, and L.M. Papouchado, Spectrum '88- Nuclear and Hazardous Waste Management International Topical Meeting, American Nuclear Society, (1986).

(13) G.G. Wicks, in Corrosion of Glass, Ceramics and Ceramic Superconductors, ed. by D.E. Clark and B.K. Zoitos, Noyes Publications, pp. 218-268 (1992).

(14) C.M. Jantzen, in Ceramic Transactions, Nuclear Waste Management IV, G.G. Wicks, D.F. Bickford and L.R. Bunnel, eds., 23, p. 37, Am. Cer. Soc., Columbus, OH (1991).

(15) C.M. Jantzen and K.G. Brown, Am. Cer. Soc. Bulletin, 72, No. 5, pp. 55-59 (1993).

(16) P.D. Soper, D.D. Walker, M.J. Plodinec, G.J. Roberts and L.F. Lightner, Bull. Am. Cer. Soc., 62, p. 1013 (1983).

(17) M.J. Plodinec, J.Non-Cryst. Solids, 84, p. 206 (1986).

(18) W.D. Kingery, H.K. Bowen, and D.R. Uhlmann, Introduction to Ceramics, 2nd Ed., Wiley, New York (1976).

(19) G.G. Wicks, "Structure of Glasses", in Encyclopedia of Materials Science and Engineering, Vol. 3, p. 2020, Pergamon, Oxford (1986).

(20) W. Lutze and R.C. Ewing, eds., Radioactive Waste Forms for the Future, Elsevier Science Publishing Co., New York, N.Y. (1988).

(21) M.J. Plodinec, G.G. Wicks and N.E. Bibler, "An Assessment of Savannah River Borosilicate Glass in the Repository Environment," DP-1629, Savannah River Laboratory, Aiken, SC (1982).

(22) M. Tomozawa, G.M. Singer, Y. Oka, and J.T. Warden,in Ceramics in Nuclear Waste Management, CONF-790420, T.D. Chikalla and J.E. Mendel, eds., p. 193, Technical Information Center, U.S. DoE, Washington, D.C. (1979).

(23) R.P. Turcotte. and J.W. Wald in Scientific Basis for Nuclear Waste Management, C. Northrup, ed., Vol. 2, p. 141 (1980).

(24) B.M. Robnett and G.G. Wicks, "Effects of Devitrification on the Leachability of High-Level Radioactive Waste Glass", DP-MS-81-60, Savannah River Laboratory, Aiken, SC (1981).

(25) D.F. Bickford and C.M. Jantzen in Scientific Basis for Nuclear Waste Management VII, G.L. McVay, ed.), 26, p. 557 (1984).

(26) T.H. Smith and W.A. Ross, "Impact Testing of Simulated High-Level Waste in Canisters", BNWL-1903, Pacific Northwest Laboratory, Richland, WA (1975).

(27) P.K. Smith and C.A. Baxter, "Fracture During Cooling of Cast Borosilicate Glass Containing Nuclear Wastes", DP-1602, Savannah River Laboratory, Aiken, SC (1981).

(28) J.M. Perez. and J.H. Westik, "Effects of Cracks on Glass Leaching", in Abstracts from ORNL Conference on the Leachability of Radioactive Solids, p. 35 (1980).

(29) L.J. Jardine, G.T. Reedy, and W.J. Mecham in Scientific Basis for Nuclear Waste Management, S.V. Topp, ed., 6, p 115 (1982).

(30) W.J. Weber and F.P. Roberts, Nucl. Technol., 60, p. 178 (1983).

(31) R.W. Turcotte and F.P. Roberts in Ceramic and Glass Radioactive Waste Forms, CONF-770102, D.W. Readey and C.R. Cooley, eds., p. 65 (1977).

(32) N.E. Bibler and J.A. Kelley, "Effect of Internal Alpha Radiation on Borosilicate Glass Containing Simulated Radioactive Waste," DP-MS-75-94, Savannah River Laboratory, Aiken, SC (1975).

(33) R.P. Turcotte, "Radiation Effects in Solidified High-Level Wastes, Part 2, Helium Behavior," BNWL-2051, Pacific Northwest Laboratory, Richland, WA (1976).

(34) F.P. Roberts, R.P. Turcotte, and W.J. Weber, "Materials Characterization Center Workshop on the Irradiation Effects in Nuclear Waste Forms, Summary Report," PNL-358, Pacific Northwest Laboratory, Richland, WA (1981).

(35) N.E. Bibler, in Scientific Basis for Nuclear Waste Management, S.V. Topp, ed., 6, p. 681 (1982).

(36) D.E. Clark, C.G. Pantano, and L.L. Hench, Corrosion of Glass, Books for Industry, New York, NY (1979).

(37) D.D. Walker, J.R. Wiley, M.D. Dukes, and J.H. Leroy, "Leach Rate Studies on Glass Containing Actual Radioactive Waste", DP-MS-80-96, Savannah River Laboratory, Aiken, SC (1980).

(38) D.M. Strachan, Scientific Basis for Nuclear Waste Management V, W. Lutze, ed., 11, p. 181 (1982).

(39) G.G. Wicks, J.A. Stone, G.T. Chandler, and S. Williams, "Long-Term Behavior of Simulated Savannah River Plant [SRP] Waste Glass Part I; MCC-1 Leachability Results, 4 Year Leaching Data", DP-1728, Savannah River Laboratory, Aiken, SC (1986).

(40) J.H. Westsik, and R.P. Turcotte, in Scientific Basis for Nuclear Waste Management, G. J. McCarthy, ed., 1, p. 341 (1979).

(41) D.F. Flynn, L.J. Jardine, and M.J. Steindler, in Scientific Basis for Nuclear Waste Management, C. Northrup, ed., 2, p. 103 (1980).

(42) H.P. Hermansson, H. Christensen, L. Werme, K. Ollila, and R. Lundqwist, in Scientific Basis for Nuclear Waste Management, S.V. Topp, ed., 6, p. 107 (1982).

(43) G.G. Wicks, P.E. O'Rourke, and P.G. Whitkop, "The Chemical Durability of SRP Waste Glass as a Function of Groundwater pH", DP-MS-81-104, Savannah River Laboratory, Aiken, SC (1981).

(44) L.O. Werme, L.L. Hench, J.L Nogues, J. Nucl. Mater., 116, p. 69 (1983).

(45) C.M. Jantzen, in Advances in Ceramics, Nuclear Waste Management, G.G. Wicks and W.A. Ross, eds., 8, p. 385, American Ceramic Society, Columbus, OH (1984).

(46) C.M. Jantzen and G.G. Wicks, in Scientific Basis for Nuclear Waste Management VIII", C.M. Jantzen, J.A. Stone, and R.C. Ewing, eds., 44, p.29 (1985).

(47) D.M. Strachan, B.O. Barnes, and R.P. Turcotte, in "Scientific Basis for Nuclear Waste Management", J.G. Moore, ed., 3, p. 347 (1981).

(48) P.B. Macedo, A. Barkett, and J.H. Simmons, in Scientific Basis for Nuclear Waste Management V, W. Lutze, ed., 11, p. 57, (1982).

(49) D.E. Clark, C.A. Maurer, A.R. Jurgensen, and L. Urwongse, in Scientific Basis for Nuclear Waste Management V, W. Lutze, ed., 11, p. 1, (1982).

(50) W.D. Rankin and G.G. Wicks, J. Am. Cer. Soc., 66, p. 417 (1983).

(51) C.Q. Buckwalter, L.R. Pederson, and G.L. McVay, J. Non-Cryst. Solids, 49, 397 (1982).

(52) G.T. Chandler, G.G. Wicks,. and R.M. Wallace, Nuclear Waste Management II, D.E. Clark, W.B. White, and A.J. Machiels, eds., 20, p. 455 (1986).

(53) G.G. Wicks, W.D. Mosley, P.G. Whitkop, and K.A. Saturday, J. Non-Cryst. Solids, 49, p. 413 (1982).

(54) J.E. Mendel,, Nucl. Technol., 32, p. 72 (1977).

(55) G.G. Wicks, W.D. Rankin, and S.L. Gore, in Scientific Basis for Nuclear Waste Management VIII, C.M. Jantzen, J.A. Stone and R.C. Ewing, eds., 44, p. 171 (1985).

(56) W. Lutze, R. Muller, and W. Montserrat, in Scientific Basis for Nuclear Waste Management XI, M.J. Apted and R.E. Westerman, eds., 112, p. 575., Materials Research Society, Pittsburgh, PA (1987).

(57) R.M. Wallace and G.G. Wicks, in Scientific Basis for Nuclear Waste Management VI, D.C. Brookins, ed., 15, p. 23, Elsevier, New York (1983).

(58) IAEA, Proceedings of the Symposium on the Oklo Phenomenon, Report IAEA-SM-204, Vienna (1975).

(59) G.G. Wicks, "WIPP/SRL In-Situ and Laboratory Testing Programs- Part I: MIIT Overview, Nonradioactive Waste Glass Studies," DP-1706, Savannah River Laboratory, Aiken, SC (1985).

(60) G.G. Wicks and M.A. Molecke,in Advances in Ceramics, Nuclear Waste Management II, Clark, D.E., White, W.B. and Machiels, A.J. eds., 20, p. 657 (1986).

(61) M.A. Molecke and G.G. Wicks, "Test Plan: WIPP Materials Interface Interactions Tests (MIIT)," 9, No. 1, Sandia National Laboratories, Albuquerque, NM (1986).

(62) G.G. Wicks and M.A. Molecke, in Waste Management '88, Post, R.G. and Wacks, M.E., eds., 2, p. 383 (1988).

(63) G.G. Wicks, "WIPP/SRL In-Situ Tests- Part II: Pictorial History of MIIT and Final MIIT Matrices, Assemblies, and Samples Listings," DP-1733, Savannah River Laboratory, Aiken, SC (1987).

(64) Proceedings from the Workshop on Testing of High-Level Waste Forms Under Repository Conditions, Commission of European Communities (CEC), EUR 12 017 EN, McMenamin, T. ed., (1989).

PROBLEMS

1- Design an evaluation strategy that can be used to assess (a) the most important technical performance features and (b) processing characteristics, for waste immobilization matrices. Include:

- Parameters to be measured and assessed
- Experiments used to measure parameters
- Weighting factors for evaluating various parameters (emphasizing most important parameters)
- Formulate a means to integrate these data in order to rank candidate waste forms
- Provide a general, generic assessment using this strategy for metals vs. polymers vs. ceramics/glasses

2- Design a characterization strategy useful to understand nuclear waste glass forms and interfaces produced during a leaching process. Include types of surface and bulk studies that can be conducted along with relevant solution analyses, etc., and define advantages and disadvantages of each tool.

3- Design a new type of in-situ or field experiment to be used to provide useful information on performance of waste glass systems in geologic repositories.

4- Design an integrated strategy that can be used to predict the performance of a 40-component waste glass system in a complex geologic repository out to time periods of thousands of years.

5- Design a strategy and experiment matrix that can be used to help extend the useful lifetime of materials in medical devices and implants, or in other fields and disciplines.

Designing Whitewares

12

D.A. Earl

DESIGNING WHITEWARES

David A. Earl
Alfred University
Ceramic Engineering & Materials Science
Alfred, NY 14802

ABSTRACT

Ceramic processing and glass science principles are applied to design glazed whiteware products including ceramic tile, dinnerware, sanitaryware, and high-voltage porcelain. The ceramic tile industry is the biggest sector of the whitewares market. This chapter summarizes several factors that are considered when designing glazed ceramic tile.

INTRODUCTION

The ceramic tile industry develops new products and processing methods to increase sales and reduce manufacturing costs. Ceramic engineers and materials scientists in R&D create products which match the aesthetics specified by ceramic artists by designing new body surface structures, glaze formulations, and glazing application techniques. Formulas and processing methods are continually modified to improve product performance (e.g. mechanical strength and wear resistance), eliminate defects, increase manufacturing speed, and utilize lower cost materials.

Ceramic tile manufacturing involves continuous automated high-volume processing of materials. Body and glaze slurries with approximately 67 wt% solids are prepared via ball milling of raw materials, chemical additives (deflocculant suspending agents and binders), and water. The body slurry is spray-dried into granules, dry pressed into greenware, and dried prior to glazing. Up to twenty different glaze coatings are applied to achieve the desired aesthetics and properties with techniques including spraying, waterfall, screen printing and roller printing. Peak firing temperatures vary from 1090°C for high-talc wall tile bodies to 1250°C for high-clay porcelain wall and floor tile bodies. During firing, some body materials melt to form silicate glass in the pores and the glaze is converted into an impervious glass-ceramic. As vitrification of the body proceeds, the proportion of glassy phase increases and the porosity decreases, resulting in higher shrinkage and fired strength. Porcelain tile bodies are normally fired to above 1200°C to facilitate the formation of mullite, whose needle-like morphology increases the strength. Mullite forms from the decomposition of kaolinite through a solid state reaction and has a range of compositions between and including $3Al_2O_3 \cdot 2SiO_4$ (3:2) and $2Al_2O_3 \cdot SiO_4$ (2:1). For all types of tile products, the firing cycle is optimized based on the firing behaviors of the specific body and glaze formulas.

Important characteristics and properties of the final product include dimensional uniformity, thermal expansion, porosity, strength, hardness, optical properties (color, gloss, and opacity), aesthetics, surface structure (resistance to slippage and staining), mechanical and chemical durability, product consistency, stability over time and a range of environmental conditions, and cost.

When designing new ceramic tile products, R&D closely considers all of the composition-processing-structure-properties relationships from the idea stage, through laboratory development and pilot line trials, to manufacturing. Once a new product is approved for production, R&D works with the Quality Control department to monitor each stage of the process, including incoming raw material testing, material checks at each stage of the process, and testing of finished products.

BODY DESIGN

New raw materials are analyzed for particle size, specific surface area and morphology, bulk density, chemistry, mineralogy, color, deflocculation behavior, properties as a function of temperature, homogeneity, and lot-to-lot variation. Differential scanning calorimetry (DSC), thermogravimetric analysis (TGA), and dilatometry are used to identify temperatures where decomposition and phase transformations occur. This information is important for designing the proper firing profile. Body materials include clay minerals such as kaolinite ($Al_2O_3 \cdot 2SiO_2 \cdot 2H_2O$) for slurry suspension stability, plasticity during forming, and green strength; a flux such as soda feldspar ($Na_2O \cdot Al_2O_3 \cdot 6SiO_2$), talc ($3MgO \cdot 4SiO_2 \cdot H_2O$), or nepheline syenite ($3Na_2O \cdot K_2O \cdot 4Al_2O_3 \cdot 8SiO_2$) to facilitate vitrification during firing; quartz due to its low cost, high thermal expansion, and role as a glass former; whiting ($CaCO_3$) or wollastonite ($CaSiO_3$) to increase the thermal expansion and lower the fired shrinkage; deflocculants (e.g. Na-hexametaphosphate) for slurry suspension stability and lower viscosity; and binders (e.g. Na-lignosulfonate) to improve green strength.

Clays are subjected to special tests due to their importance in ceramic processing and wide range of mineralogy from different sources. Clays always have some lower valency cation impurities (e.g. K^+, Na^+, Ca^{++}, etc.) adsorbed at the interfaces between crystallites in the stacked aggregates. These cations are readily desorbed when the clay is suspended in water, producing overall negatively charged particles. The surface charge over a high surface area improves slurry suspension stability and provides for high adsorption of processing additives relative to the particle volume. The cation exchange capacity (CEC) of a clay is measured to determine the maximum amount of desorbed cations per 100g of material. Clay provides plasticity for forming by the manner in which water films are held in the sheet structure. It also increases green strength through bridging of larger ceramic particles. The percent water of plasticity and the green strength of pressed samples are measured for new clay materials. Clays containing more of the colloidal mineral montmorillonite have a higher green strength due to montmorillonite's very high specific surface area and tendency to coagulate.

The tile body must have a higher thermal expansion than the glaze so the fired glaze is placed in compression at room temperature, which increases the strength and avoids glaze cracks due to tensile stresses. Quartz raises the body thermal expansion, but at high levels may cause "dunting" cracks during the kiln cooling cycle due its rapid volume contraction from the β to α phase at 573°C.

After a raw material is tested it is combined with other body materials to develop the optimum formula. New bodies are analyzed for many parameters including fired phases present, shrinkage, loss on ignition, porosity, strength, chemical resistance, repeatability, defects, and unit cost. Manufacturing quality control involves monitoring the same variables considered during the initial design stage, as well as the material condition at each stage of the process.

GLAZE DESIGN

Tile glazes are glass-ceramics formulated from glass frits, clay and other minerals, zircon opacifier ($ZrSiO_4$), colorants, deflocculants, and binders (e.g. sodium carboxymethylcellulose). Many factors are considered during the design stage, including the fired color (primarily due to specific absorption by light colorant particles), gloss (specular reflectance, effected by surface smoothness), opacity (diffuse reflectance due to crystalline particles), hardness, resistance to chemical attack, stain resistance (surface microstructure), coefficient of friction (floor applications only), susceptibility to defects, and unit cost. In many cases, multiple layers of different glaze formulas are applied to achieve the desired appearance, so reactions between formulations resulting in undesired crystallization or dissolution must be avoided.

Glaze frits generally include SiO_2 as the main glass former, alkali (Na_2O and K_2O), B_2O_3 and ZnO or SrO as the main fluxes, and CaO, MgO and Al_2O_3 to increase the coating's hardness and durability. Each company requires different formulations that are specific to their process, body formulation, and product applications (e.g. commercial vs. residential flooring). The frit composition is designed to provide the thermal expansion, softening temperature, melt viscosity, insolubility of colorants and opacifier, and crystalline phase development during firing necessary for achieving the target physical properties and avoiding defects. Compared to other batch components, frit normally has the lowest melting temperature, is the most corrosive to ceramic colorants and has the main influence on crystallization during firing. Opacity and whiteness in glossy glazes are obtained through mineral additions of zircon to the batch or by crystallization of zircon from ZrO_2 and SiO_2 in the frit. Zircon is an excellent opacifier for glazes due to its high index of refraction (2.05) relative to the glassy matrix (1.50-1.70) and low solubility in most silicate melts. Maximum light scattering and whiteness with zircon occur with a particle size range of 0.4-0.7 μm and a mass fraction of 0.16. Frits for transparent coatings ordinarily contain no ZrO_2 and include more CaO and Al_2O_3 to maintain sufficient mechanical properties. Matte glazes are designed to crystallize large particles of phases such as wollastonite or diopside ($CaMgSi_2O_6$) that significantly increase the diffuse reflectance and lower the gloss.

Zircon doped pigments are the most stable ceramic colorants up to approximately 1200°C. Zircon's tetragonal structure has the ability to accommodate vanadium, iron, and praseodymium substitutionally, and its high chemical and thermal stability make it ideal for use in ceramic coatings. The zircon triaxial system used to color industrial glazes is based on blending zircon-vanadium blue, zircon-iron coral and zircon-praseodymium yellow to obtain a wide range of colors. Zircon pigments are normally added to glaze batches in the range of 0.1% to 5.0% by weight, and their solubility during firing varies with the melt composition.

Control of crystallization and prevention of opacifier and colorant dissolution is important for optimizing the properties, appearance and reproducibility of whiteware products. X-ray diffraction is used to identify the structures and quantities of crystalline phases present in fired

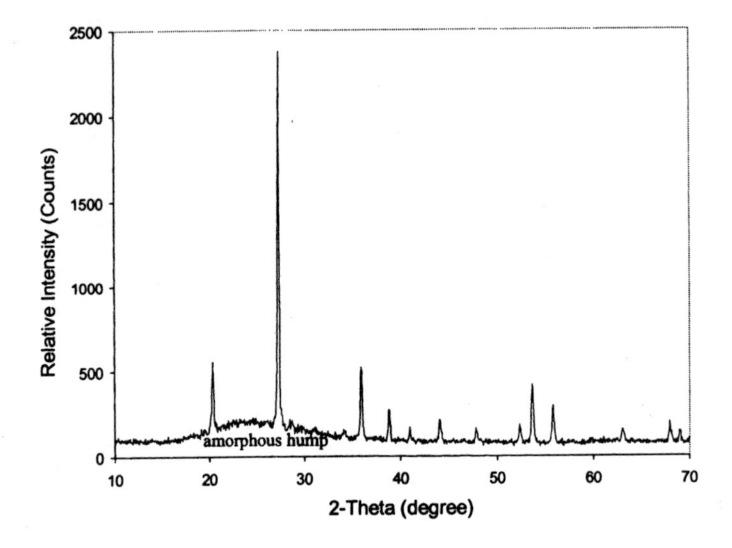

Figure 1. X-ray diffraction (XRD) pattern for a fired glossy glaze containing 5.5 wt% zircon ($ZrSiO_4$) in the glassy matrix (JCPDS # 06-0266).

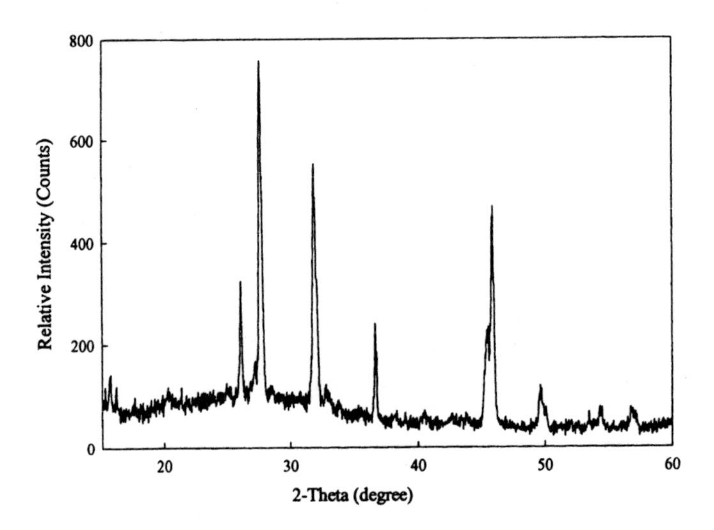

Figure 2. XRD pattern for a fired matte glaze containing 85 wt% wollastonite ($CaSiO_3$) crystalline phase (JCPDS # 31-0300).

glazes. Figures 1 and 2 show diffraction patterns for glossy opaque and matte glazes. The technique is also employed to study phase evolution in bodies during firing and to detect changes in the mineralogy of incoming raw materials.

Glaze formulas are also designed to provide consistent properties over a range of processing conditions because process variation is inherent to high-volume manufacturing. Glaze quality control in production involves monitoring each of the factors considered during the product design as well variables involved with glaze preparation and application. The glaze slurry particle size distribution, viscosity, density, application weight, and applied thickness are monitored and controlled. Incoming glaze raw materials are also subjected to the same quality control checks performed on the body materials

ENVIRONMENTAL ISSUES

Current environmental concerns in the whitewares industry include kiln stack emissions and ground water monitoring near manufacturing facilities. A growing environmental concern is the classification of silica as a carcinogen.

The clean air act of 1990 requires installation of emissions control equipment on ceramic kilns. In particular, fluorine is a major concern because it reacts with water to form the pollutant HF. Fluorine substitutes for hydroxyl in the crystal structures of some clay and talc minerals and may be released above 450°C during dehydroxylation. Wet or dry "scrubbers" installed on stacks use lime to tie up fluorine:

$$2HF + CaO \rightarrow H_2O + CaF_2 \text{ (fluorite)} \quad (1)$$

However, scrubbers are very expensive to purchase and maintain. Efforts are being made to design whitewares with minerals such as pyrophyllite which contain little or no fluorine. Other gaseous emissions from kilns can be HCl, and sulfur compounds (SO_2 and SO_3) from clays. Organic emissions from binders may also exist in small concentrations and emissions of vaporized heavy metals from glazes and trace metallic compounds from clays will attract attention as environmental regulations are tightened in the future.

Some whiteware companies are required by EPA to monitor groundwater levels of lead, barium, chromium, cadmium, and zinc if the materials are used in their process. Most companies have discontinued using lead in glazes, however, some decorations and specialty glazes still incorporate the material. Chromium and cadmium are used in small amounts

as glaze colorants for some products, while zinc is widely included in ceramic tile glazes. Although zinc in tile glazes facilitated the elimination of lead, typical zinc levels have been reduced from approximately 10 wt% to 2-4 wt% over the past ten years due to its designation as a regulated material.

ILLUSTRATIVE EXAMPLES

1. Firing Curve Design

A new tile body formula is being developed for a 50 minute firing cycle. Raw materials are a locally mined clay, Na-feldspar, wollastonite, and quartz. A chemical analysis reveals no impurities (i.e. organics, sulfides, carbonates, etc.) in the materials and DSC finds clay dehydroxylation between 450-700°C and the onset of feldspar melting and body vitrification at 1085°C. A dilatometer analysis shows that bodies made from mixtures of the materials begin to bloat above 1115°C. The desired level of porosity in lab samples is achieved after an eight-minute hold between 1085°C and 1115°C. X-ray diffraction analysis of fired and unfired bodies identifies no dissolution of quartz during firing. Based on the analytical results, the initial firing curve design is shown in Figure 1.

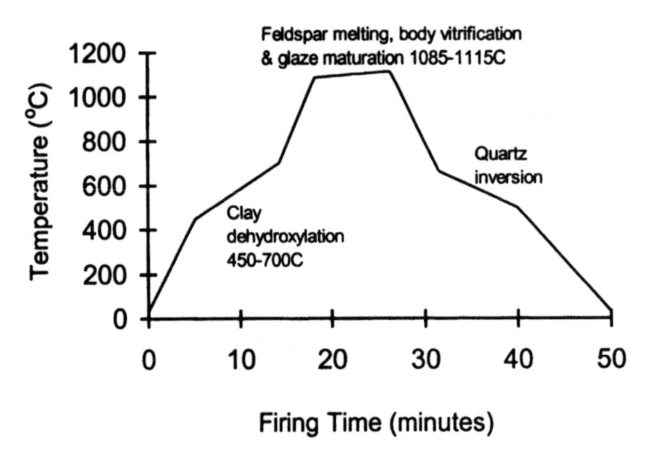

Figure 1. Experimental firing curve based on the thermal analysis and XRD results.

The heating rate is lowered at specific points to ensure complete dehydroxylation of the clay before the body begins to vitrify and a gradual transition through quartz inversion to avoid cooling cracks. After the body and firing curve have been optimized to achieve the desired quality and properties with the lowest cost and fastest production cycle, compatible glazes

must be designed. The glazes should have a lower thermal expansion than the body, begin to soften after gases have evolved from the body to avoid bubble defects, and flow and mature within the time allotted at the peak firing temperature.

2. Glaze Formula Design

A new frit composition was developed to increase the opacity (whiteness) of a glossy "fast-fire" ceramic wall tile glaze. Figure 2 shows the microstructures of glazes made with the original frit (a) and the new frit (b). Both glazes contain clay, zircon-vanadium (Zr-V) blue-green colorant, and frit comprised of SiO_2, Al_2O_3, ZrO_2, ZnO, CaO, Na_2O, and K_2O. The frit in glaze A was designed to crystallize zircon from SiO_2 and ZrO_2 in the glass to provide opacity, but the resulting particle size and shape is not optimum for scattering light. The amount of light scattered from zircon is maximized when the particles are spherical with a diameter of 0.4 to 0.7 μm, which corresponds to wavelengths in the visible spectrum. The frit composition was modified to raise the melt viscosity by increasing the Al_2O_3:Na_2O ratio in order to slow the crystal growth rate and favor the formation of spheres. The resulting microstructure is shown in Figure 2 (b). Figure 3 shows the shift in optical properties due to the glaze composition-structure changes, where the reflectance and opacity are increased due to the change in zircon precipitate size and shape.

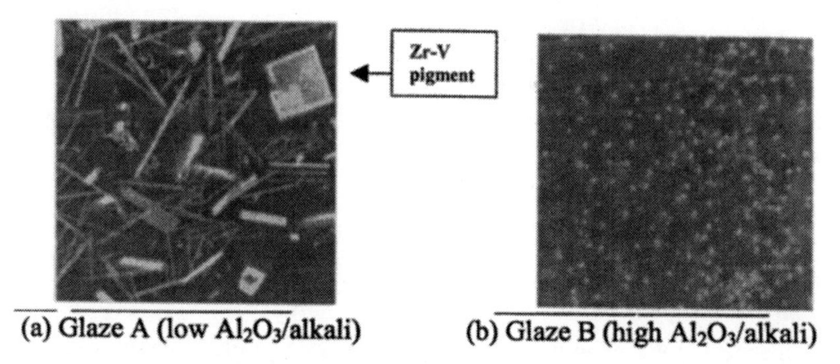

Figure 2. SEM micrographs (X4000) of glazes batched with 93% glass frit,5% clay, and 2% Zr-V pigment, and fired to a peak temperature of 1100°C in a 30 minute cycle. The glaze frit compositions differ in the Al_2O_3:alkali ratio, where A=0.3 and B=0.6, respectively. All particles shown are zircon that precipitated from the frit during firing, except for the large square particle in (a), which is Zr-V pigment with a lower side length of 4 μm. Most of the zircon particle diameters in (b) are between 0.4 and 0.7 μm.

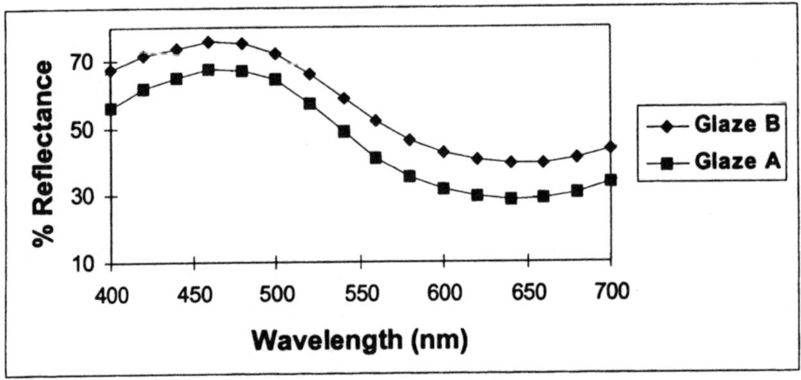

Figure 3. Spectral reflectance curves for glazes A and B, generated using a spectrophotometer with a standard D65 daylight light source. The smaller, spherical zircon particles in B increase diffuse reflectance across the visible spectrum. The Zr-V pigment in both glazes absorbs red-yellow wavelengths, producing a blue-green color.

SUMMARY

The design and manufacture of traditional ceramics such as glazed tile involves some of the same processing steps, analytical techniques, and scientific principles applied to develop advanced materials. Results from analytical tests and laboratory trials are used to quantify composition-processing-structure-properties relationships during the initial design stage and prototype development. Once a product is approved for manufacturing, material properties and performance are monitored at each stage of the process, including incoming raw materials, fabrication steps, and testing of finished products.

SUGGESTED READING

1. J. Reed, Principles of Ceramics Processing, 2nd Edition, John Wiley Interscience (1995)
2. T. Manfredini and L. Pennisi, "Recent Innovations in Fast Firing Process," in *Science of Whitewares*, eds., V. Henkes, G. Onoda, and W. Carty, American Ceramic Society, Columbus, 213-223 (1996)
3. C. Parmelee and C. Harman, Ceramic Glazes, 3rd Ed., CBLS, Marietta, OH (1973).
4. D. Earl and D. Clark, "Effects of Glaze Frit Oxides on Crystallization and Zircon Pigment Dissolution in Whiteware Coatings," J. Amer. Ceram. Soc., 83 (9), 2170-2176 (2000)

PROBLEMS

Reference the suggested reading to answer the following problems.

1. Two clays are being considered for a tile body formula; in water, one clay desorbs a small amount of Na^+ and the other desorbs a high quantity of Ca^{++}. Contrast the slurry suspension stability and viscosity resulting from the two clays. (Reference 1)

2. An experimental tile body formula contains 50% kaolin, 15% quartz, 15% whiting, and 20% Na-feldspar (assume no impurities), and must be fired using a 60 minute cycle with a peak temperature of 1100°C.
   a) Based on the expected high-temperature reactions, sketch a reasonable time vs. temperature firing profile. State your assumptions regarding the reaction temperatures and glaze maturation range. Assume the glaze will be designed to fit the body firing profile.
   b) On the same graph, sketch an approximate firing profile for the same body if the whiting is replaced with wollastonite. How would this change the glaze design requirements? (References 2 & 3)

3. a) Calculate the linear coefficient of thermal expansion for the glaze frit formula below using Hall or McLindon & West coefficients.
   b) Glaze thermal expansion is often estimated based on published linear coefficients but the calculations are not accurate for whiteware bodies. Explain.
   c) Explain how the thermal expansion of a whiteware body containing quartz in the initial batch changes with successive firings. (Reference 3 or any glass-science book.)

Frit Oxide	**Weight %**
SiO_2	55
ZrO_2	8
B_2O_3	6
Al_2O_3	4
Na_2O	2
K_2O	3
CaO	8
MgO	2
ZnO	12

4. Group exercise: Design a quality control laboratory for a ceramic tile manufacturing plant. List all of the materials properties and characteristics that should be measured and the analytical equipment required. Discuss how each measured parameter is related to product performance and/or value. Estimate the costs to start up and operate the lab.

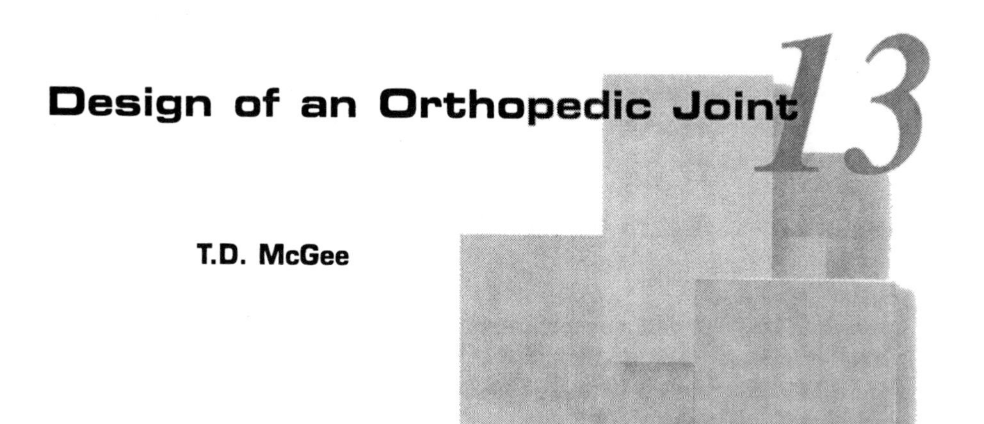

13 Design of an Orthopedic Joint

T.D. McGee

DESIGN OF AN ORTHOPEDIC JOINT

Thomas D. McGee
Materials Science and Engineering Department,
Biomedical Engineering Program, and
Collaborating Professor Veterinary Clinical Sciences
Iowa State University
Ames, IA 50011

ABSTRACT

Design of orthopedic joint replacements must satisfy the kinematic requirements of the joint. Existing designs generally do so. Ceramic components, if they are to be used, must not interfere with those requirements, but must also provide benefits over existing materials and designs. Inherent in existing designs is a limitation of implant service to about twelve years. Ceramic components that will last longer make new designs possible. Bioceramics have potential to 1) control tissue response and 2) avoid tribological failure. This can only be accomplished if the designer understands the mechanisms of failure and wear resistance of the ceramic components, the anatomy of kinematics of the joint, and the physiological processes involved in tissue reactions and bone maintenance. This paper describes the design process, including: joint function, tissue response, anatomical and physiological concerns, fracture mechanics and wear resistance. The net result is a functional design that illustrates the necessity of the designer to understand the application in order to design the ceramic components that may make possible longer years of service than are currently expected.

INTRODUCTION

Design is a unique engineering skill. In the ceramic materials area, there are three different design functions that can be considered:

1. Design of a material,
2. Design of a process,
3. Design of an application for a ceramic material.

For this discussion, design of an orthopedic implant is used as an example that requires all- three design functions. The ultimate result is a design of a new orthopedic implant that has the potential to replace existing designs with improved performance and longer life in human patients.

ENGINEERING DESIGN

Design is often considered the ultimate expression of engineering practice, partly because design has been an important component of engineering practice (e.g. bridge design, civil engineering or machine design, mechanical engineering); and partly because design represents sophisticate engineering performance in which mathematics, science, economics, esthetics and engineering skills are combined to solve a practical problem that has a useful function for society.

Design has certain features that distinguish it from engineering analysis, although engineering analysis is a part of the design process. (Many scientists teaching in engineering programs confuse analytical methods with design.) This is especially true now because theory has become such a strong component of engineering education that applications are not treated extensively. This is also the result of the attempt to give the student more exposure to industry and to technology transfer. Many projects originating in industry attempt to use university facilities to solve an industrial problem. For example, using an instrument such as a scanning electron microscope to find a soldering defect in a chip is analysis, not design.

The features that distinguish design are as follows:

1. A practical purpose, a goal that the student must reach through a creative thought process.
2. Application of mathematics and scientific principles as a part of that creative thought. Often the process is subdivided into components for better analysis.
3. Consideration of economic factors necessary to the practical purpose.
4. Consideration of social values and esthetics, including beauty and ecology.
5. Evaluation of the design in terms of the original goal. This requires analysis of the design in term of the goals.
6. Iteration, the first design almost always can be improved. Analysis usually gives new directions or emphasis that make a second or third design more effective in reaching goals. The student needs to learn when to stop, when further improvement is not expected.

Educational programs in engineering design often begin with introductory courses that are followed by a capstone design course. For example, a mechanical engineer might study design of machine elements and engineering economy before attempting to design a machine in a capstone course. This teaches the analytical principles and their application to design before attempting the capstone design course. However, before a student can create a design to solve a problem in a practical application, he or she must understand that application. This is what makes design so difficult to teach in a theory- oriented curriculum. It is rare that the design goal is completely new. For example, design of a new automotive engine must consider existing designs, their virtues and faults, their strengths and weaknesses before a new design concept is possible. The need for the necessary background in the area of application to obtain understanding and perspective in creating a new design cannot be overemphasized.

THE ORTHOPEDIC DESIGN PROBLEM

Existing orthopedic joints are wonderful improvements over the first crude attempts to replace an articulating joint, such as the hip joint. The development by J. Charnley in the late 1950's of a metal ball in a polyethylene socket, both cemented in place with polymethyl-methacrylate (PPMA) cement, was a marvelous break-through in joint design (Fig. 1).[1] Variations on the basic design are used to this day to replace diseased joints, bringing freedom from excruciating pain and return to ambulation for hundreds of thousands of patients in the USA each year.[2] Hips, knees, elbows and shoulders are replaced. These replacements have a limited lifetime, especially for younger, more-active patients. Each successive replacement is more difficult. Longer life expectancies and replacements in younger patients have created a great deal of interest in improved joint designs to benefit such patients. To satisfy these desires for improvement we must understand the existing designs, their strength and weaknesses, the functions they perform, the surgical requirements and the mechanisms of failure. Inherent in this is the need to understand the structure of bone, the anatomy of the joint and the physiological processes of wound healing, bone remodeling, and response to infection. Some of the requirements for a joint replacement are

1. Kinematics of the joint. The critical range of motions of the joint must be considered.

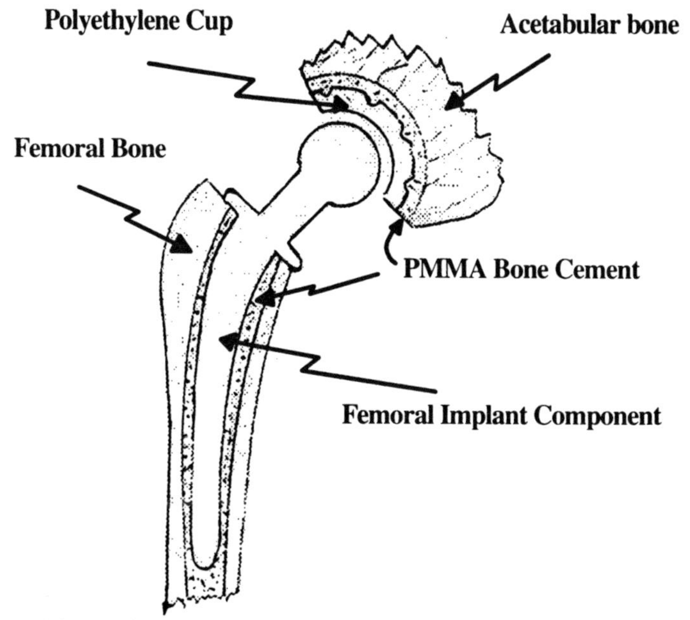

Fig.1 The general configuration of a hip joint replacement

2. Strength in the body environment. Stress corrosion and fatigue in metals are important. Brittle fracture of ceramics and deterioration of polymers are also important. Any component must withstand the repetitive forces imposed by function as long as possible.
3. Biocompatibility. Tissue response limits the choice of materials. Only about two ceramics, four metal alloys and two polymers are sufficiently biocompatible for existing designs. They are primarily bioinert, insoluble in the warm corrosive environment.
4. Wear resistance, tribology. Abrasive wear and particulate debris from wear need to be avoided. Debris can cause inflammation of adjacent tissue, and resorbtion. This is currently a weakness. Lubrication by the synovial fluid is important.
5. Physical properties such as modulus of elasticity, electrical conductivity (a source of galvanic cells involved in tissue response), thermal conductivity, etc., need to be considered.

6. Surgical compatibility. Sterilization should be easy and not change properties. Removal of joint components and placement must be easy – with jigs or fixtures as necessary. Cements must set quickly so that range of motion can be checked before closure.

MECHANISMS OF FAILURE

A new design could be an improvement if the mechanisms of failure are avoided. Replacement of hip joints has been successful for a longer time than knees, elbows and shoulders. Current designs often last 10-15 years, with some much shorter and a few lasting over 25 years. However, the usual function for adults is about 10-12 years.[3] The common mechanisms of failure usually include pain and loosening under stress caused by

1. Wear. Wear debris in a joint capsule, usually polyethelene particulates, sometimes metal particulates, cause inflammatory response, pain and loosening. In Europe, a dense polycrystalline alumina ball and an alumina socket are often used to avoid wear debris. The ball is usually supported by a metal femoral stem with close size dimensions for support of the ball.
2. Stress Shielding. Resorption of bone in areas where "stress shielding" occurs is believed to be one cause of failure. The metal femoral stem has a higher modulus of elasticity than bone. If bone supports the implant at the distal apex, stress is concentrated there and little stress is imposed nearer the ball. Bone resorbs if stress is either too high or too low. So the elastic mismatch is believed to cause local low stresses, resorption, movement, loosening and pain.
3. Fibrous capsules. The metals and polymers and some ceramics are bioinert, but bone does not bond to them. The foreign-body-response mechanism of the body attempts to wall off the foreign body from the tissue with a fibrous capsule. If this capsule is very thin the implant is stable. If movement occurs, the capsule may thicken, allowing more movement, progressive thickening, loosening and pain.
4. Excessive Temperature. Polymethylmethacrylate (PMMA) polymerizes rapidly with an exothermic reaction. The temperature rise can sometimes cauterize tissue, although manufacturers have limited the expected temperature rise. This can cause necrosis, resorption, loosening and pain.
5. Vascularity. Because bone does not bond to PMMA, it is inserted under pressure to fill the pores of porous (trabecular) bone to create

interlocking of PMMA with bone. This gives an interlocking structure and strengthens the bone. However, it interferes with the blood supply for repair. Fatigue failure of the bone / PMMA structure can lead to fracture or loosening and pain.

6. Tissue Reactions. Some patients are allergic to implant components. Leaching of metals often decolorizes adjacent tissue, possible in part by galvanic activity. More common is the breakdown of the PMMA. Both mechanisms cause inflammatory responses, loosening and pain.
7. Infection. Infection can localize at an implant, usually from a systemic infection long after the original surgery. These are extremely difficult to treat. Sometimes the appliances must be removed, the infection cured, and a replacement implanted.

RECENT MODIFICATIONS

Recent modifications to joint design include:

1. Calcium phosphates.[4] Existing implants often have calcium phosphate coatings, especially in stress-shielding locations. The calcium phosphates enhance tissue response and bonding during the first three months. Plasma sprayed coatings are porous, thin (often 30-50 μm), and poorly bonded to the metal. They enhance recovery during wound healing, but are usually gone at three months.
   All over the world, improved coatings or enamels or glass-ceramics are being studied to try to make and improve the tissue/implant interface. None are in current use.
2. Metal Surface Porosity or Roughness. Beads, slot blasting, sintered powders and various structures are used to enhance "in-growth" of tissue to improve bone integration with the implant. Some are helpful, some release particles that invoke tissue response, and, if movement occurs, the roughness can act like a file. This is an area of current research and designs are evolving in clinical practice.
3. Stabilized Zirconia. Greater toughness of stabilized zirconia could reduce fracture probability, but at the expense of wear debris when compared to alumina.
4. Calcium Phosphate, (CP), Cements. Replacement of PMMA with CP cements, or CP containing cements is probably desirable. Reaction of a basic CP with an acidic CP to make hydroxyapatite or dicalcium phosphate cement has been proposed. Modifications with $CaCO_3$ or other chemicals to improve setting time have been

proposed. None are in current use for joint fixation in the U.S.A., although some are used in Europe and trials are being conducted in the U.S.A. They are extremely weak and have very low fracture toughness.

5. Antibiotics. Antibiotics such as gentimyesin, are often included in PMMA. Research on calcium phosphates and many other compounds for controlled release rates are being conducted all over the world. These do not help with delayed, deep infections because they are gone when the infection occurs.
6. Cementless prostheses. Because of the PMMA problems uncemented joints have been in use. For this to be successful the bone must be reamed to a close fit with the prosthesis with minimum tissue damage. Robot machine devices have been used to improve fit. This method was introduced to clinical practice but has become less popular recently because longer lifetime has not been achieved.

REQUIREMENT OF A NEW JOINT DESIGN

From an analysis of current designs, the mechanisms of failure, and recent trends the following requirements seem appropriate.

1. Minimum tissue removal for surgery.
2. Satisfy kinematic requirements.
3. Tissue interface strong as cortical bone, not subject to fatigue failure. Provision for vascular supply to support the bonding tissue.
4. Biocompatible ceramic interface containing calcium phosphate to enhance tissue bonding.
5. Free of wear debris.
6. Metal components not in contact with tissue and bioinert.
7. Cementless or with a strong calcium phosphate cement.
8. Minimum cost.
9. Reasonable stress on components.

To satisfy these requirements, the following are needed:

1. New bioceramic interface that tissue will bond to and support the implant without PMMA cement.
2. Joint design incorporating the interface and suitable wear components.
3. Process for producing the ceramic interface and the joint.

THE BIOCERAMIC MATERIAL

Because the calcium phosphates are modified by the tissue interactions, they do not have reliable, enduring strength. A material is needed that has the bioactivity of the calcium phosphate without the loss in strength resulting from its bioactivity. This can be accomplished with a composite.[5,6] Metals and polymers do not bond well to the calcium phosphates. A ceramic / CP composite must include a strong ceramic component, but it must not react with the calcium phosphate when the composite is fired. In general, acidic oxides react with basic oxides so acid/base considerations are important. Tricalcium phosphate (TCP) $Ca_3(PO_4)_2$, is the salt of a strong base and a weak acid. Spinel ($MgAl_2O_4$) is the salt of a strong base and a weak acid. The Mg^{++} ion does not proxy well for the Ca^{++} ion in crystal structures, nor Al^{+++} for P^{5+}. Therefore, reactions between them should be limited. About 50 volume percent of each should be present to have interconnected phases. Spinel is bioinert, TCP is bioactive. The melting point of TCP is 1670°C. That of spinel is 2135°C. Therefore, the coarser grain size should belong to the TCP. If so, removal of the TCP will not change the flaw structure of the spinel, so enduring strength should be achieved.[7]

Using this reasoning, a material shown in Table I called an Osteoceramic was produced. Its properties are

Table I. Properties of the Composite

Composition:	$\alpha Ca_3(PO)_2$ and $MgAl_2O_4$
Compressive strength	199 Mpa
Tensile strength	70 Mpa
Youngs Modulus	114 Gpa
Reversible Thermal Expansion	10.7×10^{-6}/°C
Bulk Density	3.09 gm/cc
True Density	3.37 gm/cc
Ca:P ratio	1.62

PROCESSING THE CERAMIC COMPOSITES

For the composite to be accurately formed it should be economically machined. Near shape forming and diamond grinding are possible, but diamond machining is expensive. (The European alumina components are dense, sintered Al_2O_3, diamond machined, and expensive, requiring aspherical machining processes.) Experiments with the partially fired

composite revealed it can be readily machined with ordinary machine tools after firing at 1200°C. Allowance for shrinkage for final firing at 1450°C (about 10% linear) must be included. Because the additional shrinkage makes dimensional tolerances less accurate, and because the composite should be bonded to metal components, a cemented or pre-stressed metal structure should be used to support the composite in metal components.

The wear interface in the joint design should be alumina because of its lubricity and wear resistance. The geometry should accommodate machining to circular symmetry.

A NEW JOINT DESIGN

Tissue Removal and Design Considerations

The femoral ball is dense at its surface and the trabecular walls are oriented to deliver surface forces on the ball to the cortical bone at the distal diaphysis. The center of the ball is the center of rotation of the joint. A replacement should have a center of rotation in the same location. This could be done if the replacement is a sphere with its center in the appropriate position. This is possible if the "socket" is recessed into the femoral ball and has a smaller radius of curvature. A small radius means higher stress, and less displacement needed to dislocate the joint. The higher stress can be tolerated if the ball is made of polished sapphire. Small polished sapphire balls are less expensive than large polycrystalline alumina components. Other means can be used to accommodate dislocation. So a sapphire ball at the center of rotation seems feasible. Supporting this with a metal shaft, with the ball cemented to the shaft to prevent rotation allows the ball to be supported by a metal bracket that is cemented to the femoral interface composed of the osteoceramic.

The acetabular component is made with the same reasoning. Reaming the cartilage and a minimum of dense bone to fit an osteoceramic interface reduces tissue removal. A metal assembly to hold two sapphire rings contoured to fit the sapphire ball can be mounted to allow the rings to rotate freely on the sapphire. They form a cage that limits dislocation displacement and the rings can accommodate limited lateral and medial rotation (abduction and addiction), so the kinematics of the normal joint can be simulated.

Geometry of the Joint

The basic geometry of the joint defines the geometry of the new joint replacement, Fig. 2

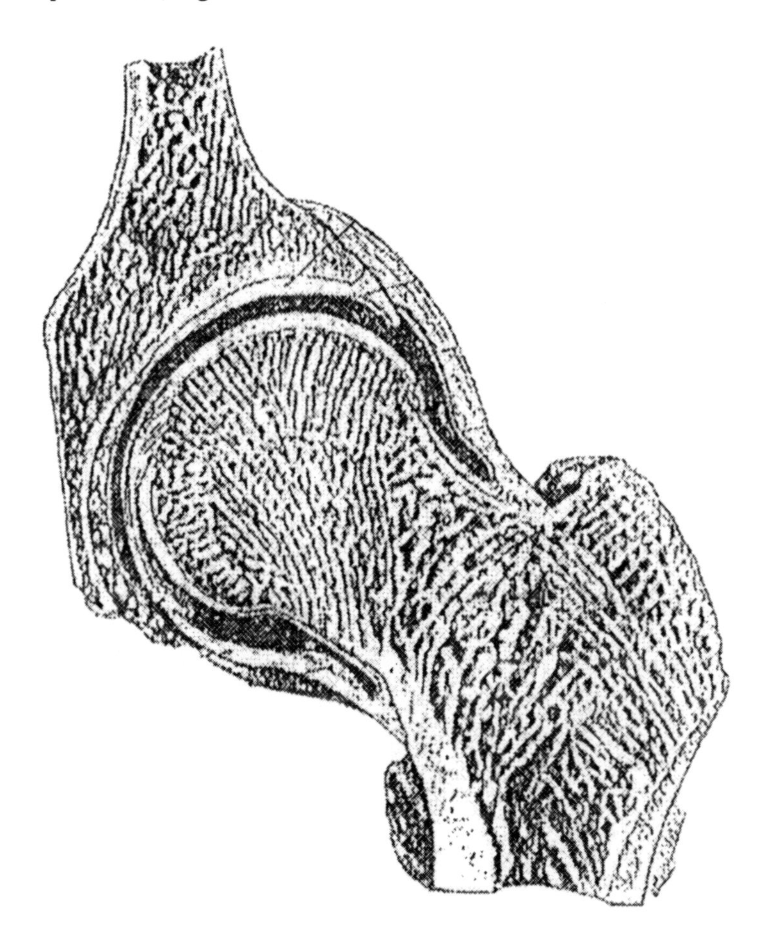

Fig. 2. Cross section of a typical hip joint, full scale.

The First Concept Design

Based on the above logic, a first concept of the new joint can be visualized, Fig. 3.

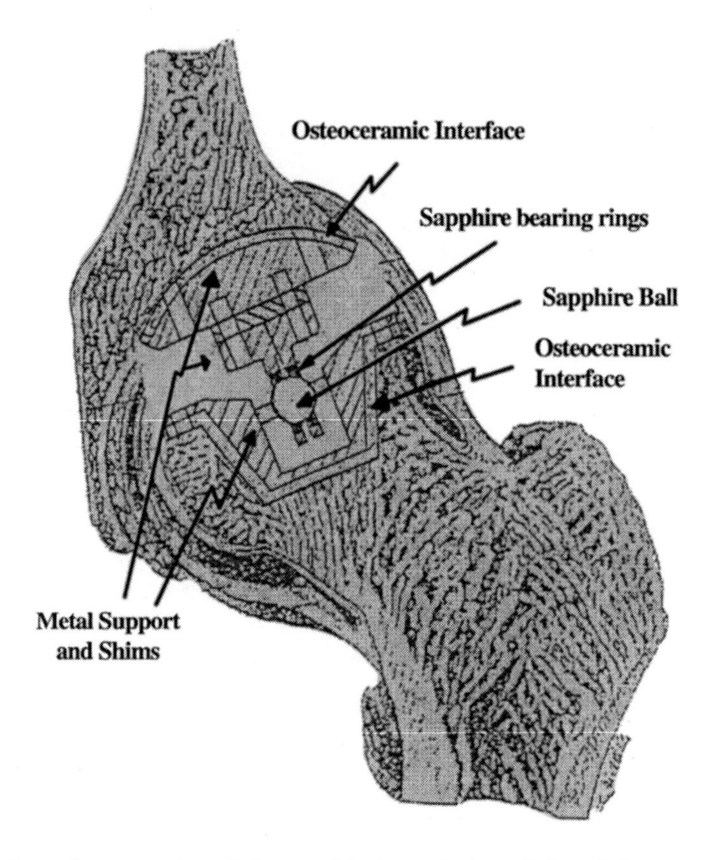

Fig. 3 Cross section of the new hip joint design, fit into joint structure.

Evaluation of the Design Concept

The design concept can be compared with the requirements to determine if this idea is compatible with the requirements:

1. Minimum surgical removal. The usual hip joint replacement removes the entire femoral head just above the greater trocanther and inserts a stem into the femoral shaft. This new design has much

less removal of tissue from the femur. The acetabular component also has less removal. However, the structure of the acetabulum will dictate sometimes that deeper removal is necessary. This can be accommodated by the shims of the design.

2. Satisfy kinematic requirements. The rotation of the two sapphire rings on the ball is uninhibited in anterior/posterior movement. The free space on the ball normal to that plane is enough for normal adduction/abduction function. The kinematics are satisfied.
3. Tissue interface strong as cortical bone, etc. The osteoceramic is strong as cortical bone and not subject to fatigue by corrosion. This is an uncemented joint with temporary fixation using bone screws to stabilize the implant until the tissue bonds to the joint. The blood supply to the bonding tissue is not compromised, except for the removal of the ligament to the head of the femur. At the time of removal that ligament could be reattached to the femur near the bottom of the joint capsule. Whether or not this is necessary or practical is unknown.
4. Biocompatible ceramic interface. The osteoceramic is biocompatible.
5. Free of wear debris. Hard polished sapphire should have a coefficient of friction that is very low in the synovial fluid because single crystal sapphire can be smoother than polycrystalline alumina. Based on the excellent wear resistance of polycrystalline alumina, this requirement should be satisfied. However, a weakness in design is that the normal force on the small diameter ball will be higher by about 25 times greater than on a normal hip.
6. Metal components not in contact with the tissue. This requirement is satisfied.
7. Cementless or with a strong calcium phosphate cement. This design is cementless.
8. Minimum cost. The simple rotational nature and small size of the sapphire components makes their cost very low. The osteoceramic cost is believed low also because it can be machined with ordinary machine tools. Because the metal components are not in contact with tissue they can be machined from 316L stainless steel, so they should also be inexpensive when compared with the other biocompatible metals.
9. Reasonable stress on components. The stresses on the tissue contacts, and the metal supporting them, are low because the joint is

large. The only critical areas are the shaft supporting the ball and the ring support assembly. These are simple bearing stresses that can be calculated accurately. The highest stress is on the shaft supporting the ball. If the patient weighs180 pounds and the dynamic force can be 5 times the patients weight, the shaft must support 900 pounds. This is distributed across both ends of the shaft so the shearing force is 450 pounds. Taking the shear stress of 316L SS as 45000 PSI (605 MPa) this requires and area of 0.01 in^2. This requires a minimum radius of 0.056 inches, or 1.42 mm, easily compatible with the design.

EXAMPLE DESIGN PROBLEMS

1. Dislocation Resistance

Displasia of the hip occurs, sometimes, from trauma, but most commonly, it is the result of instability produced by aging. Resistance to displasia is the result of the ligaments and the tendons and the joint capsule surrounding the joint. (Fig. 1) For the design presented in Fig. 3 the resistance is determined by the sapphire bearing rings that ride on the sapphire ball.

For the natural hip in Fig. 2 determine the vertical displacement needed before lateral movement would bring about displasia. Then, assuming the bearing rings are spring loaded so that they could be separated on their axis design a ball with dimensions required to produce the same vertical displacement. Fig. 4 shows the proportions of bearing rings and ball in greater detail.

Answer:

From Fig. 2 the vertical movement can be determined directly from the drawing as the vertical height x, between the bottom of the acetabular lip and the top of the ball.

Measuring x as 3 mm we can now calculate the diameter of the ball:

Let D = diameter ball.

Let d = diameter shaft supporting the ball.

The vertical movement must stop when the bearing ring reaches the surface of the supporting shaft. That vertical displacement, then is $(D/2) \sin 40° - (d/2) = x$

Then $D = 2[x+d/2)] \div \tan 40°$

$$= 2[3+1,25] \div 0.643$$
$$= 2(4.25) \div 0.643$$
$$= \frac{8.5}{0.643} = 13.2\text{mm}$$

2. Bearing Stress

If the sapphire ball in Fig. 4 has the diameter calculated in problem 1 of 13.2 mm calculate the bearing stress on the sapphire rings and compare it with the bearing stress of the natural joint shown in Fig. 2.

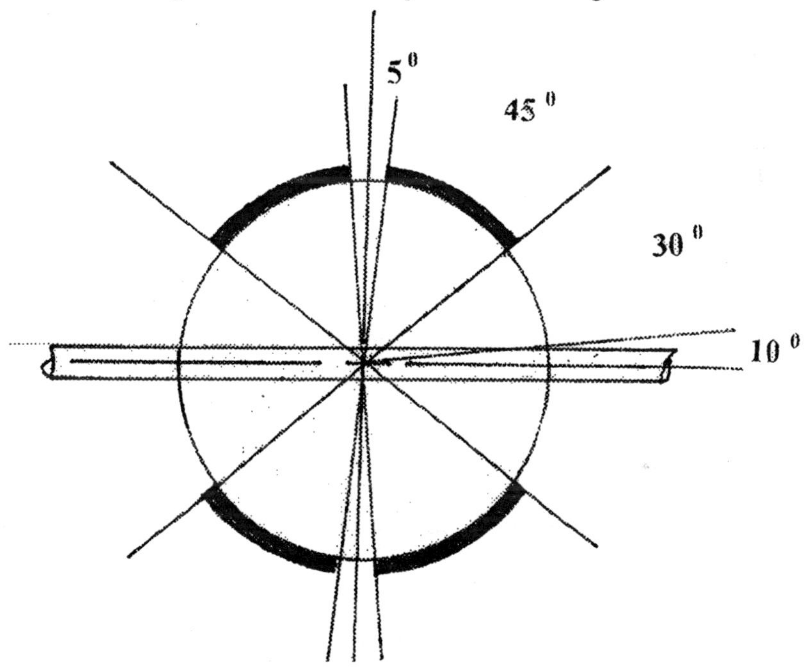

Fig. 4 Angular relationships of the bearing ring to the sapphire ball

To make this analysis it is necessary to know the structure of the acetabulum represented by Fig. 2. From the figure we see that a ligament is attached at the center of the ball of the femur. Also it is apparent that the load-bearing portion of the ball is only about 60°, and the diameter D_B is 20 mm. The head of a spherical bearing cap has a surface, from the geometry of a sphere,

$$A_1 = \frac{\pi D_B^2}{4}(1 - \text{Cos}\theta)$$
$$= \pi\frac{(20)^2}{4}(1 - \text{Cos}60°)$$
$$= 157 \text{ mm}^2$$

The attached ligament at the end of the ball covers an angle of about 10°, but the acetabulum must provide space on rotation to accommodate it at other positions, so the space in the acetabulum is larger (not visible in Fig. 2) by about a factor of two. This reduces the area by:

$$A_2 = \frac{\pi D_B^2}{4}(1 - \text{Cos}20°)$$
$$= 20 \text{ mm}^2$$

Then $A_1 - A_2 = 157 - 20 = 137 \text{ mm}^2$

In the plane perpendicular to the plane of Fig. 2 there is also a channel for the ligament attached to the head of the ball, so the acetabulum seat is reduced. Often this is about 10% of the partial hemispherical seat, so the bearing area is estimated at:

$A_B = 0.90\ (A_1\text{-}A_2) = 0.90 \times 137 = 123 \text{ mm}^2$

Then the bearing stress in the natural joint is

$$\sigma = \frac{F}{a} = \frac{1000}{123 \times 10^{-6} \text{m}^2} = 8.1\text{MPa}$$

The sapphire rings represent segments of a sphere with a solid angle of 45°, the difference between a solid angle of 85° and a solid angle of 50°. Then, using the same equation as in Prob. 1 for the surface of a portion of a sphere, the bearing area A_3 is:

$$A_3 = \frac{\pi D^2}{2}[(1 - \text{Cos}85°) - (1 - \text{Cos}40°)]$$

$$= \frac{\pi(13.2)^2}{2}\left[(1-0.087)(1-0.766)\right]$$
$$= 87.1\pi\left[(0.913)(0.234)\right]$$
$$= 58.3\text{x}10^{-6}\,\text{m}^2$$

Then the bearing stress is:

$$\sigma = \frac{F}{A_3} = \frac{1000}{58.3\text{x}10^{-6}} = 17.1\text{MPa}$$

This is a low and acceptable stress.

OTHER PRACTICE PROBLEMS

The design presented in Fig. 3 is the first step in design, the initial concept. This example illustrates the reasoning in bringing forth a new design, but many details have been skipped over to arrive at the concept presented. For this concept to be practical, some of those details must be fully explored. Some of the questions are

1. Bone bonding – acetabulum and femoral cap. Interlocking geometry is essential to bonding. Bone projections or supports exposed to stress must be formed and repaired by the tissue reactions. This is possible if, for example, concentric grooves and radial grooves are provided at the surfaces – so that bone growing into the grooves has sufficient strength to provide for the loading forces, including rotation. What size, location and depths of grooves should be provided? What thickness of interface is required?
2. Range of Motion. Ideally the limit of the range of motion is the ability of the muscles to contract and the ligaments to stretch, not a mechanical barrier in the joint. This new design seems to have this feature, but kinematic studies and modification of the implant geometry may be needed. What is the angle of adduction and abduction provided by this design? Adduction is toward the body centerline in the plane of the drawing and abduction is away from it.
3. Sapphire Bearing Assembly. This is a semi-constrained design because at least five degrees of freedom are included. The bearing loads, degree of polish, static and dynamic friction coefficients, and directional forces in the bearing in response to expected loads have not been presented. Some of this is easily determined from engineering mechanics, but experimental verification of friction in

synovial fluid is needed. What is the unit stress at the bearing interface? If the coefficient of friction is 0.05 what is the resistance to movement accompanying that stress?

4. Detailed Analysis of Metal Stresses. In the example the stresses believed to be most critical, the shear stress in the metal shaft supporting the bearing, were calculated. However, all metal components must be strong enough for their function. Detailed stress calculations and modifications, if necessary, are required. This includes the horizontal forces on the sapphire bearing cups and their containment. What is the horizontal force separating the cups for a 6 mm dia. ball with the geometry of problem 2?
5. Metal/interface Cement. A permanent cement is needed between the bioceramic interface and the metal components. Dental cements are probably suitable. But the problems of accurate fit, cement selection, and cement thickness require detailed analysis. What cement properties would you require in selecting a cement from internet specifications?
6. Minimum surgical removal. Standard surgical screws are used for temporary fixation of the implant. Only one cross-sectional plane has been given in Fig. 3. Where would you place the screws considering rotation perpendicular to the circular axis?
7. Surgical Tools. For a new design to be practical, the surgical approach, the exposure of the surfaces to be altered, and the alignment of fixtures for preparing the bone must be fully explored. What fixtures would you need for the surgical procedures that must be completed?
8. Adjustment of Center of Rotation. Shims, rotation of the components, and other features must be analyzed to be sure the surgeon can control the location of the center of rotation. The surgeon must be able to verify range of motion and proper kinematic function before closing. What shapes of shims would you recommend? What fixtures would you need to obtain the necessary combination before actual placement of the implant?
9. Modifications. This initial design is intended to replace the hip joint with a minimum of tissue removal. If the femoral neck were fractured, how would you extend the same concept to a femoral stem component?

10. Another concern is the possibility of displacement of the ball from the bearing by excessive loads. How much load is required? Could elastic limits in the metal components be controlled to allow dislocation, but non-surgical restoration?

CONCLUSION

A preliminary new hip joint design has been completed. Improvements are possible, and testing of the design may show important factors that have not been considered, but this is a reasonable first attempt at an improved design. This is just the initial stage of design. Most of the details of the material design, the process design and the joint design have been deleted to save space. It is obvious that the success of any such design is dependent on the details of the design process, an understanding of existing designs, the judgment used in making the decisions about important elements of the design, and the degree to which the designer really understands the requirements for success. It is that last feature that is most difficult to acquire in the usual teaching situation. Usually the design problem should be much simpler than the one used here as an illustration.

PATENT STATUS

Concepts disclosed here are subject to existing patents and applications[8,9].

REFERENCES

1. J. Charnley Low Friction Arthroplasty of the Hip. Springer-Verlag 1979
2. Yoon B. Park "Orthopedic Prosthesis Fixation" Chapt. 48 The Biomecial Engineering Handbook J. D. Bronzino, ed. in chief CRC Press 1995.
3. The National Center for Health Statistics 1994.
4. K. deGroot Bioceramics of Calcium Phosphates CRC Press 1981
5. T.D. McGee, "Artificial Bone and Tooth Material" US Patent # 3,787,900. 1974.
6. T.D. McGee, A. Graves, K.S. Tweden, and G.G. Niederauer, "A Biologically Active Ceramic Material with Enduring Strength" Chapt. 42 Encyclopedia of Biomaterials 1995.
7. T.D. McGee & C.E. Olson, "General Requirements for a Successful Orthopedic Implant" Chapt. 13 Encyclopedia for Biomaterials 1995.

8. T.D. McGee, “Method of Restructuring Bone”, US Patent No. 6,312,467 (Nov. 6 2001).
9. T.D. McGee, “Method of Restructuring Bone” US Patent No. 6,364,909 (April 2, 2002).

14

Design Optimization of Ceramic-to-Metal Joints

J.H. Selverian

DESIGN OPTIMIZATION OF CERAMIC–TO-METAL JOINTS

J.H. Selverian
OSRAM SYLVANIA Inc.
71 Cherry Hill Drive
Beverly, MA 01915-1068

ABSTRACT

A method of designing ceramic-to-metal seals using optimization techniques and a probability of failure criterion is described. A generic ceramic-to-metal lamp seal is taken as an example. The effect of various design parameters is discussed.

INTRODUCTION

Ceramic-to-metal and glass-to-metal joints are critical in many lighting applications. Almost every lamp made has at least one ceramic-to-metal or glass-to-metal seal and several billions lamps are made each year. Examples include standard incandescent lamps where the seal is between lead glass and dumet, quartz halogen lamps where the seal is between quartz and molybdenum, high-pressure sodium vapor lamps where the seal is between alumina and niobium, Figure 1.

In the previously mentioned seals with alumina the actual seal is accomplished by a glass frit. In metal halide high intensity discharge (HID) lamps the chemical fills required to provide the desired photometry can react with the frit and the niobium leading to seal failure. These types of chemical compatibility problems frequently dictate the materials used in making a seal and can lead to a compromised design from a mechanical design point-of-view. For these reasons it is desirable to have a seal without frit or niobium.

These joints must last the life of the lamp, up to 20 000 hours, withstand operating temperatures on the order of 600°C and withstand thousands of thermal cycles due to power-on/power-off. To aid in the design of these joints finite element analysis is frequently used. However, when designing with brittle materials, which exhibit volume dependent mechanical properties, the selection of a failure criterion is not straightforward. In addition, brittle materials typically display large variability in strength due to the presence of flaws. It can be a very tedious process to determine the best design if the typical trial-and-error approach is used. Also, fabricating ceramic parts of various dimensions is an expensive and time-consuming process.

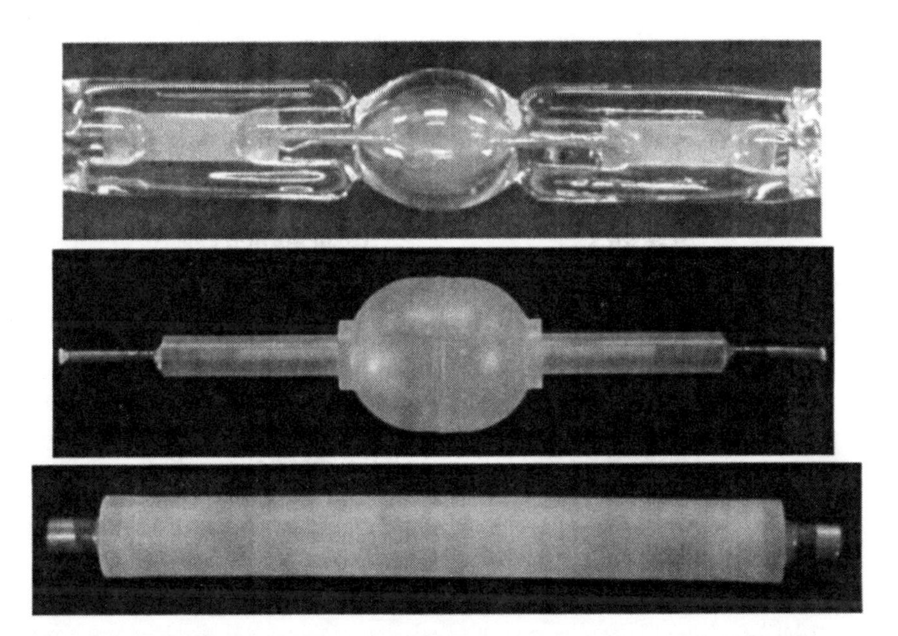

Figure 1. Three types of commercial lamps involving glass- or ceramic-to-metal seals. Top: Automotive Xenarc® high-pressure Xe metal halide HID lamp, 7 MPa operating pressure. Press seal between molybdenum and quartz. 6 mm diameter of arc tube. Center: Metal halide HID lamp, 3 MPa operating pressure. Frit seal between niobium and alumina. 11 mm diameter arc tube. Bottom: Sodium vapor HID lamp, 13 kPa operating pressure. Frit seal between niobium and alumina. 10 mm diameter arc tube.

The goal was to develop an automated method of determining the "best" joint geometry and to use this design as the starting point for a more focused experimental study. Due to uncertainties in materials properties and other variables in the manufacturing processes, experimental verification and fine-tuning of the design with test lamps will be needed for the foreseeable future. In this work finite element stress analysis is linked with probabilistic failure analysis of brittle materials to evaluate the robustness of a given joint design. Numeric optimization was used to determine the "best" design. An example of the design process follows.

Design Variables

A generic joint is used as an example, Figure 2. The variables which were allowed to vary in this study were; the diameter of the molybdenum rod, the outer diameter of the alumina tube, the overhang of the alumina tube and cermet on the outside of the joint, the length of the cermet and the weight fraction of the tungsten in the cermet. This gives a total of five design variables. The ranges of the design variables were limited to those dictated by tooling and other manufacturing considerations, Table I. This was accomplished using constraints on the design variables in the optimization routine.

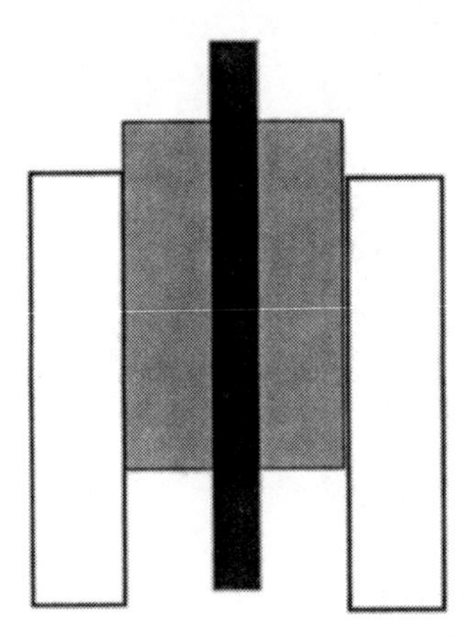

Figure 2. Schematic of the test joint (only end portion is shown) composed of an alumina outer tube (white), and tungsten/alumina cermet (gray) and a molybdenum electrical feedthrough (black). This drawing is not to scale. A negative outside overhang (tube does not overhang the cermet) and a positive inside overhang (the tube does overhang the cermet) of the alumina tube and the cermet is shown here.

Table I. List of allowable values for the design variables. All dimensions are in millimeters except for the weight fraction of metal in cermet.

Design Variable	Allowable Values	
	Minimum	Maximum
Outer radius of Mo rod	0.125	0.2
Outer radius of alumina tube end	2.2	3.0
Outside overhang of alumina & cermet	-2.0	2.0
Length of cermet plug	2.0	5.0
Weight fraction of tungsten in cermet	0.1	0.9

Stress Analysis of the Joint

The residual stresses at room temperature after the joining process were used to calculate the probability of failure of the joint. The finite element analysis was done using ABAQUS [1] with quadratic reduced integration elements; a mesh is shown in Figure 3. The alumina and cermet were treated as elastic solids and the molybdenum feedthrough was treated as an elasto-plastic material, i.e., the molybdenum was allowed to deform plastically. The joint was considered to be stress-free at 800°C and was then cooled to room temperature. The stress state at room temperature was used in all of the following analyses. Only static fracture was considered as a failure mechanism.

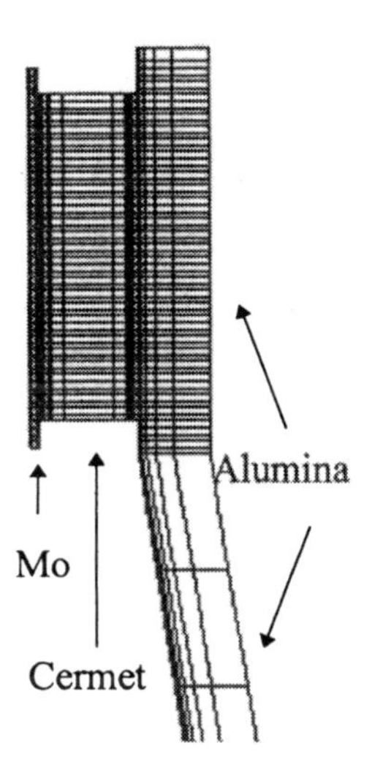

Figure 3. Axisymmetric mesh used for FEA. Only 1/4 of the joint needed to be modeled because of symmetry (only the end is shown here). The size of the mesh in the cermet was kept constant in all of the analysis to eliminate problems caused by the singularity due to the cermet/molybdenum interface.

Material Properties

The Weibull distribution is typically used to express the mechanical properties of brittle materials. Full details on constructing a Weibull plot are given in reference 2. A Weibull plot for the alumina used in this work is shown in Figure 4. Diametrical compression ring tests were used to load the samples. While greater than 40 samples are desirable for a Weibull analysis, only 20 were used in this study due to the limited availability of material. Two parameters are obtained from a Weibull plot which describe the material. The first is the Weibull modulus (m), also called the shape parameter. This is the slope of the Weibull plot. It describes the dispersion in the strength of the material. A low Weibull modulus means that there is a significant amount of scatter in the strength, a high Weibull modulus means that there is a small amount of scatter in the strength. As the Weibull modulus approaches infinity the strength of the material approaches a single value. The Weibull modulus for a ductile metal is typically in the range of 40. The second parameter obtained from a Weibull plot is the characteristic strength (σ_θ). This is the strength value of the material with a probability of failure of 63.2%, where the y-axis crosses zero.

Each set of Weibull parameters is only valid for the failure mode of the samples tested. Ceramics can fail by volume flaws (voids) or by surface defects (pits/scratches). If more than one flaw type is active in a population of samples the data on a Weibull plot will fall on two lines, one for each flaw type. The data in Figure 4 all fall on one line, which indicates that only one flaw type is active. This flaw type was identified by fractography as a volume flaw. It is important, after the fact, to make sure that the test samples and real samples have failed by the same failure mode, i.e., both by surface flaws of both by volume flaws. If this is not the case then the Weibull parameters that were developed do not apply to the real samples. In this situation the test should be redesigned to measure the properties of the ceramic based on the active flaw type and the analysis should be re-run.

One limitation with the characteristic strength is that the measured strength of a brittle material depends on the volume tested. This is due to the increased probability of having a large flaw in a large test volume. The relationship between the strength and stressed volume is given by [3]:

$$\frac{\sigma_{V1}}{\sigma_{V2}} = \left(\frac{V_2}{V_1}\right)^{1/m} \tag{1}$$

where $V_{1,2}$ is the stressed volume of sample 1 and 2, $\sigma_{V1,V2}$ are the measures strengths of samples 1 and 2, and m is the Weibull modulus. From this equation it is seen that the characteristic strength of a sample will depend on the volume of the sample tested. This makes comparing the results of tests of different sized samples difficult. One way around this is to use a volume weighted characteristic strength, this parameter is called the scale parameter (σ_o) [3]:

$$\sigma_{oV} = \sigma_{\theta V} V_e^{1/m_V} \qquad \sigma_{oS} = \sigma_{\theta S} A_e^{1/m_S} \tag{2}$$

where the subscripts "V" and "S" refer to volume and surface properties, V_e is the effective volume, A_e is the effective area. The effective area and volume depend on the stressed region of the sample, which depends on the test configuration used. As an example, for a tensile test V_e equals the volume of the gauge section and V_a equals the surface area of the gauge section. For simple geometries such as a tensile test or a 4-pt bend test V_e and V_a are easy to calculate. However, for more complicated test configurations FEA of the test configuration is needed to determine V_e and V_a [3].

Fracture parameters used in the probability of failure analysis for the alumina and cermet were measured and are listed in Table II.

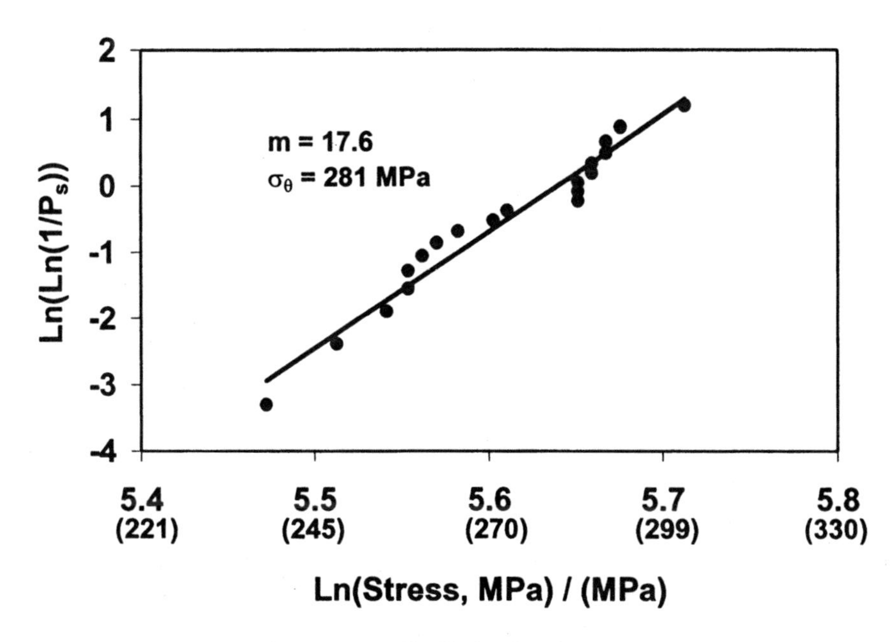

Figure 4. Weibull plot for the PCA.

Table II. Typical fracture parameters for the alumina and a 30 wt% tungsten-alumina cermet.

Material	Weibull Modulus	Characteristic Strength (MPa)	Scale Parameter (σ_oV)
30 wt% tungsten-alumina cermet	2.7	235	83 MPa(m)$^{3/2.74}$
Alumina arc tube	17.6	281	184 MPa(m)$^{3/17.6}$

Optimization

In order to use an optimization routine a cost function must be defined. This cost function refers to the score or quantitative rating given to a particular design and is the parameter that is minimized. A single number is needed to represent how good a design will perform. The probability of failure was calculated with CARES/Life [3, 4] and was used as the cost function. The NASA developed CARES/Life code predicts the probability of failure of a ceramic component as a function of its time in service. CARES/Life couples results from finite element analysis with probabilistic models describing material strength as a function of time (strength degradation due to slow crack growth) and stress state.

The optimization code ASA [5] was used. A main program was written to generate the finite element mesh and to tie together the finite element, failure analysis and optimization routines.

A starting set of design parameters was used to initiate the optimization run. Once the finite element analysis was finished and the cost function evaluated, the optimization routine would then proceed to define a new set of design parameters and initiate a new finite element analysis. This would continue until a predefined number of designs were examined, Figure 5. In this study the results converged after approximately 600 designs.

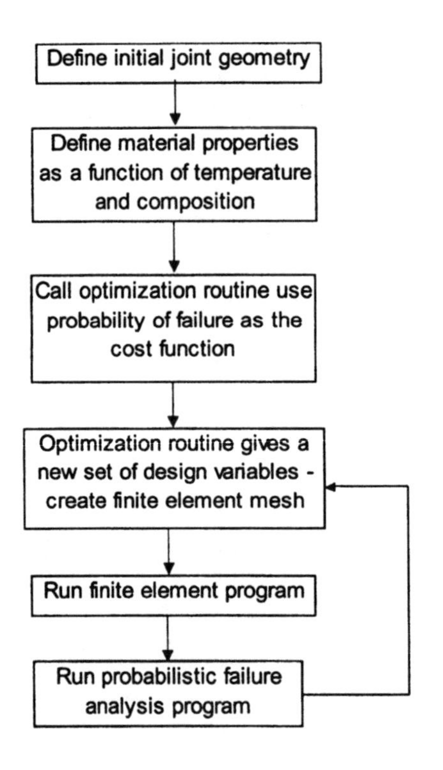

Figure 5. Flow chart of process. An initial design is defined and the optimization routine is called. The probability of failure of the joint is used as the cost function for the optimization routine. The optimization routine adjusts the design variables until the lowest "cost" is found. The entire procedure is automated.

RESULTS

The outer radius of the cermet was kept constant at 1.70 mm due to manufacturing constraints. The inside overhang of the alumina tube and cermet was kept constant at 0.5 mm. Initially, the inside overhang was varied but initial

runs showed that this dimension has essentially no effect on the joint quality and was therefore kept constant to speed-up the optimization process.

The cermet undergoes compressive strains at the interface with the alumina and tensile strains at the interface with the molybdenum, Figure 6. These strains are due to the mismatch in the thermal expansion coefficients and are the cause of joint failure. The highest stresses and highest probability of failure are at the ends of the cermet at the molybdenum/cermet interface, Figures 7 and 8. In real joints similar to the one studied here, the joint fails at this interface (Figure 9), in agreement with the calculations.

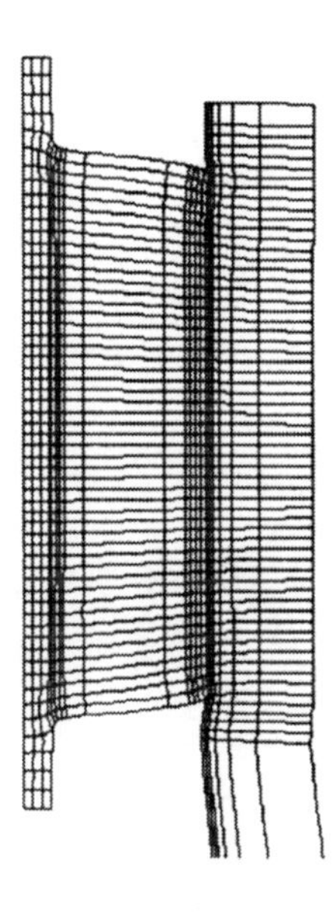

Figure 6. Deformed mesh. Close-up of the deformations in the joint region. Deformations were magnified by a factor of 100 to make them easily visible. The strains in the cermet near the alumina are compressive and near the molybdenum they are tensile.

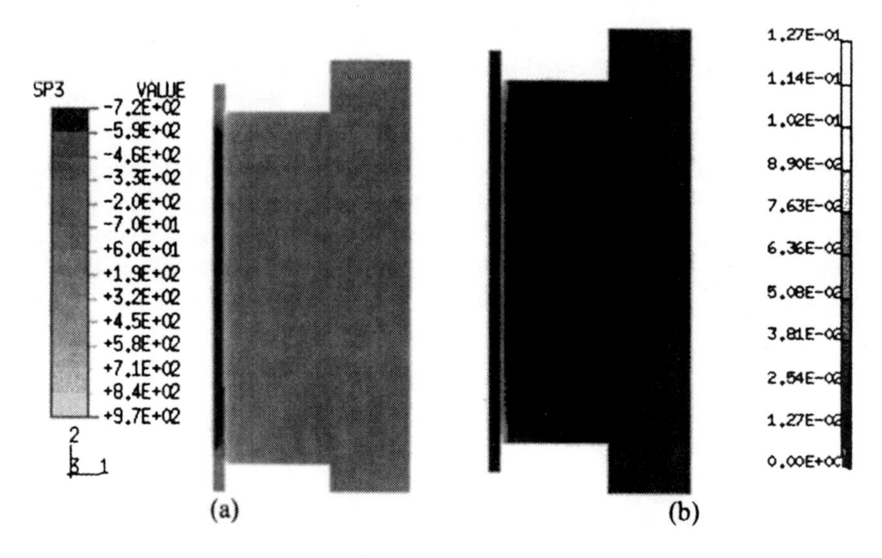

Figure 7. Contours of (a) maximum principal stress (MPa) and (b) probability of failure for the initial design. The joint is predicted to fail along the molybdenum/cermet interface.

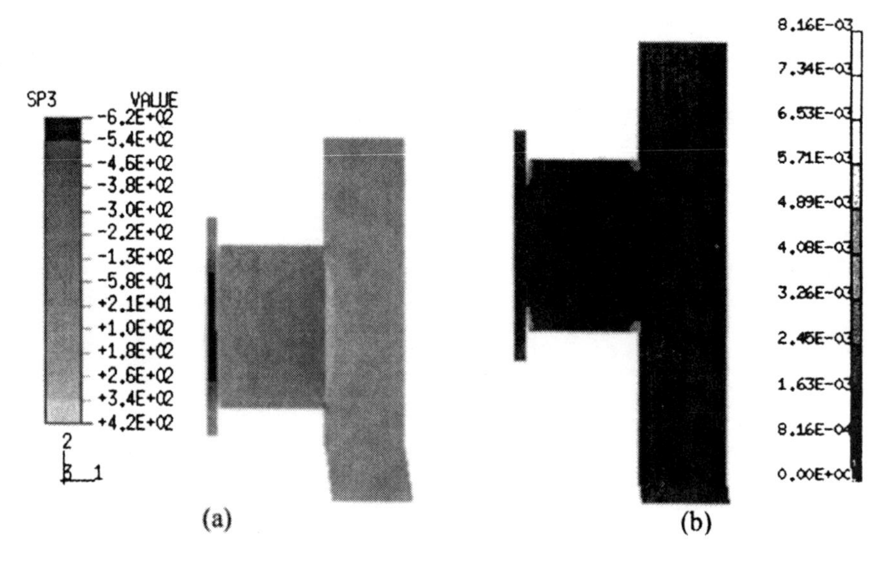

Figure 8. Contours of (a) maximum principal stress (MPa) and (b) probability of failure for the optimized design.

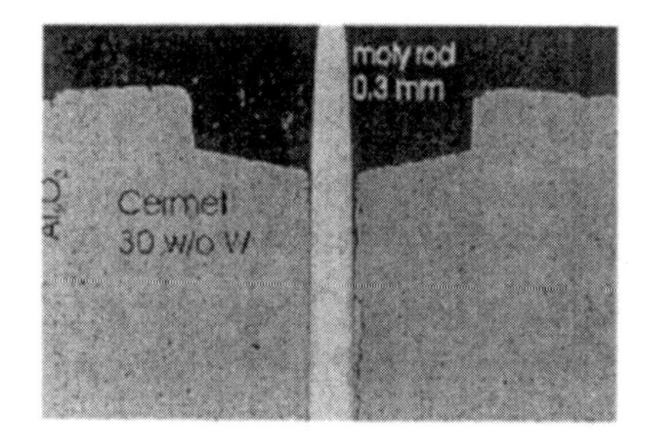

Figure 9. Cross-section of a joint. Cracks are seen at the cermet/molybdenum interface.

Since the alumina has a higher Weibull modulus and scale parameter than the cermet, the design was changed (Table III) by the optimization routine so that the stresses in the cermet were reduced and those in the alumina were increased. This shows the weakness of the typical method of reducing the global maximum tensile stress for parts made from more than one material. In actuality, the maximum stress should not be uniform over the entire joint, but higher in the stronger material and subsequently lower in the weaker material. This type of a non-uniform stress distribution will give a more robust design than will a uniform stress distribution.

Table III. Results from the optimization run. The main changes were the molybdenum feedthrough was reduced in diameter, the outside overhang was increased, the cermet was shorted and the tungsten content of the cermet was increased. The probability of failure was significantly reduced, from 0.9183 to 0.0520.

Dimension	Initial Design	Optimized Design
Molybdenum outer radius (mm)	0.150	0.128
Cermet outer radius (mm)	1.70	1.70
Alumina outer radius (mm)	2.85	2.88
Outside overhang (mm)	0.70	1.59
Length of cermet (mm)	5.0	2.3
Outside overhang (mm)	0.5	0.5
Weight fraction of tungsten in cermet	0.30	0.76
Probability of failure (cost)	0.9183	0.0520

Importance of Individual Design Variables

A regression analysis was performed on all of the designs from the optimization run to identify the most influential design variables, Table IV. All of the variables were found to be statistically significant at 95% confidence.

Table IV. Linear regression analysis of designs from the optimization runs.

Normalized Coefficients from Regression Analysis for				
Mo Outer Radius	Alumina Outer Radius	Outside Overhang	Cermet Length	Weight Fraction of Tungsten in Cermet
0.24	-0.14	-0.31	0.13	-0.55

Each number given in Table IV is the normalized coefficient. The R^2 coefficient was 0.88 indicating a good fit. The relative importance of a variable and the sensitivity of the design to variations in its value are indicated by the normalized coefficient. A negative value indicates that the cost decreases as this variable increases; a positive value indicates that the cost increases as this variable increases. Also, the higher the magnitude of the coefficient the more significant the variable is and the greater the sensitivity of the design is variations in its value (tolerance). The R^2 value of 0.88 for regression analysis suggests that the correlation between the five variables studied is well represented by a linear relationship. Note however, that the actual relationship is not linear. Using the cermet as an example, as the amount of metal in the cermet increases from its initial value of 30%, the joints get better, however, as the optimum value is reached and exceeded the joints get worse - this is a parabolic type relationship. The relationship will always be nonlinear if the optimum value lies within the allowable design range. The variables listed in order of decreasing importance are: weight fraction of tungsten in cermet, outside overhang, outer diameter of molybdenum feedthrough, outer diameter of alumina tube and the length of the cermet.

CONCLUSION

Such an approach to metal-to-ceramic joint design can be used to give a starting point for experimental testing of joints. In practice, because of the complicated nature of the fabrication process and difficulties in obtaining accurate material property data, the design space around the optimum found by the approach outlined above must be studied experimentally. Additionally, the method indicates which of the several design variables are the most critical and can be used to help set tolerances for the final manufacturing process.

BIBLIOGRAPHY

[1]ABAQUS, Hibbit, Karlsson and Sorensen Inc., 1080 Main Street, Pawtucket, Rhode Island, 02860.

[2]R.W. Davidge, *Mechanical Behaviour of Ceramics*, 1st edition, pp. 133-139, Cambridge University Press, New York, 1980.

[3]CARES/Life - Ceramics Analysis and Reliability Evaluation of Structures Life Prediction Program, N.N Nemeth, L.M. Powers, L.A. Janosik, and J.P. Gyekenyesi, NASA - Lewis Research Center, 1993.

[4]Ceramic-to-Metal Joints: Part II - Performance Testing and Strength Prediction, J.H. Selverian and S. Kang, Amer. Cer. Soc. Bull., **71** [10] 1511-1520, (1992).

[5]ASA - Adaptive Simulated Annealing, L. Ingber, the source code for this software is available free at; http://www.alumni.caltech.edu/~ingber.

Additional Information:

1) CARES software at http://www.grc.nasa.gov/WWW/LPB/cares/
2) Nemeth, N. N., Powers, L. M., Janosik, L. A., and Gyekenyesi, J. P.: "Designing Ceramic Components for Durability," Amer. Cer. Soc. Bull., **72** [12] 59-69 (1993).
3) IIT Research Institute / Reliability Analysis Center at http://www.rac.iitri.org/

Questions for Discussion:

1) How is designing with brittle materials inherently different than designing with ductile materials?
2) What are the important material properties for ceramic-to-metal and/or glass-to-metal sealing? What are the important characteristics of the service environment?
3) When designing structures, data on the fracture properties of the materials are required. Typically a 4-pt bend test is used to characterize these properties for a ceramic. Is this test method acceptable? Is it always the best? What are the potential drawbacks of this method?
4) Typically Weibull plots are used to determine the required properties of a ceramic. Most structures are designed to operate where the probability of failure is very low. However, in the low probability of failure region of the Weibull plot there are, by definition, very few data points. What does this mean in terms of obtaining accurate and applicable Weibull values?

5) When using optimization techniques, often the most critical and difficult part is developing a good cost function. In the previous sections the cost function consisted of only one variable (probability of failure). In other applications the cost function may have two or more terms and a weighting function is needed to describe their relative importance. An example would be in a light source where you want the maximum light output from the smallest sized source. Consider cases where there is more than one term in the cost function and discuss how to determine what weighting function to use.
6) Design a ceramic-to-metal joint that will serve as an electrical feedthrough capable of carrying 10 Amps at 120V. The insulating ceramic will be made from alumina. Select a metal to join with the alumina using either a frit or a metal braze alloy. It must withstand 300°C in air for a minimum of 1000 hours and remain hermetic. The feedthrough must be able to withstand a force of 5N applied to the metal conductor. Be sure to explain your selection process and support your design with the required simplified calculations.

15 Designing for Thermochemical Applications

E.D. Wachsman

DESIGNING FOR THERMOCHEMICAL APPLICATIONS

Eric D. Wachsman
University of Florida
Gainesville, FL 32611-6400

ABSTRACT

Fabrication and operating environments can play a dramatic role in the thermochemistry of ceramics. This effect is most evident in the defect chemistry of oxide-ion conductors. These materials are used in, and being investigated for, a wide variety of technological applications including fuel cells, membranes and sensors. Their required functionality in these applications goes far beyond mechanical strength and toughness, to include such properties as ionic and electronic conductivity, thermochemical stability, and catalytic activity. These properties arise out of their defect chemistry which is directly dependent on their oxygen stoichiometry and thus their environment. Thermochemical considerations in the selection of materials for these applications will be reviewed.

INTRODUCTION

Research in functional materials has progressed from those materials exhibiting structural to electronic functionality. The study of ion conducting ceramics ushers in a new era of "chemically functional materials." This chemical functionality arises out of the defect equilibria of these materials, and results in the ability to transport chemical species and actively participate in chemical reactions at their surface. Moreover, this chemical functionality provides a promise for the future whereby the harnessing of our natural hydrocarbon energy resources can shift from inefficient and polluting combustion - mechanical methods to direct chemical conversion.

There are numerous ceramic materials that conduct cations (e.g., Li^+ and Na^+) and anions (e.g., F^- and O^{2-}) and are therefore chemically functional. For the purposes of this paper we will focus on oxide-ion conducting ceramics. Depending on their defect chemistry and environment, these materials can exhibit exclusively ionic (electrolyte) or electronic conduction, or mixed (both ionic and electronic) conduction.

In the first part of this paper we will review the unique materials properties associated with chemical functionality, as these properties are not generally considered in the design of ceramic devices. In the second part we will summarize how these properties influence the selection of materials for solid oxide fuel cells (SOFCs), and then present a case study - the development of a low temperature SOFC electrolyte.

CHEMICAL FUNCTIONALITY

Defect Equilibria

As stated above, the chemical functionality arises out of the defect equilibria. (For a review of defect equilibria the reader is refered to ref.'s 1-4, and the numerous references therein.) This can be seen by the heterogeneous equlibria between gaseous oxygen and vacant ($V_O^{\bullet\bullet}$) and occupied (O_O^x) oxygen-sites and electrons (e') in the solid (Kröger-Vink defect notation):

$$O_O^x = 1/2O_2 + V_O^{\bullet\bullet} + 2e' \quad (1)$$

The resultant law of mass action for the external equilibria, assuming $[O_O^x]$ is high (site fraction ~1), can be written as:

$$K_e(T) = P_{O_2}^{1/2}\,[V_O^{\bullet\bullet}]\,n^2 \quad (2)$$

where $K_e(T)$ is the temperature dependent equilibrium constant. There are two things to note when using this relationship. The first is that one needs to be careful of units (is $K_e(T)$ defined in terms of site fraction or concentration). The second is that the equilibria could have just as easily been written in terms of oxygen interstitials (O_i'') and electron-holes ($h^\bullet$).

Solid-state defect chemistry involves formation of a pair of defects (positive $^\bullet$ and negative ') to preserve charge neutrality, as above. The formation of a pair of defects is termed disorder, and is a direct result of the entropic contribution to the free energy, which is always greater in a disordered (high concentration of defects) system. Thus, the equilibrium concentration increases with temperature.

Solid oxide electrolytes typically exhibit Anti-Frenkel Disorder in which the predominant native defects are anion vacancies and interstitial anions, and are related by:

$$O_O^x = O_i'' + V_O^{\bullet\bullet} \quad (3)$$

The resultant law of mass action is:

$$K_d(T) = [O_i'']\,[V_O^{\bullet\bullet}] \quad (4)$$

The equilibrium constant $K_d(T)$ similarly assumes $[O_O^x]$ is high (site fraction ~1).

The relation between the concentrations of electrons (n) and holes (p) is:

$$K_i(T) = n\,p \quad (5)$$

where $K_i(T)$ is the equilibrium constant for the formation of intrinsic electron-hole pairs.

Thus for a pure oxide exhibiting Anti-Frenkel Disorder there are 4 species and 3 equilibrium relationships. The final relationship is that of charge neutrality, which for this system is:

$$2[V_O^{\bullet\bullet}] + p = 2[O_i''] + n \quad (6)$$

The defect equilibria as a function of P_{O_2} is entirely defined by these 4 relationships.

The simultaneous solution of these 4 relationships results in an analytical expression for the concentration of each of the defects as a function of P_{O_2}. This solution for the defect concentrations is shown graphically in a typical electrolyte defect equilibrium diagram (DED), Figure 1. This diagram shows three regions, low, intermediate, and high P_{O_2}, in which a pair of defects dominate the neutrality relationship due to their relative high concentration.

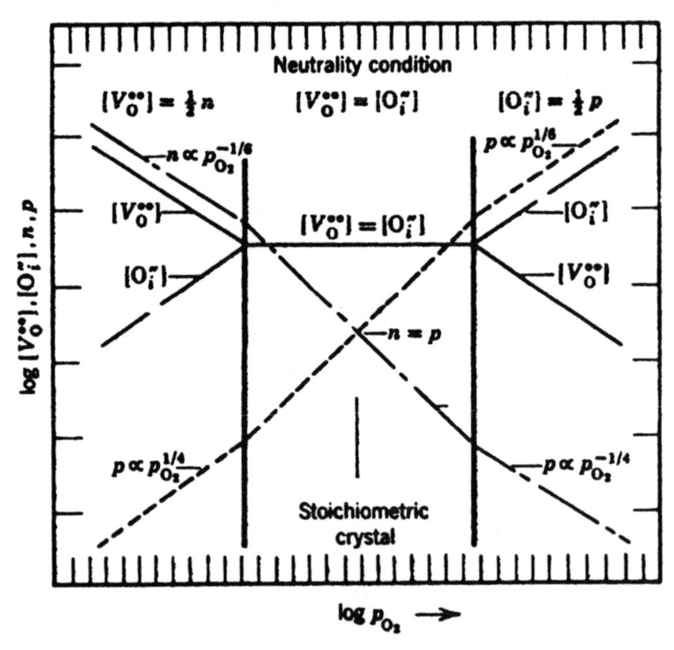

Figure 1. Representative defect equilibrium diagram for solid oxide electrolytes, from Kingery *et al* [1].

Over the intermediate region $K_d(T)$ dominates and $[O_i''] \sim [V_O^{\bullet\bullet}] = K_d(T)^{1/2}$. Therefore, for solid oxide electrolytes the concentration of oxygen vacancies and interstitials is independent of P_{O_2} over a range of oxygen partial pressure. At the stoichiometric point ($[O_i''] = [V_O^{\bullet\bullet}]$) the concentration of electrons and holes is $n = p = K_i(T)^{1/2}$.

At low P_{O_2} the charge neutrality relationship is dominated by $2[V_O^{\bullet\bullet}] \sim n$. Substitution of this into Eqtn. 2 results in a $P_{O_2}^{-1/6}$ dependence for the concentration of the dominant species ($V_O^{\bullet\bullet}$ and e') in this region. At high P_{O_2} the charge neutrality relationship is dominated by $2[O_i''] \sim p$. This results in a $P_{O_2}^{1/6}$ dependence for the concentration of the dominant species (O_i'' and $h^\bullet$) in this region.

As can be seen, the concentration of these intrinsic defects at any temperature is directly related to P_{O_2} by the equilibrium constants for defect formation. In addition, the presence of extrinsic charged defects (either as intentional dopants or impurities) will also affect their concentration.

For example, zirconia is phase stabilized in the fluorite structure by the addition of lower valent cations such as Ca^{2+} or Y^{3+} (acceptor dopants). These dopants are charged defects, Ca_{Zr}'' and Y_{Zr}', respectively, which not only stabilize the fluorite phase, they also create additional $V_O^{\bullet\bullet}$ in order to preserve charge neutrality. The typical dopant concentration (~10%) and the concentration of $V_O^{\bullet\bullet}$ formed are greater than the intrinsic O_i'' and $V_O^{\bullet\bullet}$ concentrations. This results in an additional intermediate region in the DED where the dominant pair of defects are the acceptor dopant and $V_O^{\bullet\bullet}$. (The representative DEDs for this system can be seen in ref. 4).

As the dopant concentration increases, dopant-vacancy associates form (e.g., Ca_{Zr}''-$V_O^{\bullet\bullet}$) both due to coulombic attraction and lattice strain energy. In addition, multivalent dopants or impurities (e.g., $Tb^{3+/4+}$, $Co^{3+/2+}$) can also be present. Due to the ability to change oxidation state these impurities can further add to the concentration of electronic species. The resultant dependence of defect concentrations on P_{O_2} (DEDs) are much more complicated.

Ionic and Electronic Conductivity

All of the defects in the DED above (O_i'', $V_O^{\bullet\bullet}$, e', and $h^\bullet$) are mobile. The conductivity (σ_i) of each mobile species is proportional to its charge (z_i q), mobility (u_i) and concentration ([i]):

$$\sigma_i = z_i \, q \, u_i \, [i] \tag{7}$$

The total conductivity (σ_t) is the sum of the conductivity of each species-i. The mobility of electronic species is more than an order of magnitude greater than ionic species. Therefore, for a material to be considered primarily an ionic conductor the concentration of ionic defects needs to be more than an order of magnitude greater than the concentration of e' and $h^\bullet$.

The total bulk conductivity of three oxide electrolytes is plotted in Figure 2. The observed Arrhenius behaviour is due primarily to the thermally activated mobility. However, the temperature dependence of mobile defect concentration (e.g., K_d, K_e, and K_i) and defect association can also contribute to the apparent activation energy.

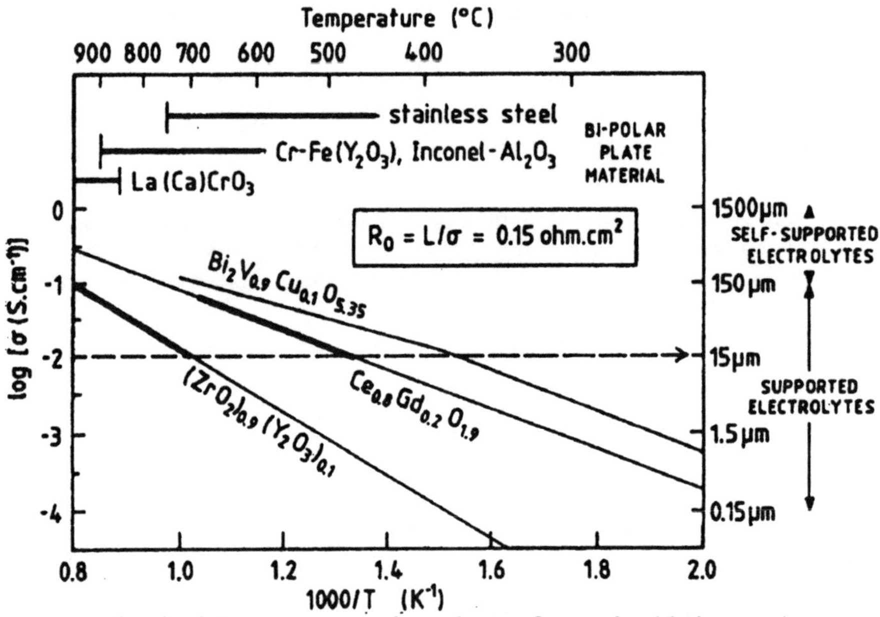

Figure 2. Conductivity temperature dependence of several oxide-ion conductors showing effect of materials selection for a designed performance target (R_0) on materials compatibility and fabrication considerations, from Steele *et al* [5].

The observed conductivity is also dependent on the P_{O_2} as indicated by the DED, Fig. 1. Thus for a material with this DED, at low P_{O_2} the conductivity is n-type, with $\sigma_t \sim P_{O_2}^{-1/6}$; at high P_{O_2} the conductivity is p-type, with $\sigma_t \sim P_{O_2}^{1/6}$; and at intermediate P_{O_2} σ_t is independent of P_{O_2}.

The transference number of any species-i is defined as the ratio of the conductivity of that species to σ_t.

$$t_i \equiv \sigma_i / \sigma_t \tag{8}$$

Solid oxide electrolytes are almost exclusively ionically conducting ($t_{ionic} > 0.99$) over a large range of oxygen partial pressure. This is the intermediate range of the DED where σ_t is independent of P_{O_2}. In contrast, for mixed conductors $t_{ionic} \sim 0.5$ and this can occur over a region where σ_t is a function of P_{O_2}.

Catalytic Activity

Catalytic activity is the ability of a substance to promote the rate of a chemical reaction. Catalytic selectivity is the propensity of that catalyst to promote a desired reaction over any potential undesired reactions. Heterogeneous catalysts are typically composed of noble metals or transition metals/metal oxides dispersed on a high surface area support. The support can play an active role in catalysis, and

much has been made of the role of $V_O^{\bullet\bullet}$ in oxide catalysts as well as several surface oxygen species (e.g., O^-, etc).

Perovskites can be catalytically active and relatively selective depending on the nature of the B-site cation and the degree of aliovalent substitution on the A-site. This can be seen in the numerous publications on the use of these materials as oxidation and reduction catalysts. In these studies the B-site is typically a transition metal and doping with lower valent cations results in $V_O^{\bullet\bullet}$ formation. In general, the more readily reducible the catalyst, the more active it is. For example, in a series of B-site substitution of first row transition metals in $LaBO_3$, we observed that those metals with near half full d-orbitals are most active for NO reduction (to N_2 and O_2), Figure 3. Therefore, the catalytic properties of these materials are related to their defect equilibria.

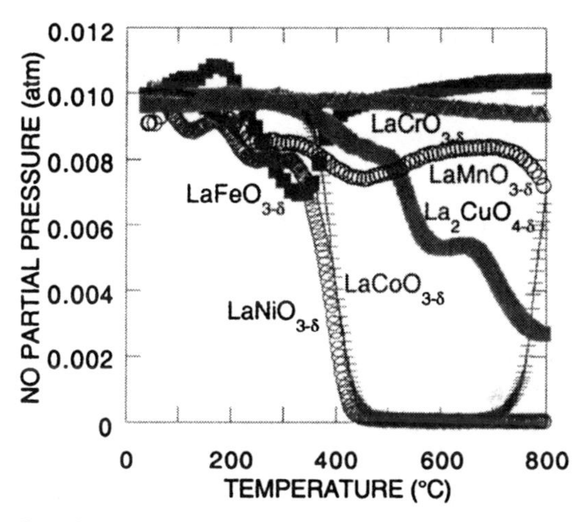

Figure 3. Effect of B-site transition metal, in $LaBO_3$ catalysts, on NO reduction activity. Temperature programmed reaction of NO over partially reduced catalysts showing reduction in NO concentration. Oxygen species formed enters lattice and quantitative N_2 formation not shown for clarity [6].

Since catalysis is a surface phenomenon the reaction rate is proprtional to surface area. This leads to a significant fabrication issue for ceramic electrochemical devices, in that typical sintering temperatures will take a highly active catalyst powder with a specific surface area of ~$100 m^2/g$ and reduce it to $1\text{-}10 m^2/g$.

As materials are developed that are more conductive, and/or the fabrication of current materials results in thinner films, it is the heterogeneous kinetics at the materials surface that dominates the overall performance of the resultant device. Much more work needs to be done to understand the catalytic and electrocatalytic properties of ion-conducting ceramics. A further understanding in this regard would significantly minimize the electrode polarization of fuel cells and batteries, maximize the surface exchange coefficient of gas separation membranes, and enhance the signal and selectivity of electrochemical sensors.

Thermochemical Stability

Cells based on ion conducting ceramics typically consist of several layers. Each layer has a different functional requirement and, therefore, a different composition and microstructure. Thermochemical concerns during fabrication/sintering are the formation of secondary non-conductive phases at the interfaces between layers. Electrodes are typically porous and can be dual phase. Therefore, maintaining the desired microstructure and phase distribution during sintering is also important.

The above concerns during sintering are common to ceramic devices. What is different about the thermochemical stability issues of ceramic electrochemical cells is that they are operated at elevated temperature, in oxidizing to reducing conditions, under applied potentials, and that species within the solid are mobile and will move in response to the potential/environment. Thus while two oxides may form a stable interface when sintered in air, subsequent operation under an applied potential will change the oxygen activity at that interface and may cause undesired reactions to occur. If the resulting phase is non-conductive it can have a serious impact on the device performance.

Thermochemical Expansion

Thermal expansion is a well known phenomenon, taught in almost every engineering major, involving the change in dimensions with a change in temperature. In contrast, thermochemical expansion is a much less recognized phenomenon involving the change in dimensions with a change in chemical potential. The chemical potential of an ion conducting ceramic depends on the concentration of defects. As shown above, the concentration of defects depends on P_{O_2}. Therefore, the dimensions (expansion) of these materials depends on P_{O_2}.

An example of a material that exhibits this phenomenon is $CeO_{2-\delta}$. If $CeO_{2-\delta}$ is exposed to a low P_{O_2} atmosphere, at temperatures sufficient for ionic conduction, the lattice will lose oxygen (δ increases) consistent with the defect equilibria for that material. The result of this loss of lattice oxygen is in fact an increase (not decrease) in dimensions. One explanation for this phenomena is that the radius of the Ce^{3+} cation, that is formed upon reduction, is larger than Ce^{4+}. Typical values for the thermochemical expansion are on the order of a 1% expansion for a $\Delta\delta = 0.1$.

Potential Gradient

The defect equilibria, described above, results in a concentration of species that is directly related to the equilibrium P_{O_2}. More accurately, the *activity* of the defect species is directly related to the prevalent oxygen *activity*. Therefore, under a potential gradient the activity of the defect species at any point in the solid is directly related to the local oxygen activity. (For simplicity we will continue to use P_{O_2} even though partial pressure may be a difficult concept in the bulk of a solid).

The majority of cells employing ion conducting ceramics operate under a potential gradient (chemical and/or electric). For example a fuel cell typically operates with air ($P_{O_2}{}^L = 0.21$ atm) on one side and fuel ($P_{O_2}{}^0 = 10^{-21}$ atm) on the other side, see Figure 4.

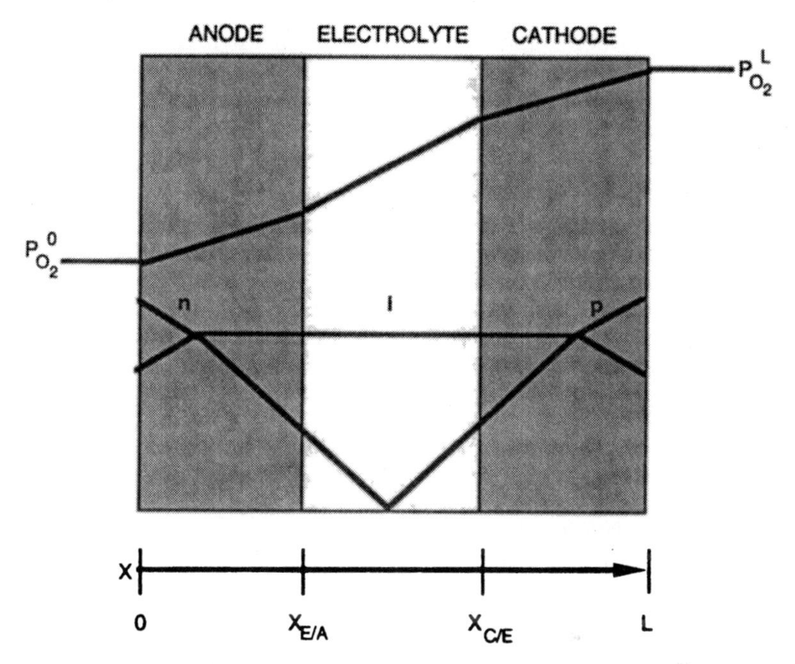

Figure 4. Conceptual representation of a solid oxide fuel cell, showing both the oxygen partial pressure gradient and the resulting change in the majority carrier (linear potential gradients shown for simplicity).

Chemical potential (μ_{O_2}) is proportional to $\ln(P_{O_2})$:

$$\mu_{O_2} - \mu_{O_2}{}^{\circ} = RT\ln(P_{O_2}) \tag{9}$$

The electric voltage or open circuit potential (V_{OCP}) that is created in a cell under a chemical potential gradient is:

$$V_{OCP} = (\mu_{O_2}{}'' - \mu_{O_2}{}')/nF \tag{10}$$

where ' and " indicate the different sides of the cell, n is the number of electrons in the reaction (= 4 for O_2), and F is Faraday's constant.

Therefore, the x-axis of the DED (Fig. 1) is proportional to the chemical potential (minus the standard state $\mu_{O_2}{}^{\circ}$) and is a function of distance through the cell thickness as shown in the bottom half of Figure 4. The difference between two points on the axis is proportional to the electric potential. Given the 20 orders of magnitude difference in P_{O_2} between the two sides of a fuel cell, one can see how at one side a material can exhibit n-type conductivity and on the other side of the same material be either an electrolyte or exhibit p-type conductivity (asuming it is stable in both environments). However, this 20 orders of magnitude in P_{O_2} corresponds to an V_{OCP} of only ~1 V.

Since all of the properties we are discussing depend on the concentration of defects, and the concentration of these defects can change so dramatically from one side of the cell to the other (both at the surface and within the material), understanding the effect of these gradients in potential is crucial to the selection of materials.

CELL DESIGN CONSIDERATIONS

Solid state electrochemical cells may have one overriding function, however, they are typically multicomponent devices with each component having multiple functions and therefore multiple material property requirements. This is exemplified by the various functional requirements of an SOFC. While the overall device function is to produce electricity from a fuel source, the process requires a series of electrochemical steps each carried out by a particular component of the cell.

Solid Oxide Fuel Cells

SOFCs offer great promise as a clean and efficient process for directly converting chemical energy to electricity while providing significant environmental benefits. Since electrochemical cells do not involve a Carnot cycle, the efficiency of an SOFC is about three times that of an internal combustion engine. Therefore, they produce about one-third less CO_2 per kilowatt hour. In addition, they produce negligible CO, HC, or NO_x.

One of the key performance parameters of a fuel cell is its power density (P W/cm^2). The power generated is the cell voltage (V_{cell}) multiplied by the current density (i A/cm^2) generated at that voltage. At open circuit conditions ($i \equiv 0$) V_{cell} is equal to the V_{OCP}, which is set thermodynamically by the chemical potential gradient across the cell. The voltage drop from V_{OCP} is the cell overpotential (η_{cell}). When the cell exhibits ohmic behavior $\eta_{cell} = i \cdot R_{cell}$, where R_{cell} is the area specific cell resistance (Ω cm^2), resulting in

$$V_{cell} = V_{OCP} - i \cdot R_{cell} \tag{11}$$

There are three electrochemically active components in the operation of a fuel cell: the electrolyte, cathode, and anode. R_{cell} is the sum of the resistances of each of these components. In addition, a forth component, the interconnect, serves to provide an electric path between cells. The resistance of this component shows up in the stack performance. The function and considerations for materials selection of each of these components are described below.

Electrolyte

The selection of materials and fabrication methods for the entire cell are predicated by the selection of the electrolyte. First, the majority of the potential drop

is across the electrolyte, so the material has to be thermochemically stable from the oxidizing (air) to the reducing (fuel) conditions of operation. Second, the electrolyte needs to have a high conductivity and that conductivity is ideally exclusively ionic, $t_{ionic} > 0.99$, over the entire potential gradient. Any electronic conductivity in the electrolyte results in an internal electric short, reducing the cell efficiency. Since the electrolyte physically separates oxygen from a combustible fuel it is imperative that the electrolyte is stable in each of these environments, and whatever fabrication method is used results in a dense layer.

The majority of solid oxide electrolytes investigated have been those based on one of the quadravalent-cation oxides ZrO_2, HfO_2, CeO_2 or ThO_2 doped with either a divalent-cation oxide, such as CaO, or a trivalent-cation oxide, such as Sc_2O_3, Y_2O_3 or any of the rare earth oxides. These solid oxide electrolytes have very high oxygen ion conductivities, on the order of 0.1 S cm^{-1} at temperatures in the range of 1000 °C. Other solid oxide electrolytes, such as those based on Bi_2O_3 doped with similar rare earth or alkaline earth oxides, have comparable oxygen ion conductivities to the group IVB oxides but at much lower temperatures. Yttria-stabilized zirconia (YSZ) has been the electrolyte of choice in SOFCs because it represented the most favorable compromise in chemical and thermal stability, ionic conductivity, and cost.

The fluorite structure is common to all of these systems. It is the structure stable at high temperatures for the pure oxides and it is the phase associated with high conductivity. The presence of the dopant oxide serves two purposes: stabilization of this high temperature phase down to room temperature; and in the case of the ZrO_2 based electrolytes, the formation of anion vacancies in order to preserve electroneutrality. It is these oxygen vacancies that are the mobile species in solid oxide electrolytes.

Cathode

The primary function of the cathode is to provide the electrocatalytic sites for the reduction of oxygen from the oxidant gas stream.

$$1/2O_2 + V_O^{\bullet\bullet} + 2e' \rightarrow O_O^x \quad (12)$$

This reaction involves three species and therefore up to three phases: gas (O_2), an ionic conductor ($V_O^{\bullet\bullet}$, O_O^x) and an electronic conductor (e'). For example a porous Pt layer meets the functional requirements of a cathode. At the interface between the Pt and the oxide electrolyte e' and $V_O^{\bullet\bullet}$, respectively, are present, and since the layer is porous O_2 can diffuse to this interface. This region is called the triple phase boundary (TPB).

The other material requirements of the cathode are that it has a high electronic conductivity for transport of electrons to (or holes from) the electrocatalytic sites; that it is thermochemically stable in an oxidizing environment and in contact with the electrolyte and interconnect materials; and that it has a matching thermal coefficient of expansion (TCE).

Because of the electrical conductivity and thermochemical stability requirements candidate materials are essentially limited to conducting oxides (while noble metals may also meet these requirements they are too expensive to be practical). Electronic or mixed conducting oxides exist in numerous structures, with variations on the perovskite ABO_3 lattice as one of the more common group of structures (e.g., $LaMnO_3$, $LaFeO_3$, $LaCoO_3$, etc.). These oxides are typically p-type semiconductors in which substitution of lower valent cations produces additional mobile anion vacancies (e.g., $La_{1-x}Sr_xCoO_3$). These materials are active catalysts due to the mobility and thermodynamic activity of their oxygen-ions/vacancies.

Among these candidate materials, compostions based on $La_{1-x}Sr_x(Co/Fe/Ni)O_3$ have the highest conductivity, both electronically and ionically. The ionic contribution to the conductivity is important because it spreads the reaction region over the entire cathode surface thus relaxing the porosity requirement and increasing the available surface area for the reaction to occur. Further, these materials are catalytically active (see Fig. 3). However, the cathode material that is used in current generation SOFCs is $La_{1-x}Sr_xMnO_3$. Even though this material has a lower electronic conductivity, negligible ionic conductivity, and a lower catalytic acivity it has one important property that has made it the material of choice, its TCE can be tailored to match that of YSZ. Thus in designing a device one may have to compromise on functional performance in order to meet other operating/fabrication requirements.

Anode

The primary function of the anode is to provide the electrocatalytic sites for the oxidation of the fuel. For a H_2 fuel the reaction would be

$$H_2 + O_O^x \rightarrow H_2O + V_O^{\bullet\bullet} + 2e' \quad (13)$$

Similar to the cathode, this reaction involves up to three phases and typically occurs at the TPB. Current SOFCs use a porous Ni-YSZ cermet for the anode with the TPB occuring at the Ni/YSZ/gas interface. The other material requirements of the anode are that it has a high electronic conductivity for transport of electrons from the electrocatlytic sites; that it is thermochemically stable in a reducing environment and in contact with the electrolyte and interconnect materials; and that it has a matching TCE.

Interconnect

The primary function of the interconnect is to provide a series electrical path between adjacent cells. Therefore, the interconnect needs to have a high conductivity that is ideally exclusively electronic. Any ionic conductivity in the interconnect results in an internal ionic short, reducing the stack efficiency. Current SOFCs use an interconnect based on $LaCrO_3$. Since the interconnect physically separates oxygen from a combustible fuel it is imperative that the interconnect is stable in each of these environments, and whatever fabrication method is used results in a dense layer.

CASE STUDY - LOW TEMPERATURE SOFCs

The current SOFC technology must operate in the region of 1000°C to avoid unacceptably high ohmic losses. High temperatures limit the availability of compatible materials and demand (a) specialized (expensive) materials for the fuel cell interconnects and insulation, (b) time to heat up to the operating temperature and (c) energy input to arrive at the operating temperature. Therefore, if fuel cells could be designed to give a reasonable power output at lower temperatures tremendous benefits may be accrued, not the least of which is reduced cost. The problem is, at lower temperatures the conductivity of the conventional YSZ electrolyte decreases to the point where it cannot supply electrical current efficiently to an external load. Three approaches have surfaced to address this issue, as described below.

Thin Electrolytes - The area specific electrolyte resistance (R_0) is linearly proportional to the thickness (L)

$$R_0 = L/\sigma \tag{14}$$

Therefore, one approach to reduce the operating temperature has been the use of thin film ($\leq 10\mu m$) YSZ supported on a porous electrode. This has resulted in a significant improvement in power density at 800°C [7,8].

However, sintering of a pin-hole free thin film on large area porous supports is technologically challenging. Further, thin film fabrication methods tend to be more expensive than the more conventional tape casting. More importantly, since one of the functions of the electrolyte is to separate a combustible fuel from air, thin films on a porous support have long term reliability concerns. Finally, if one is sucessful at thin film fabrication of electrolytes, then a thin film of a more conductive electrolyte would have even better performance.

Higher Conductivity Materials - Another aproach to reduce R_0 is to use electrolyte materials with $\sigma > \sigma_{YSZ}$. There are a variety of materials that have $\sigma > \sigma_{YSZ}$ and are being investigated as potential electrolytes for SOFCs. Two examples of this are ceria and bismuth oxide based electrolytes. As can be seen in Figure 2 these materials have 1-2 orders of magnitude greater conductivity than YSZ.

B. Steele *et al* [5, 9] compared these electrolyte materials based on a design R_0 of 0.15 Ω cm^2. (The reader is referred to his ref.'s 5 and 9 for further details on the selection of this target R_0 as well as comparisons of other electrolyte materials.) Based on a design R_0 one can readily calculated the required L based on the electrolyte σ as shown on the right Y-axis of Figure 2. This figure then shows the tradeoff between σ and film thickness (and fabrication method) for a desired operating temperature.

An other way of comparing the merits of these different electrolytes is to calculate their theoretical performance based completely on their respective σ and ignoring all other factors (note electrode resistances can not be ignored in a real cell, this comparison is just to show the effect of electrolyte selection only and the reader is referred to problems 2 and 3 to see the effect of electrode resistance). This is done in Figure 5 for an SOFC operating at 800 °C using a tape cast 100 µm YSZ, samaria doped ceria (SDC), and a yttria stabilized bismuth oxide (YSB) electrolyte.

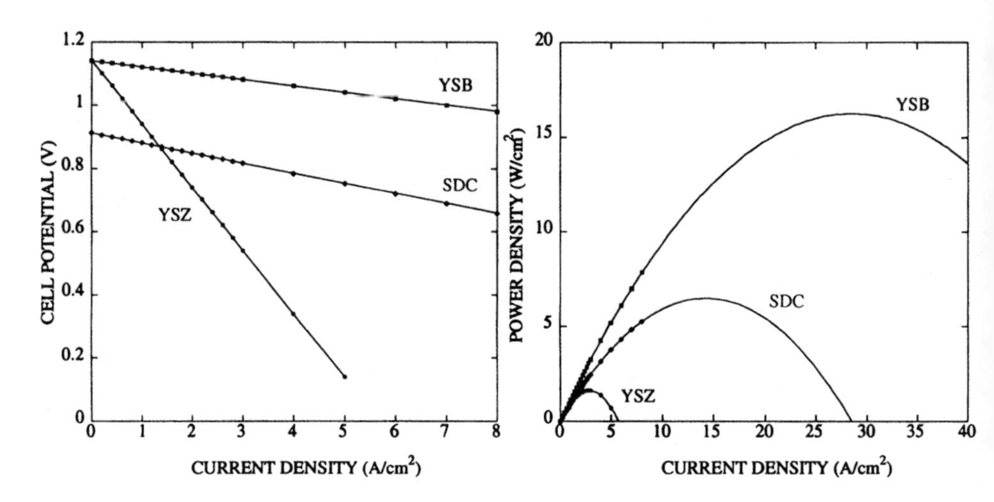

Figure 5. Theoretical performance for an SOFC operating at 800 °C, with air cathode and H_2/3%H_2O anode, using a 100 μm YSZ, SDC, and YSB electrolyte based completely on their respective σ (ignoring all other factors); cell voltage (a) and power density (b) as a function of current density.

The theoretical V_{OCP} at 800 °C, with air on the cathode side and H_2/3%H_2O on the anode side is about 1.14 V. YSZ and YSB show a linear drop in V_{cell} with i, consistent with their respective R_0 of 0.20 and 0.02 Ω cm^2 (Figure 5a). This corresponds to a P of 1.6 W/cm^2 for the YSZ SOFC (satisfying current power density goals of 1 W/cm^2) and an incredible P of 16 W/cm^2 for the YSB SOFC (Figure 5b).

Since σ_{YSB} is so high, resulting in the potential for an SOFC to exceed targeted power density goals by an order of magnitude, why is it not the electrolyte of choice by SOFC developers. The reason is that it fails with respect to one of the other key requirements, thermochemical stability. Unfortunately, while Bi_2O_3 electrolytes exhibit purely ionic conduction down to 10^{-20} atm P_{O_2} (in O_2/Ar mixtures) they are readily reduced to metallic Bi at this oxygen partial pressure in H_2/H_2O mixtures [10]. Another concern with bismuth oxide electrolytes has been their phase/structural stability [11].

SDC also has a high conductivity, resulting in a R_0 of 0.03 Ω cm^2, still a significant improvement over YSZ. It does not have the decomposition and phase problems of YSB. However, the defect equilibria of SDC is such that while in air it is predominantly an ionic conductor, under a fuel atmosphere it has significant n-type electronic conduction. Thus, in a fuel cell arrangement the overall t_{ionic} is only ~ 0.8. The result of this can be seen in Figure 5a.

Under open circuit conditions the actual measured V_{OCP} is related the the theoretical thermodynamic V_{OCP} by:

$$t_{ionic} = V_{OCP}(\text{measured})/V_{OCP}(\text{theoretical}) \qquad (13)$$

Therefore, under open circuit conditions the combination of electronic and ionic (mixed) conduction in SDC will result in O_2 permeating through the electrolyte and oxidizing fuel without producing electricity (efficiency loss). Thus, at low i the V_{cell} is less than that obtained with an YSZ SOFC (Figure 5a).

However, the i-V_{cell} characteristics of these two electrolytes cross, and as can be seen in Figure 5b, the result is that an SDC SOFC theoretically has a much higher P than YSZ. It is because of this that many groups are actively investigating and developing ceria electrolyte SOFCs [12-13].

One other issue that needs to be addressed with ceria electrolyte SOFCs, however, is thermochemical expansion. As described above, under reducing conditions ceria expands. Further, an SOFC electroyte is exposed on one side to air and on the other to a fuel environment. Therefore, there will be significant dimensional differences (strain) from one side of the electrolyte to the other that can potentially result in cracking and catastrophic failure.

Bilayer Electrolytes - Bilayered electrolytes have been investigated in order to overcome the inherent limited thermochemical stability of highly conductive oxides such as doped ceria and stabilized bismuth oxide [14-15]. The rationale for the earlier investigations [14] was that a thin, dense layer of a more stable oxide could protect these high conductivity materials from reduction by preventing contact with the reducing gas. For example, Yahiro et al. [14] demonstrated that a thin protective layer of YSZ on the fuel side of a ceria SOFC increased the V_{OCP} and resultant power density.

In general, however a layered YSZ-ceria type composite electrolyte has no intrinsic advantage over just a thin YSZ electrolyte itself, other than providing a nonporous substrate for YSZ deposition. Moreover, the rationale of preventing contact of, for instance, a ceria electrolyte with a reducing gas is conceptually an over simplification of the use of a bilayer electrolyte. Rather one should consider the functional gradient, in terms of oxygen activity (or effective local P_{O_2}), through the electrolyte and use this to modify or block electronic transport.

We demonstrated this approach using bilayer ceria-bismuth oxide electrolytes [15]. Bismuth oxide is thermochemically stable on the oxidizing side, and while ceria has electronic conductivity on the reducing side it does not decompose. By taking in to consideration the ionic and electronic conductivity of each of these materials and adjusting their relative thicknesses, as shown in Figure 6, we are obtaining an interfacial P_{O_2} that maintains the thermochemical stablity of bismuth oxide. Further, the transport properties of bismuth oxide are such that it blocks electronic conduction originating in the ceria layer. Thus, this bilayer approach allows us to optimize (design) an electrolyte that has the conductivity advantages of these materials without their thermochemical disadvantages.

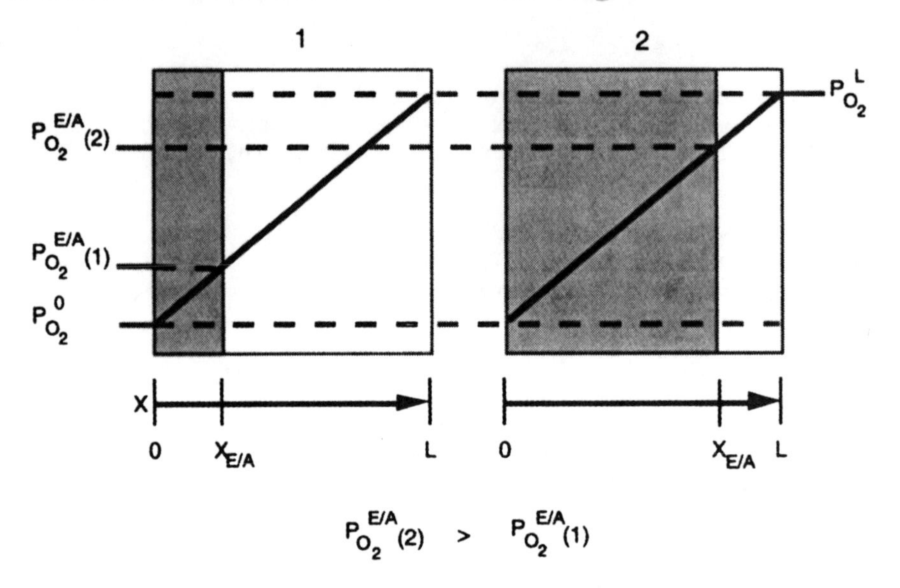

Figure 6. Conceptual representation of a bilayered solid-oxide electrochemical cell showing effect of relative thickness on interfacial P_{O_2} (shown for simplicity with a linear P_{O_2} gradient).

SUGGESTED PROBLEMS

1. A major goal of SOFC developers is to reduce the operating temperature to a range where stainless steel can be used for the interconnect. This substitution of materials has been estimated to reduce the cost of SOFC stacks by an order of magnitude [16]. If the targeted R_0 for the electrolyte is of 0.10 Ω cm^2, what electrolytes are suitable and at what thickness? Assuming your company has tape casting (but not thin film deposition) capabilities, how will this influence your selection?

2. Assuming that with an $La_{1-x}Sr_xMnO_3$ cathode and Ni-YSZ anode R_{anode} = $R_{cathode}$ = 0.15 Ω cm^2 for an SOFC operating at 800 °C, with air cathode and H_2/3%H_2O anode. (a) Plot i-V_{cell} and i-P_{cell} for a 100 μm thick YSZ, SDC, and YSB electrolyte. (b) Plot i-V_{cell} and i-P_{cell} for a 10 μm thick YSZ, SDC, and YSB electrolyte. State your assumptions.

3. Assuming that with an $La_{1-x}Sr_xCoO_3$ cathode and Ni-SDC anode R_{anode} = $R_{cathode}$ = 0.05 Ω cm^2 for an SOFC operating at 800 °C, with air cathode and H_2/3%H_2O anode. (a) Plot i-V_{cell} and i-P_{cell} for a 100 μm thick YSZ, SDC, and

YSB electrolyte. (b) Plot i-V_{cell} and i-P_{cell} for a 10 μm thick YSZ, SDC, and YSB electrolyte. State your assumptions.

REFERENCES

[1]W.D. Kingery, H.K. Bowen and D.R. Uhlman, "Introduction to Ceramics," John-Wiley & Sons, 2nd Ed., Chapter 4, 1976.

[2]H. Rickert, "Electrochemistry of Solids," Springer-Verlag, 1982.

[3]P.J. Gellings and H.J. Bouwmeester Ed.'s, "Solid State Electrochemistry," CRC Press, 1997.

[4]N.Q. Minh and T. Takahashi, "Science and Technology of Ceramic Fuel Cells," Elsevier, 1995.

[5]B.C.H. Steele, K. Zheng, N. Kiratzis, R. Rudkin, and G.M. Christie, "Preliminary Investigations on Direct Methanol Ceramic Fuel Cells for Electric Vehicles," in 1994 Fuel Cell Seminar Abstracts, San Diego, CA, 479-482 (1994).

[6]E. D. Wachsman, P. Jayaweera, G. Krishnan, and A. Sanjurjo, "Electrocatalytic Reduction of NO_x on $La_{1-x}A_xB_{1-y}B'_yO_{3-\delta}$; Evidence of Electrically Enhanced Activity," *Solid State Ionics*, **136-137**, 775-82 (2000).

[7]S. DeSouza, S.J. Visco, and L.C. De Jonghe, "Thin-Film Solid Oxide Fuel Cell with High Performance at Low-Temperature," *Solid State Ionics*, **98**, 57-61 (1997).

[8]J.W. Kim, A.V. Virkar, K.Z. Fung, K. Mehta, and S.C. Singhal, "Polarization Effects in Intermediate Temperature, Anode-Supported Solid OxideFuel Cells," *J. Electrochem. Soc.*, **146**, 69-78 (1999).

[9]B.C.H. Steele, *J. Power Sources*, **49**, 1 (1994).

[10]E. D. Wachsman, G. R. Ball, N. Jiang, and D. A. Stevenson, *Solid State Ionics*, **52**, 213 (1992).

[11]N. Jiang and E. D. Wachsman, "Structural Stability and Conductivity of Phase-Stabilized Cubic Bismuth Oxides," *Journal of the American Ceramic Society*, **82** [11], 3057-64 (1999).

[12]B.C.H. Steele, "Appraisal of $Ce_{1-y}Gd_yO_{2-y/2}$ Electrolytes for IT-SOFC Operation at 500°C," *Solid State Ionics*, **129**, 95-110 (2000).

[13]R. Doshi, V.L. Richards, J.D. Carter, X. Wang, and M. Krumpelt, "Development of Solid-Oxide Fuel Cells that Operate at 500°C," *J. Electrochem. Soc.*, **146**, 1273-78 (1999).

[14]H. Yahiro, Y. Baba, K. Eguchi, and H. Arai, *J. Electrochem. Soc.*, **135**, 2077 (1988).

[15]E. D. Wachsman, P. Jayaweera, N. Jiang, D. M. Lowe, and B.G. Pound, "Stable High Conductivity Ceria/Bismuth Oxide Bilayered Electrolytes," *Journal of the Electrochemical Society*, **144-1**, 233-236 (1997).

[16]K. Krist and J.D. Wright, "Fabrication Methods for Reduced Temperature Solid Oxide Fuel Cells," Proc. SOFC III, Electrochem. Soc., **93-4**, 782-791 (1993).

ACKNOWLEDGMENT

I would like to thank the US Department of Energy (Contract # DE-AC26-99FT40712) for supporting this work.

16

Designing with Piezoelectric Devices

K. Uchino

DESIGNING WITH PIEZOELECTRIC DEVICES

Kenji Uchino
International Center for Actuators and Transducers, Materials Research Institute
The Pennsylvania State University, University Park, PA 16802

ABSTRACT

The performance of piezoelectric ceramic devices is dependent on complex factors, which are divided into three major categories: the properties of the ceramic itself, and coupled issues with the device design and drive technique. This paper reviews their designing issues from these viewpoints. Optimization of the piezo-ceramic composition, doping, and the cut angle of a single crystal sample is necessary to enhance the induced strains under a high stroke level drive, and to stabilize temperature and external stress dependence. The device design affects considerably its durability and lifetime, as well as its improved performance. Failure detection or "health" monitoring methods of ceramic actuators will increase the reliability remarkably. Regarding drive techniques, pulse drive and ac drive require special attention; the vibration overshoot after applying a sharp-rise step/pulse voltage to the actuator causes a large tensile force and a long-term application of AC voltage generates considerable heat.

INTRODUCTION

The application area of piezoelectric ceramics has become remarkably broad particularly in actuators and transformers [1,2]. Accordingly, their design algorithm is now being established. The device design has two major roles: improvement in the device characteristics and in the reliability/lifetime. The performance of ceramic actuators is dependent on complex factors, which are divided into three major categories: the properties of the ceramic itself, and the coupled issues of the device design and drive technique. This paper considers the piezoelectric device design issues from these viewpoints for improving both performance and reliability.

MATERIALS IMPROVEMENTS

Performance

Higher strain or stress is the primary key factor to the actuator/transducer material selection. Because lead zirconate titanate (PZT) used to be obtained only in a polycrytalline phase, previous researchers used the sample geometry with the applied electric field only parallel to the spontaneous polarization (i.e., poling direction). Figure 1 shows the typical strain curves for a piezoelectric lanthanum-doped PZT (PLZT) and an electrostrictive $Pb(Mg_{1/3}Nb_{2/3})O_3$-$PbTiO_3$ (PMN) [1]. Note the maximum strain level of 0.1%.

However, after the discovery of superior characteristics in single crystals $Pb(Zn_{1/3}Nb_{2/3})O_3$-$PbTiO_3$ [3], the crystal orientation dependence of piezoelectricity has been a focus of investigation, and strain up to 1.3% has been reported [4,5]. Figure 2 demonstrates the crystal orientation dependence of the piezoelectric d_{33} constant [defined by the relation: d = (induced strain x)/(applied electric field E)] calculated for PZT 60/40, using the Devonshire phenomenological theory for a solid solution system [6]. As is shown, a maximum of the apparent piezoelectric constant can be obtained by canting the electric field direction from the spontaneous polarization (perovskite <111> axis) by 57 degree. This suggests that the [100] epitaxially grown PZT film with a rhombohedral composition should exhibit 3 times larger strain under the same electric field level, as compared with the case of the [111] film (this is the conventional configuration!). Furthermore, due to the simpler domain reversal process, the ideal square-like hysteresis loops can be expected in the polarization or strain versus electric field for the same [100] films (Fig. 3), which is very essential for designing MEMS devices.

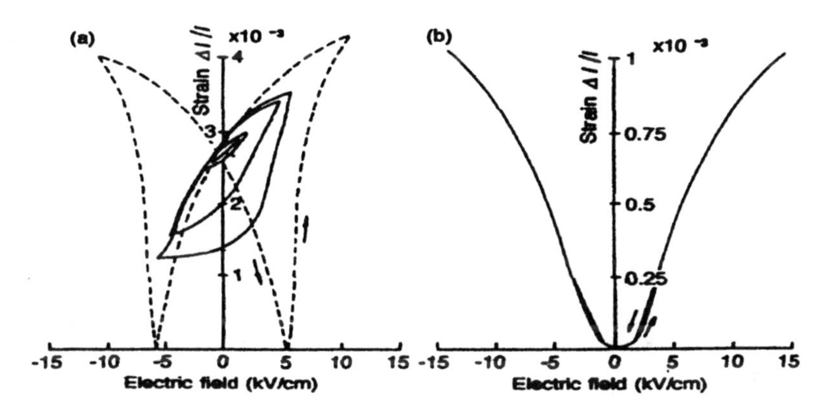

Fig.1 Field induced strain curves observed in a piezoelectric PLZT (a) and an electrostrictive PMN-PT ceramics.

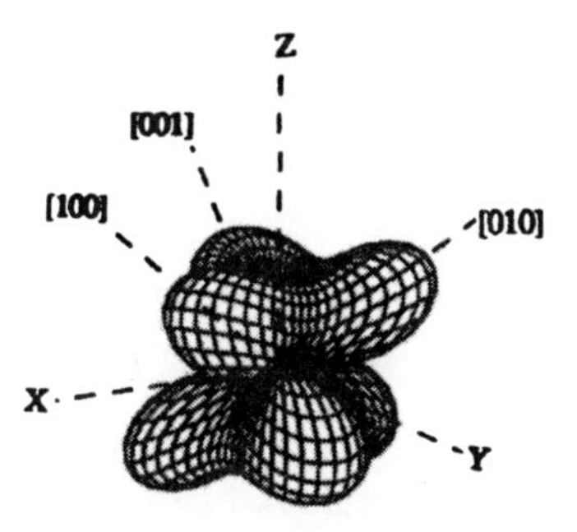

Fig.2 Crystal orientation dependence of the effective piezoelectric constant d_{33}^{eff} for a rhombohedral PZT 60/40, calculated from the Devonshire phenomenological theory.

Reliability

The reproducibility of the strain characteristics depends on grain size, porosity and impurity content, in general. Increasing the grain size enhances the magnitude of the field-induced strain, but degrades the fracture toughness and increases the hysteresis [7]. Thus, the grain size should be optimized for each application. On the other hand, porosity does not affect the strain behavior significantly. The tip deflection of unimorphs made from a $Pb(Mg_{1/3}Nb_{2/3})O_3$-based material with various sample porosities did not show variation below 8% porosity [8]. Hence, fine powders made from wet chemical processes such as coprecipitation and sol-gel will be required, and a suitable ceramic preparation process should be designed to optimize grain size and porosity.

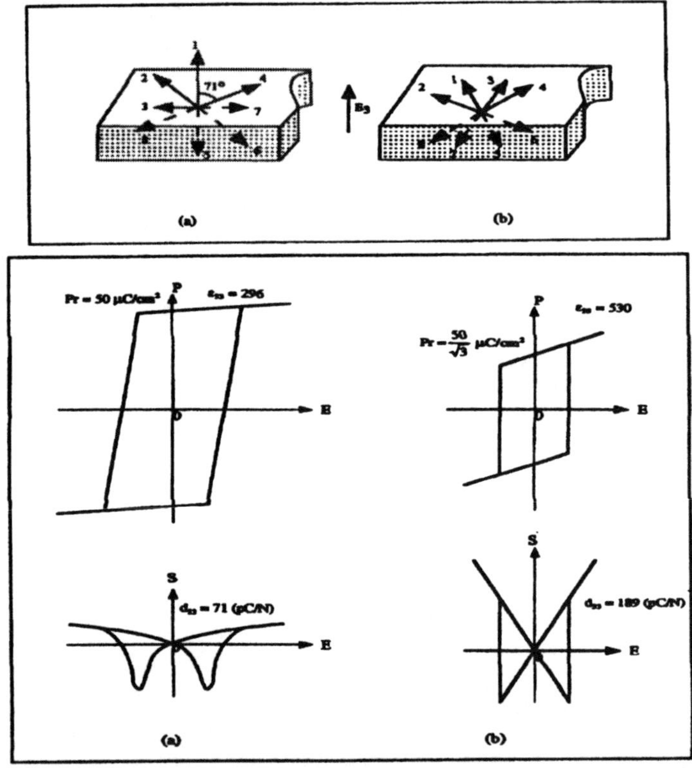

Fig.3 Possible domain states (top) and the expected polarization and strain hysteresis curves (bottom) for the [111] (left) and [100] (right) plates of a rhombohedral PZT.

The impurity (donor- or acceptor-type) level is another key material-design parameter which provides remarkable changes in strain and hysteresis/loss. Figure 4 shows dopant effects on the quasi-static field-induced strain in $(Pb_{0.73}\ Ba_{0.27})(Zr_{0.75}Ti_{0.25})O_3$ [9]. Since donor (valence > +4) doping provides "soft" characteristics, the sample exhibits larger strains and less hysteresis when driven under a high pseudo-DC electric field (1 kV/mm).

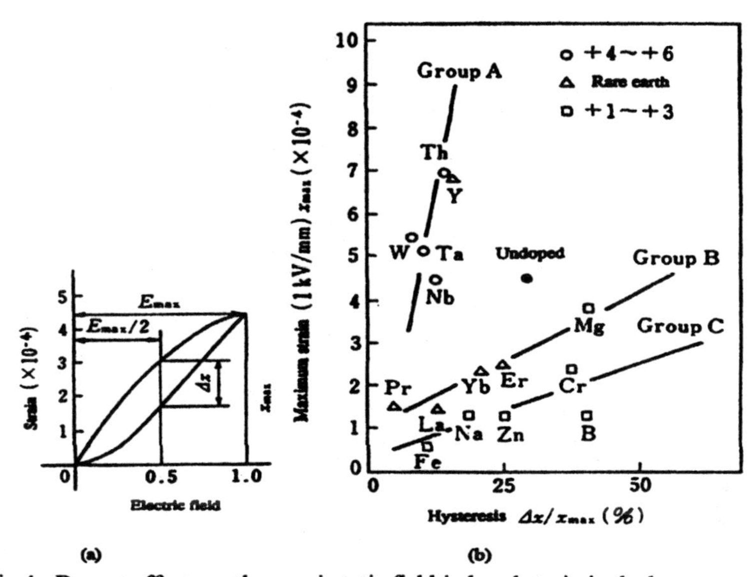

Fig.4 Dopant effects on the quasi-static field-induced strain in the base ceramic $(Pb_{0.73}Ba_{0.27})(Zr_{0.75}Ti_{0.25})O_3$. Donor (valence > +4) doping provides larger strains and less hysteresis.

In contrast, the acceptor doping provides "hard" characteristics, leading to a small hysteretic loss and a large mechanical quality factor when driven under a small AC electric field, suitable to ultrasonic motor and piezo-transformer applications. Figure 5 shows the temperature rise versus vibration velocity for undoped, Nb-doped and Fe-doped $Pb(Zr,Ti)O_3$ samples driven at a resonance mode. The suppression of heat generation is remarkable in the Fe-doped (acceptor-doped) ceramic [10].

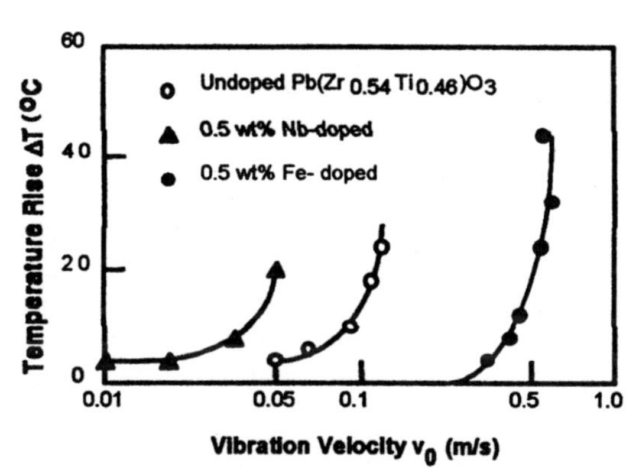

Fig.5 Temperature rise versus vibration velocity for undoped, Nb-doped and Fe-doped $Pb(Zr,Ti)O_3$ samples. Heat generation is remarkably suppressed in the Fe-doped ceramic.

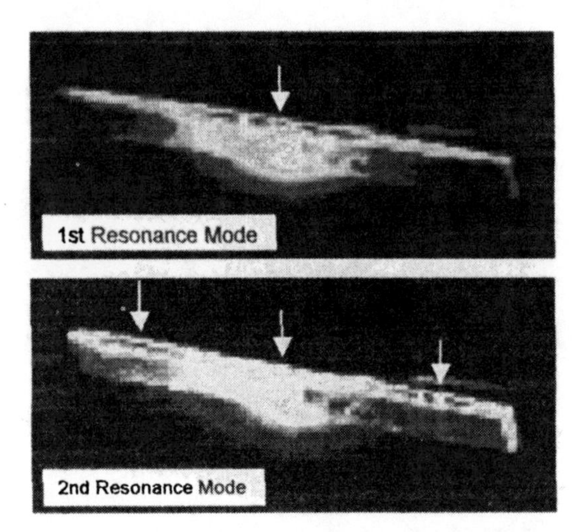

Fig.6 Temperature distribution on a rectangular PZT plate driven at the first and second extensional k_{31} resonance modes. Note the highest temperature parts by arrows.

Figure 6 shows the temperature distribution on a rectangular PZT plate driven at the first and second extensional k_{31} resonance modes. It is important to note that the maximum temperature is observed at the nodal points where the maximum stress and strain are induced when the sample is driven at the resonance mode. The major loss factor contributing to this temperature rise is considered as the (intensive) mechanical loss of the PZT.

Figure 7 shows the mechanical quality factor Q_m (inverse value of mechanical loss tan ϕ') versus composition x at two effective vibration velocities v_0=0.05 m/s and 0.5 m/s for $Pb(Zr_xTi_{1-x})O_3$ doped with 2.1 at.% of Fe [11]. The decrease in mechanical Q_m with an increase of vibration level is minimum around the rhombohedral-tetragonal morphotropic phase boundary (52/48). Thus, the morphotropic boundary composition should be adopted for designing high power piezoelectric devices.

As shown above, the conventional piezo-ceramics have the limitation in the maximum vibration velocity (v_{max}), more than which all the additionally input electrical energy converts into heat, rather than into mechanical energy. The typical rms value of v_{max} for commercialized materials, defined by the temperature rise of 20 °C from room temperature, is around 0.3 m/sec for rectangular k_{31} samples [11]. $Pb(Mn,Sb)O_3$ (PMS) - lead zirconate tatanate (PZT) ceramics with v_{max} of 0.62 m/sec have been reported [12]. By doping rare-earth ions such as Yb, Eu and Ce additionally into the PMS-PZT, we recently developed high power piezoelectrics, which can exhibit v_{max} up to 0.9 m/sec. Compared with commercially available piezoelectrics, one order of magnitude higher input electrical energy and output mechanical energy can be expected from this new material without generating significant temperature rise. This demonstrates the importance of ion doping from the material design viewpoint.

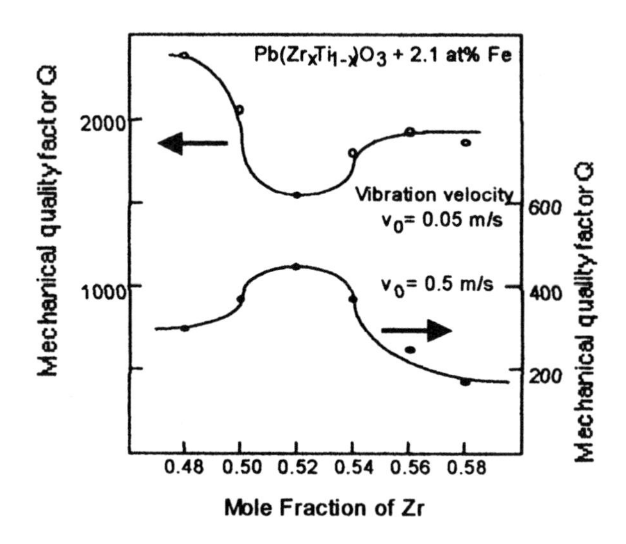

Fig.7 Mechanical Q_m versus composition x at two effective vibration velocities v_0=0.05 m/s and 0.5 m/s for $Pb(Zr_xTi_{1-x})O_3$ doped with 2.1 at.% of Fe.

The temperature dependence of the strain characteristics must be stabilized using either composite or solid solution techniques [13]. Recent new trends are found in developing high temperature actuators for engine surroundings and cryogenic actuators for laboratory equipment and space structures. Ceramic actuators are recommended to use under bias compressive stress, because the ceramic is, in general, relatively weak under externally applied tensile stress. The compressive uniaxial stress dependence of the weak-field piezoelectric constants d in various PZT ceramics indicated the significant enhancement in the d values for hard piezoelectric ceramics [14]. Systematic studies on the stress dependence of induced strains are eagerly awaited, including the composition dependence of mechanical strength, which will help in designing the stress bias mechanism for actuator application.

DEVICE IMPROVEMENTS

Performance

One of the significant drawbacks in piezoelectrics is their small strains (only about 0.1%). Thus, significant effort has been devoted to amplify the displacement. There are two categories of devices: amplification with respect to space and to time. The first is demonstrated by bimorphs and moonies/cymbals, and the latter is exemplified by inchworms and ultrasonic motors.

Moonies/cymbals: Two of the most popular actuator designs are multilayers and bimorphs [15] (see Fig. 8). The multilayer, in which roughly 100 thin piezoelectric/electro-strictive ceramic sheets are stacked together, has the advantages of low driving voltage (100 V), quick response (10 μs), high generative force (1000 N), and high electromechanical coupling. But the displacement, on the order of 10 μm, is not sufficient for some applications. This contrasts with the characteristics of the bimorph which consists of multiple piezoelectric and elastic plates bonded together to generate a large bending displacement of several hundred μm, but has relatively slow response time (1 ms) and low generative force (1 N).

A composite actuator structure called the "moonie" (or "cymbal") has been designed to provide characteristics intermediate between the multilayer and bimorph actuators; this transducer exhibits an order of magnitude larger displacement than the multilayer, and much larger generative force with quicker response than the bimorph [16]. The device consists of a thin multilayer piezoelectric element (or a single disk) and two metal plates with narrow moon-shaped cavities bonded together as shown in Fig. 8. The moonie with a size of 5 x 5 x 2.5 mm^3 can generate a 20 μm displacement under 60 V, eight times as large as the generative displacement produced by a multilayer of the same size [17]. This new compact actuator has been utilized in a miniaturized laser beam scanner.

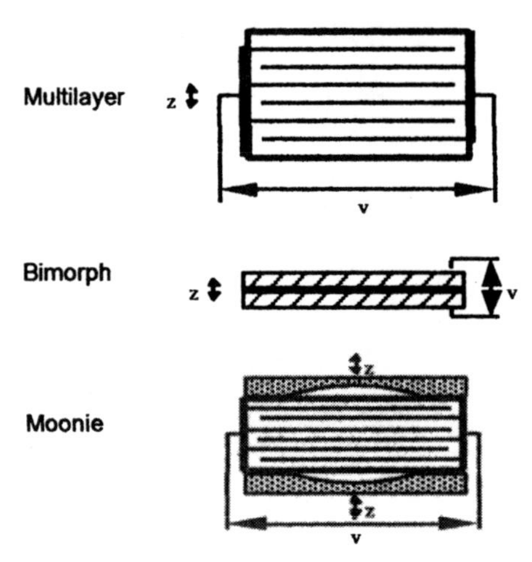

Fig. 8 Typical designs for ceramic actuators: multilayer, bimorph and moonie.

Compact Ultrasonic Motors: Figure 9 shows the famous Sashida motor [18]. By means of the traveling elastic wave induced by a thin piezoelectric ring, a ring-type slider in contact with the "rippled" surface of the elastic body bonded onto the piezoelectric is driven in both directions by exchanging the sine and cosine voltage inputs. Even though the ripple displacement is not large, the repetition of displacement at a very high frequency (15-200 kHz) provides relatively high speed to the slider. The PZT piezoelectric ring is divided into 16 positively and negatively poled regions and two asymmetric electrode gap regions so as to generate a 9th bending mode propagating wave at 44 kHz. A prototype was composed of a brass ring of 60 mm in outer diameter, 45 mm in inner diameter and 2.5 mm in thickness, bonded onto a PZT ceramic ring of 0.5 mm in thickness with divided electrodes on the back-side. The rotor was made of a polymer coated with hard rubber or polyurethane.

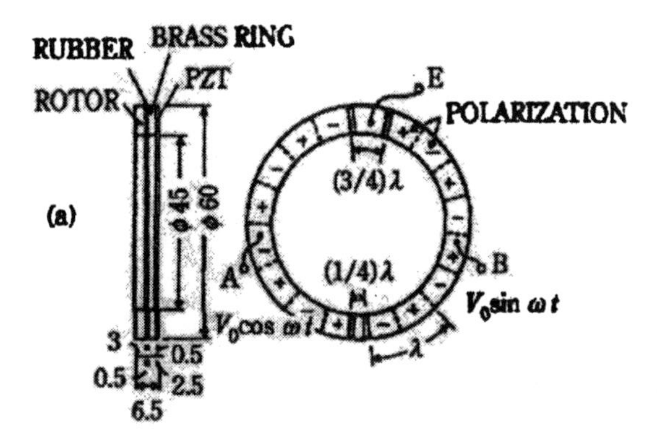

Fig. 9 Stator structure of Sashida's motor.

Canon utilized the "surfing" motor for a camera automatic focusing mechanism, installing this ring-type motor compactly in the lens frame. Using basically the same principle, Seiko Instruments miniaturized the ultrasonic motor to a diameter as small as 10 mm [19]. A driving voltage of 3 V provides torque of 0.1 mN·m. Seiko installed this tiny motor into a wrist watch as a silent alarm. AlliedSignal developed ultrasonic motors similar to Shinsei's, which would be utilized as mechanical switches for launching missiles [20].

A significant problem in miniaturizing this sort of traveling wave motor can be found in the ceramic manufacturing process; without providing a sufficient buffer gap between the adjacent electrodes, the electrical poling process (upward and downward) easily initiates a crack on the electrode gap due to the residual stress concentration. This may restrict the further miniaturization of traveling wave type motors. In contrast, standing wave type motors, the structure of which is less complicated, are more suitable for miniaturization as we will discuss in the following. They require only one uniformly poled piezo-element, less electric lead wires and one power supply. Another problem encountered in these traveling wave type motors is the support of the stator. In the case of a standing wave motor, the nodal points or lines are generally supported; this causes minimum effects on the resonance vibration. To the contrary, a traveling wave does not have such steady nodal points or lines. Thus, special considerations are necessary. In Fig. 9, the stator is basically fixed very gently along the axial direction through felt so as not to suppress the bending vibration.

We adopted the following concepts for designing new compact ultrasonic motors: (a) to simplify the structure and reduce the number of component, (b) to use simple (i.e., uniform) poling configuration, and (c) to use the standing-wave type for reducing the number of drive circuit components.

A *Windmill* motor design with basically a flat and wide configuration, using a metal-ceramic composite structure is a good example to explain [21,22]. The motor is composed of four components: stator, rotor, ball-bearing and housing unit [Fig. 10(a)]. The piezoelectric part has a simple structure of a ring electroded on its top and bottom surfaces (ϕ 3.0mm) poled uniformly in the thickness direction. The metal ring machined by Electric Discharge Machining has four inward arms placed 90° apart on its inner circumference. The metal and piezoelectric rings are bonded together, but the arms remain free; they thus behave like cantilever beams [Fig. 10(b)]. The length and cross-sectional area of each arm were selected such that the resonance frequency of the second bending mode of the arms is close to the resonance frequency of the radial mode of the stator. The rotor is placed at the center of the stator and rotates when an electric field is applied at a frequency between the radial and bending resonance modes. The truncated cone shape at the rotor end guarantees a permanent contact with the tips of the arms.

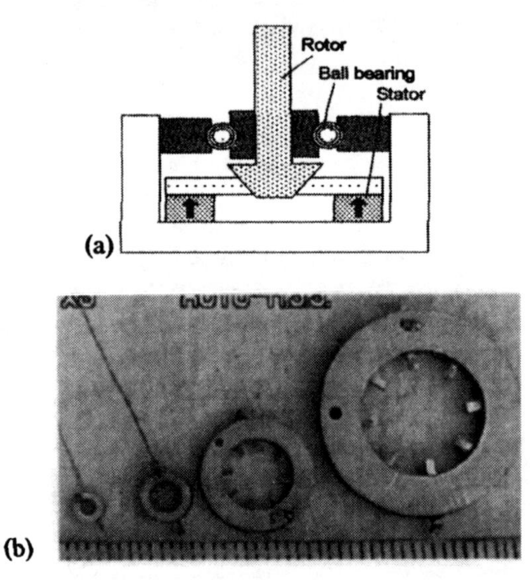

Fig. 10 "Windmill" motor using a metal coupler with multiple inward arms. (a) Cross sectional view, and (b) photos of various size stators (3-20 mmϕ).

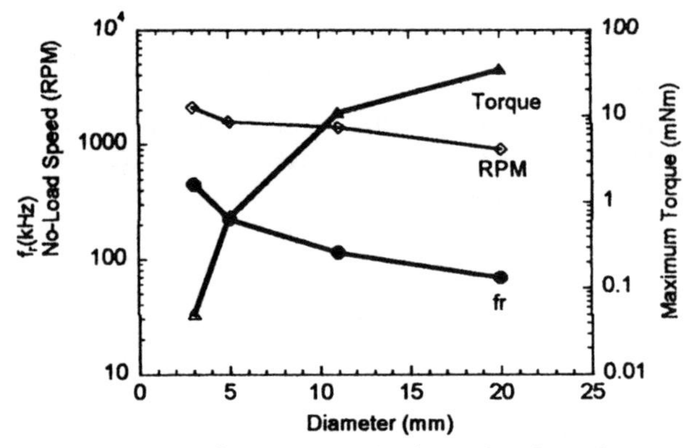

Fig.11 Radial mode resonance frequency, no-load speed and starting torque vs. diameter of the stator, measured at 15.7 V.

The operating principle of this motor is as follows: in the contraction cycle of the stator, the four arms at the center of the metal ring clamp the rotor and push it in the tangential direction. Since the radial mode frequency of the stator is close to the second bending mode frequency of the arms, the respective deformations are added and the tips of the arms bend down. In the expansion cycle, the arms release the rotor from a different path such that their tips describe an elliptical trajectory on the surface of the rotor. This motion seems to be a human finger's grasping-and-rotating action.

Figure 11 shows the size dependence of the motor characteristics. This sort of scaling principle is useful for designing the ultrasonic motors. When driven at 160 kHz, the maximum revolution 2000 rpm and the maximum torque 0.8 mNm were obtained for a 5 mm ϕ motor [22].

Another example is a *metal tube* type with a thin and long configuration, as shown in Fig. 12(a) [23]. We utilized a metal hollow cylinder, bonded with two PZT rectangular plates uniformly poled. Both can be easily found/prepared and are inexpensive. When we drive one of the PZT plates, Plate X, a bending vibration is excited along x axis. However, because of an asymmetrical mass (Plate Y), another hybridized bending mode is excited with some phase lag along the y axis, leading to an elliptical locus in a clockwise direction. On the other hand, when Plate Y is driven, a counterclockwise wobble motion is excited. Also note that only a single-phase power supply is required. The motion is analogous to a "dish-spinning" performance, with two rotors made to contact the wobbling tube ends to achieve rotation [Fig. 12(b)].

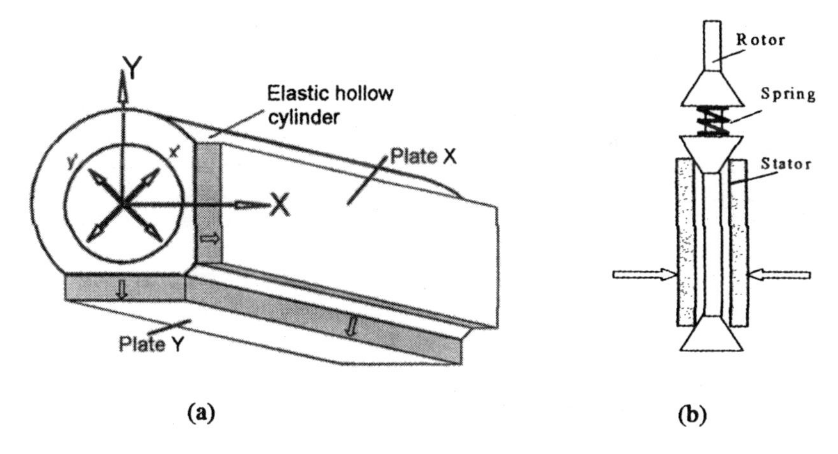

Fig.12 (a) Structure of a metal tube stator, and (b) assembly of the motor.

Finite Element Analysis (FEA) is a very useful tool to obtain realistic vibration modes in a stator. We used an ATILA software code (Magsoft Corporation) for the metal tube stator calculation. The result is shown in Fig. 13, where (b) shows the deformation of the tube slightly after the time for (a), when only Y plate is excited (X plate is electrically short-circuited in the calculation). These figures clearly show a counterclockwise rotation of the metal tube.

The rotor of this motor was a brass cylindrical rod with a pair of stainless ferrule pressed with a spring. The assembly is shown in Fig. 12(b). The no-load speed in the clockwise and counterclockwise directions as a function of input rms voltage was measured for a motor with 2.4 mm ϕ in diameter and 12 mm in length. The motor was driven at 62.1 kHz for both directions, just by exchanging the drive PZT plate. The no-load speed of 1800 rpm and the output torque of more than 1.0 mNm were obtained at 80 V for both directions. This significantly high torque was obtained due the dual stator configuration and the high pressing force between the stator and rotors made of metal.

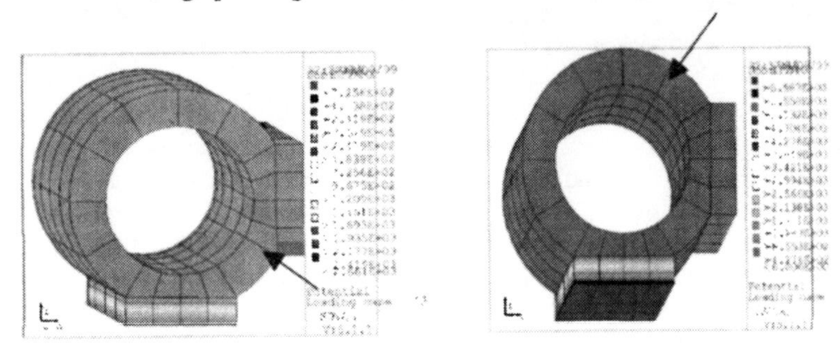

(a) At one time (b) Slightly later

Fig.13 FEA calculation results for a metal tube stator, when only Y plate (the bottom plate) is excited (Y plate is short-circuited). Notice the counterclockwise rotation.

MEMS devices: PZT thin films are deposited on a silicon membrane, which can be micro-machined as total for fabricating micro actuators and sensors, i.e., micro electromechanical systems. Figure 14 illustrates a blood tester developed in collaboration with OMRON Corporation in Japan. Applying a voltage to two surface interdigital electrodes, the surface PZT film generates surface membrane waves, which soak up blood and the test chemical from the two inlets, then mix them in the center part, and send the mixture to the monitor part through the outlet. The FEA calculation was conducted to evaluate the flow rate of the liquid by changing the thickness of the PZT or the Si membrane, inlet and outlet nozzle size, cavity thickness etc.

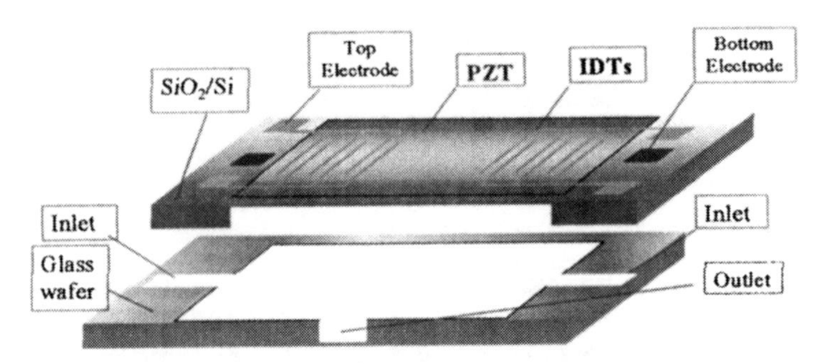

Fig.14 Structure of a PZT/silicon MEMS device, blood tester.

Reliability

Silver electrodes have a serious problem of migration under a high electric field and high humidity. This problem can be overcome with use of a silver-palladium alloy (or more expensive Pt). To achieve inexpensive ceramic actuators, we need to introduce Cu or Ni electrodes, which require a sintering temperature as low as 900°C. Low temperature sinterable actuator ceramic is a target of research.

Delamination of the electrode layer is another problem in multilayers as well as bimorphs. To enhance adhesion, composite electrode materials with metal and ceramic powder colloid, ceramic electrodes, and electrode configurations with via holes are recommended [24]. Internal stress concentration during electric field application sometimes initiates the crack in the actuator device. We calculate the stress concentration for a multilayer (8 layers) type piezo-actuator (Fig. 15) with an ATILA software code. Large tensile stress (x-x) concentration (which may cause a crack) can be clearly observed just outside of the internal electrode edge. In principle, we need to design the electrode pattern so as not to generate a maximum tensile stress more than the critical ceramic mechanical strength. From this sense, several electrode configurations have been proposed, as shown in Fig. 16: plate-through type, slit-insert type, and float-electrode-insert type [25]. These new three types can enhance the actuator lifetime. However, the plate-through type requires a special expensive task to make an external insulative coating, and the maintenance of very thin slits is a major drawback for the slit-insert type. The reason why the lifetime is extended with decreasing layer thickness has not yet been clarified.

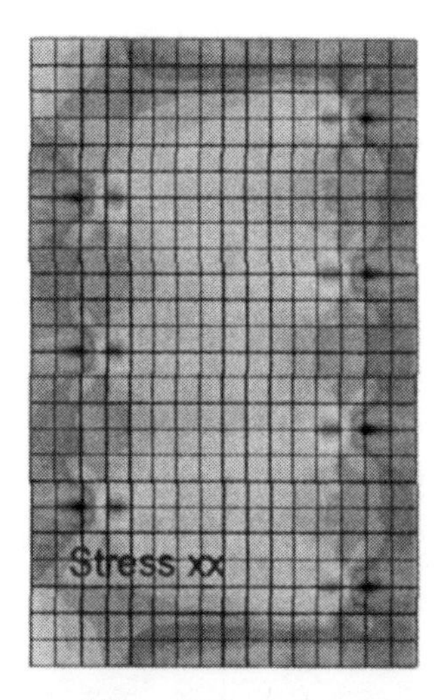

Fig. 15 Stress concentration for a multilayer (8 layers) type piezo-actuator calculated with an ATILA software code. Notice the largest tensile and compressive stresses at just outside and inside of the internal electrode edge.

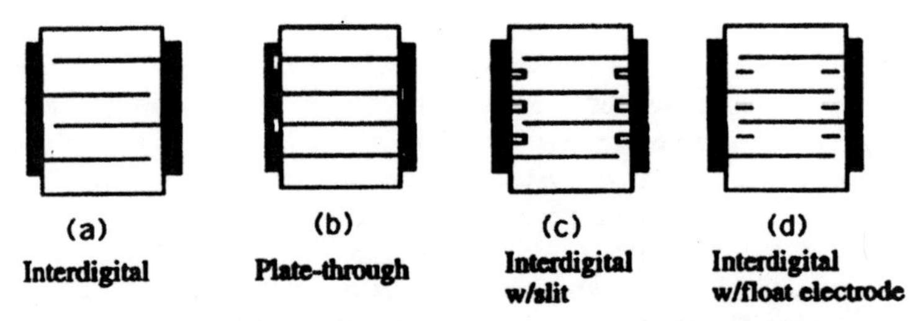

Fig.16 New electrode patterns to suppress the internal stress concentration: plate-through type (b), slit-insert type (c), and float-electrode-insert (d) types.

A similar internal stress consideration is required when we design piezoelectric transformers. The initial Rosen type (a rectangular PZT plate design as shown in Fig. 17) had a reliability problem; that is, easy mechanical breakdown at the center portion due to the coincidence of the residual stress concentration (through the poling process) and the vibration nodal point (highest induced stress). In addition to improved mechanically tough ceramic materials, by using a multilayer type without generating any poling direction mismatch (Philips Components, NEC [26]), or by redesigning the electrode configuration and exciting a 3rd longitudinal vibration mode of the rectangular plate (NEC) [27], the piezo-transformers shown in Fig. 17 have been commercialized as a back-light inverter for the liquid crystal displays.

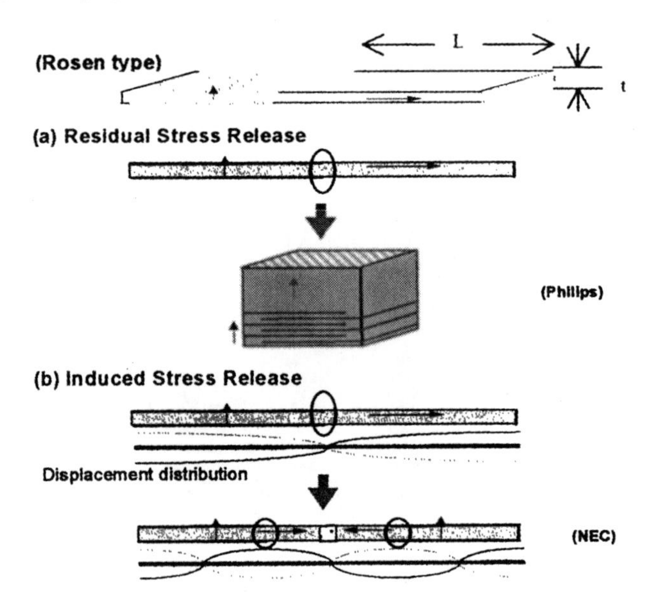

Fig.17 New types of piezoelectric transformer with a consideration of residual stress due to the polarization (a), or of induced stress (b).

Heat generation also provides additional key requirement in designing the multilayer actuator. Zheng et al. reported the heat generation from various sizes of multilayer type piezoelectric ceramic actuators [28]. Figure 18 shows the temperature change in the actuators when driven at 3 kV/mm and 300 Hz, and Fig. 19 plots the saturated temperature as a function of V_e/A, where V_e is the effective volume (electrode overlapped part) and A is the surface area. This linear relation is reasonable because the volume V_e generates the heat and this heat is dissipated through the area A. Thus, if we need to suppress the temperature rise, a small V_e/A design is preferred. For example, a hollow cylinder actuator is preferred to a solid rod actuator, when we consider the heat suppression. It is notable that in this off-resonance case, the loss contributing to the heat generation is not the intensive mechanical loss like in the resonance, but the intensive dielectric loss (i.e., P-E hysteresis loss) [28].

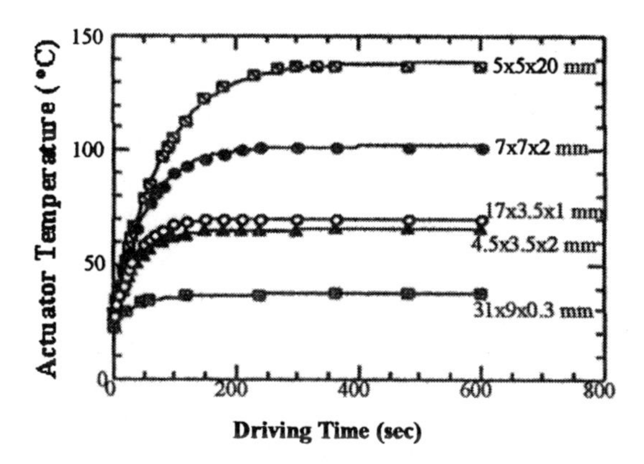

Fig.18 Temperature rise for various actuators while driven at 300 Hz and 3 kV/mm.

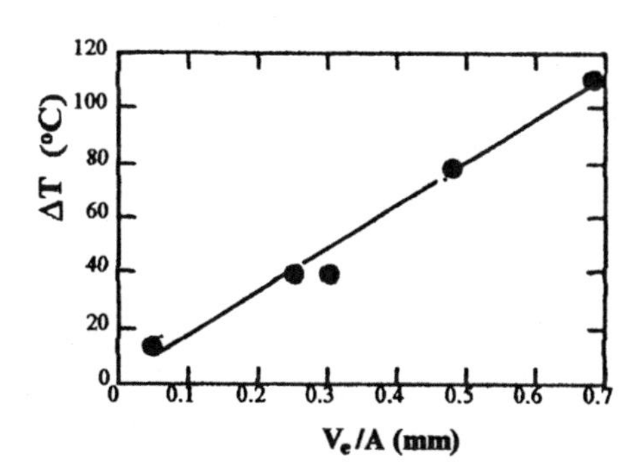

Fig.19 Temperature rise versus V_e/A (3 kV/mm, 300 Hz), where V_e is the effective volume generating the heat and A is the surface area dissipating the heat.

Although the aging effect is very important, not many investigations have been done so far. The aging effect arises from two factors: depoling and destruction. Creep and zero-point drift of the displacement are caused by the depoling of the ceramic. Another serious degradation of the strain is produced by a very high electric field under an elevated temperature, humidity and mechanical stress. Change in lifetime of a multilayer piezoelectric actuator with temperature and DC bias voltage has been reported by Nagata [29]. The lifetime t_{DC} under DC bias field E obeys an empirical rule:

$$t_{DC} = A\,E^{-n} \exp(W_{DC}/\,kT), \qquad (1)$$

where n is about 3 and W_{DC} is an activation energy ranging from 0.99 - 1.04 eV. This is another important issue in designing the device.

Lifetime prediction or "health" monitoring systems have been proposed using failure detection techniques [30]. Figure 20 shows such an "intelligent" actuator system with acoustic emission (AE) monitoring. The actuator is controlled by two feedback mechanisms: position feedback, which can compensate the position drift and the hysteresis, and breakdown detection feedback which can stop the actuator system safely without causing any serious damages to the work, e.g. in a lathe machine. Acoustic emission measurements of a piezo-actuator under a cyclic electric field provides a good predictor for lifetime. AE is detected largely when a crack propagates in the ceramic actuator at the maximum speed. During normal drive of a 100-layer piezoelectric actuator, the number of AE was counted and a drastic increase by three orders of magnitude was detected just before complete destruction. Note that part of the piezo-device can be utilized as an AE sensor.

A recent new electrode configuration with a strain gauge type (Fig. 21) is another intriguing alternative for health monitoring [31]. By measuring the resistance of the strain-gauge shaped electrode embedded in the ceramic actuator, we can monitor both electric-field induced strain and the symptom of cracks in the ceramic.

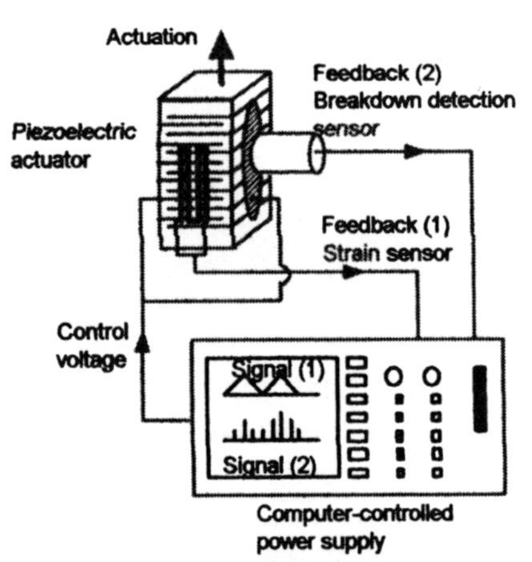

Fig.20 Intelligent actuator system with both position feedback and breakdown detection feedback mechanisms.

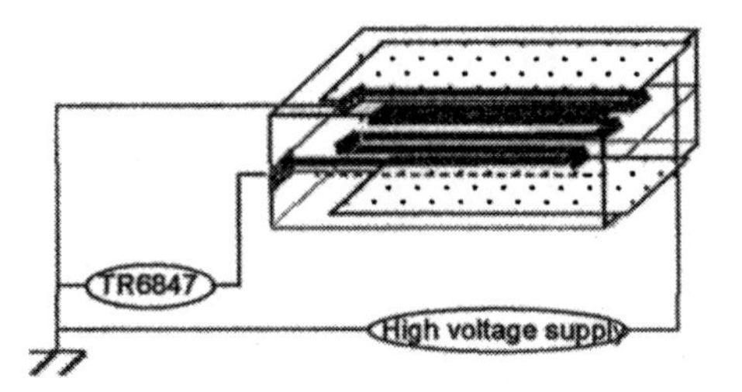

Fig.21 Strain gauge configuration of the internal electrode for an intelligent actuator.

DRIVE/CONTROL TECHNIQUES

Performance

Piezoelectric/electrostrictive actuators are classified into two categories, based on the type of driving voltage applied to the device and the nature of the strain induced by the voltage (Fig. 22): (1) rigid displacement devices for which the strain is induced unidirectionally along the direction of the applied DC field, and (2) resonating displacement devices for which the alternating strain is excited by an AC field at the mechanical resonance frequency (ultrasonic motors). The first can be further divided into two types: servo displacement transducers (positioners) controlled by a feedback system through a position-detection signal, and pulse drive motors operated in a simple on/off switching mode, exemplified by dot-matrix printers.

The material requirements for these classes of devices are somewhat different, and certain compounds will be better suited to particular applications. The ultrasonic motor, for instance, requires a very hard piezoelectric with a high mechanical quality factor Q_m, to suppress heat generation. The servo displacement transducer suffers most from strain hysteresis and, therefore, a PMN electrostrictor is used for this purpose. The pulse drive motor requires a low permittivity material aimed at quick response with a certain power supply rather than a small hysteresis, so soft PZT piezoelectrics are preferred rather than the high-permittivity PMN for this application.

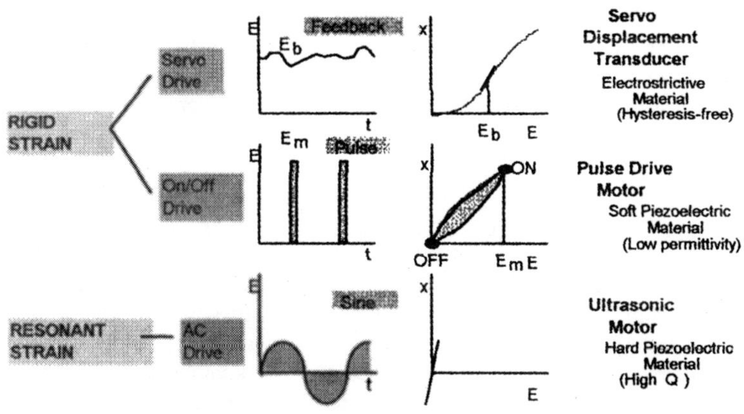

Fig.22 Classification of piezoelectric/electrostrictive actuators.

Recently, we proposed a new application of a piezoelectric transformer, that is, driving piezoelectric actuators. Because of advantages of the piezo-transformer to the conventional electromagnetic types in (a) small size/weight, (b) high efficiency, (c) no magnetic noise generation, and (d) non-flammable structure, the piezoelectric types will expand the application area rapidly, including the piezoelectric actuators. Notice that because most of the piezo-actuators are driven under 100-1000V, a transformer is inevitably required when we use a battery driven circuitry. The basic concept is illustrated in Fig. 23. When we tune the transformer operating frequency exactly to the same frequency of an ultrasonic motor's resonance (for example, designing a ring transformer, using the same size piezo-ring as the ultrasonic motor), we may use it as a drive system, or fabricate a transformer-integrated motor. When we couple a rectifier with the transformer, we can drive a multilayer or bimorph piezoelectric actuator.

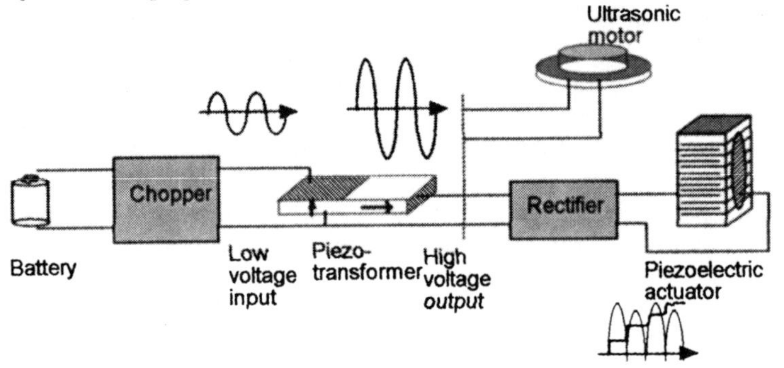

Fig.23 Basic concept of piezo-actuator drive systems (AC or pseudo DC) using a piezoelectric transformer.

In the following, we introduce a compact drive system designing with piezoelectric transformers for a multilayer piezoelectric actuator, aiming at active vibration control on a helicopter. In this sort of military application, we need to realize a compact, light-weight and electromagnetic-noise free system with keeping quick response (minimum 200Hz). For this sake, we have chosen a multilayer piezoelectric device as an actuator, and piezo-transformers (rather than electromagnetic transformers) as drive system components. Figure 24 summarizes our drive system for piezoelectric actuator control, using piezoelectric transformers [32].

We have targeted to develop two kinds of power supplies with piezo-transformers driven by a helicopter battery $24V_{DC}$: one is a high voltage DC power supply (100-1000V, 90W) for driving the piezoelectric actuator, and the other is an AC adapter ($\pm 15V_{DC}$, 0.1-0.5W) for driving the supporting circuitry. We utilized large and small multi-stacked piezo-transformer elements both with an insulative glass layer between the input and output parts for ensuring the complete floating condition, for the high power supply and the adapter, respectively. The used actuator was supplied from Tokin Corp., which has the size of $10 \times 10 \times 20 mm^3$, and is capable to generate a 16μm displacement under the maximum voltage of 100V.

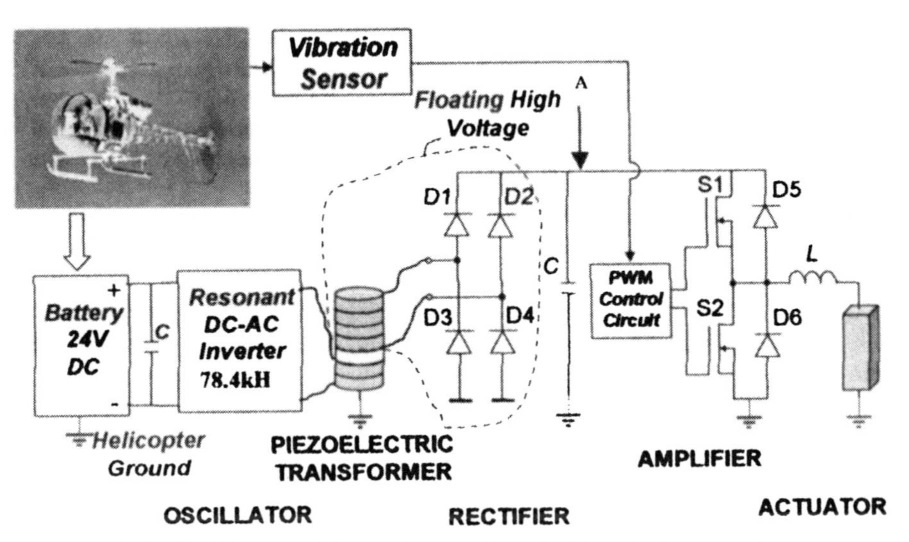

Fig. 24 Compact drive system for piezoelectric actuator control.

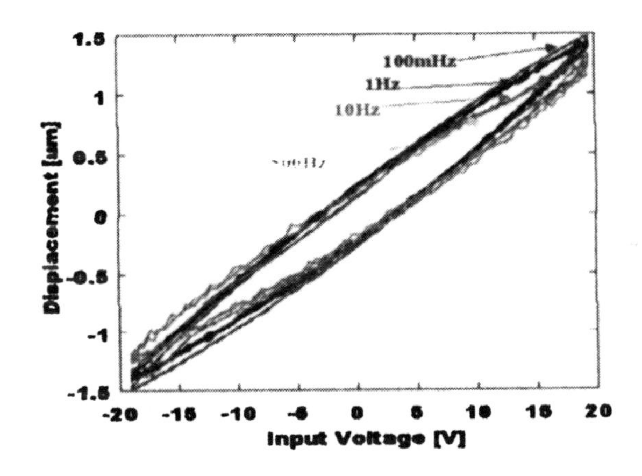

Fig.25 Displacement curves of the piezoelectric actuator driven by the newly developed power amplifier with piezoelectric transformers.

We initially tried to drive the actuator by applying the rectified high voltage directly after the power transformer (i.e., voltage at the point A in Fig. 24). However, because of a very slow response during a rising or falling voltage process (~1 sec), we decided to use this rectified voltage as a constant high DC voltage, and additionally use a power amplifier to control the voltage applied to the actuator. Among various power amplifier families [33], we have chosen a Class-D switching amplifier (see the inverter in Fig. 8), because it has some advantages over the other switching and linear amplifiers, such as permitting amplitude and frequency control and realizing fast actuation response of the actuator by chopping a DC voltage. The voltage level generated in the transformer was 300V for the input voltage of 75V applied through a timer 555. The signal from a PWM (pulse width modulation) driving circuit was applied to two power MOSFET's of the half bridge. Thus, the constant 300V_{DC} voltage from the piezo-transformer was chopped, and control of frequency and magnitude was achieved. This output voltage was applied to the piezo-actuator through a filtering inductance of 100mH [34]. The PWM carrier frequency was chosen as 40kHz, which is below the mechanical resonance frequency of the used piezo-actuator (~60kHz).

Figure 25 shows the displacement curves of the actuator driven by the newly developed power amplifier. The displacement was directly measured with an eddy current sensor. As seen in this figure, the displacement ±1.5μm was controlled by ±20V applied voltage. This drive system could be used at least up to 500Hz, which is sufficient for the active vibration control on a helicopter. Note that the new drive system is compact, light-weight by 1/10 in comparison with conventional electromagnetic transformer circuits, and magnetic-noiseless.

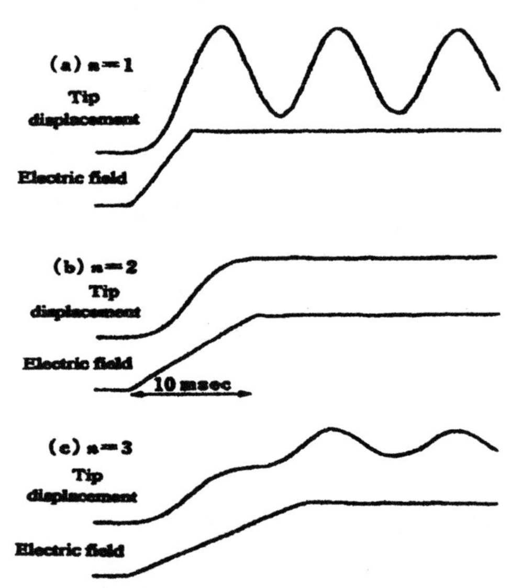

Fig. 26 Transient vibration of a bimorph excited after a pseudo-step voltage applied. n is a time scale with a unit of half of the resonance period, i.e., 2n = resonance period.

Reliability

Pulse Drive: Pulse drive of the piezoelectric / electrostrictive actuator generates very large tensile stress in the device, sometimes large enough to initiate cracks. In such cases, compressive bias stress should be employed on the device through clamping mechanisms such as a helical spring and a plate spring. Figure 26 shows another solution to suppress this problem; that is, suppression of the transient vibrations of a bimorph by choosing a suitable pseudo-step voltage applied. In Fig. 26, the rise time is varied around the resonance period (n is the time scale with a unit of $T_0/2$, where T_0 stands for the resonance period). It is concluded that the overshoot and ringing of the tip displacement is completely suppressed when the rise time is precisely adjusted to the resonance period of the piezo-device (i.e., for n = 2) [35].

Temperature rise is occasionally observed, particularly when the actuator is driven cyclically in the pulse drive. The rise time adjustment is also very important from the heat generation viewpoint. In consideration of the pulse drive using a trapezoidal wave, like an application to diesel engine injection systems, we examined the temperature rise. The trapezoidal wave with a maximum voltage 100 V, 60 Hz, and a duty ratio 50 % was applied to a commercialized multilayer actuator (product by Ceramtech). The results are summarized in Table I. The temperature rise is surprisingly dependent on the step rise time, which is basically due to the actuator's vibration overshoot and ringing. As shown in Fig. 27, for the shorter rise time, the actuator shows significant vibration ringing, which provides the additional heat generation through a sort of mechanical resonance. Depending on the degree of the overshoot, the temperature rise was increased by more than 50 % than in the case without any vibration ringing (this corresponds to roughly the off-resonance condition). Thus, when the piezoelectric actuator is used under a pulse drive condition, in addition to the off-resonance heat generation due to the intensive dielectric (P-E hysteresis) loss, the resonance type loss due to the intensive mechanical loss is superposed. Care must be taken for reducing the mechanical vibration overshoot and ringing by changing the step rise time, in order to suppress the heat generation minimum. The rise time should be chosen as a resonance period of the actuator.

Table I. Heat generation of a commercial actuator (Ceramtech) under various rise time driving (trapezoidal waveform, 50% duty cycle, 60 Hz, 100V unipolar voltage).

Rise time (μs)	72	85	97	102	146	240	800
Temperature rise (C)	23.7	21.6	20.5	20.2	20.3	18.6	16.6
V_{rms} (V)	72.9	72.8	72.7	72.7	72.6	72.1	71.3

Antiresonance Drive: Regarding ultrasonic motors, the usage of the antiresonance mode has been proposed [36]. Quality factor Q and temperature rise have been investigated on a PZT ceramic rectangular bar, and the results for the fundamental resonance (A-type) and antiresonance (B-type) modes are illustrated in Fig. 28 as a function of vibration velocity. It is recognized that Q_B is higher than Q_A over the whole vibration velocity range. In other words, the antiresonance mode can provide the same mechanical vibration level without generating heat. In a typical piezoelectric material with k_{31} around 30 %, the plate edge is not a vibration nodal point and can generate a large vibration velocity.

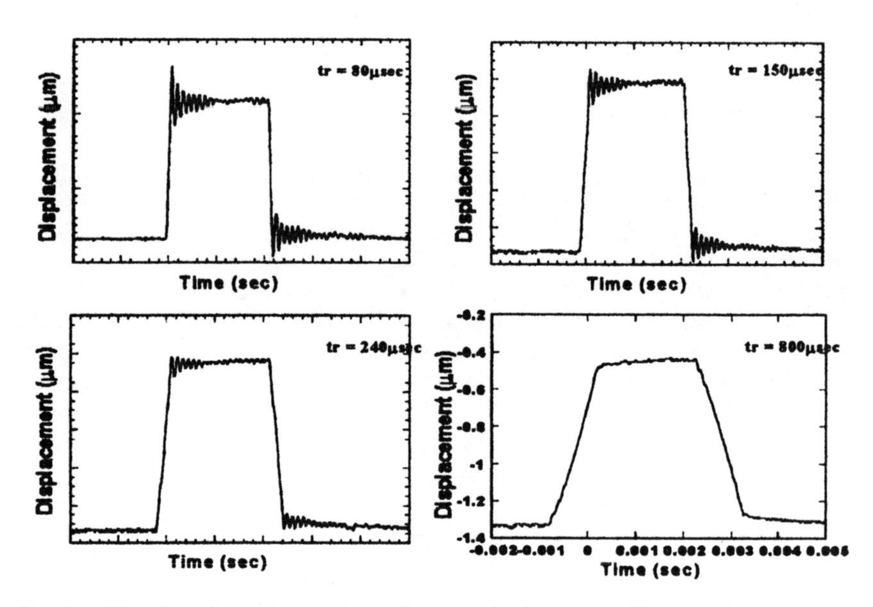

Fig. 27 Rise time dependence of the vibration ringing for the Ceramtech multilayer actuator driven under 100 V (unipolar) trapezoid waveform at frequency of 60 Hz.

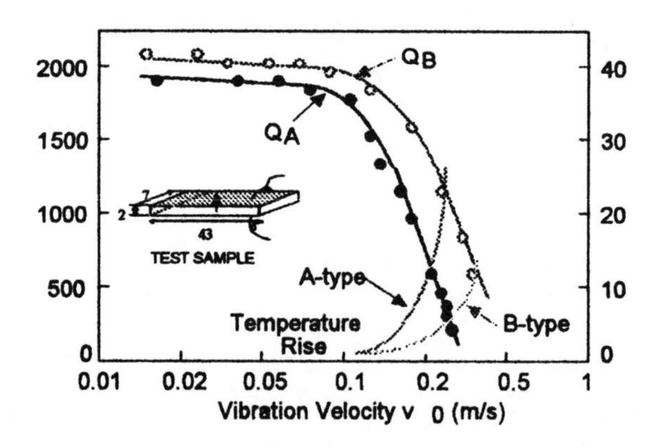

Fig.28 Vibration velocity dependence of the quality factor Q and temperature rise for both A (resonance) and B (antiresonance) type resonances of a longitudinally vibrating PZT ceramic transducer through transverse piezoelectric effect d_{31}.

All the previous ultrasonic motors have utilized the mechanical resonance mode at the so-called "resonance" frequency. However, the mechanical resonant mode at the "antiresonance" frequency reveals higher Q and efficiency than the "resonance" state.

Moreover, the usage of "antiresonance," whose admittance is very low, requires low current and high voltage for driving, in contrast to high current and low voltage for the "resonance." This means that a conventional inexpensive power supply may be utilized for driving the ultrasonic motor.

SUMMARY AND FUTURE

We discussed the design issues of piezoelectric devices, particularly of piezoelectric actuators. The design is considered from two aspects: improvement of performance and reliability. Also because the actuator is really an interdisciplinary device, we need to consider materials, device designs and drive/control techniques in total.

As an example, we will present a typical flow of an ultrasonic motor design.

Materials:(1) Using a phenomenological theory for a solid solution system, the maximum electromechanical coupling factor k and d composition is derived in the PZT system (e.g., around 52-48 morphotropic phase boundary).

(2) When the sample is a single crystal, the optimum crystal cut angle should be selected (e.g., [001] plates for a rhombohedral composition).

(3) Doping an acceptor-type ion, a larger Q material is seeked.

(4) In order to increase the fracture toughness, smaller grain samples are prepared, designing the ceramic fabrication process.

Devices: (5) In order to increase the lifetime, a simple motor design with uniformly poled PZT specimen is considered.

(6) The vibration modal analysis will be conducted using an FEM program (e.g., ATILA software code).

(7) In order to suppress the heat generation, the surface area of the device should be increased, or a air circulation should be considered.

(8) Cost-efficient fabrication process is considered. Particular focus is put on supporting techniques such as resin and frictional coating materials.

Drive/Control Systems:

(9) Since the ultrasonic motor is pure capacitive, we need to design an energy efficient drive circuit (e.g., the application of piezoelectric transformers is one of a promising approach).

(10) The antiresonance mode application (rather than the conventional resonance mode) may provide a breakthough in developing drive/control systems for high power ultrasonic motors.

High power piezoelectric actuators and transformers are very promising 21st Century technologies. In order to expand the application fields, we need to establish the designing principle of these devices as soon as possible, as well as the standardization of the evaluation methods.

CHAPTER PROBLEMS

1. A piezoelectric multilayer actuator under a certain applied voltage, V, will exhibit an amplified displacement as compared with the displacement generated in a single disk of the active material. An even more pronounced displacement amplification is expected from an electrostrictive device. Verify this using the following equations for the piezoelectric and electrostriction effects: $x = d\,E$ and $x = M\,E^2$.

2. Using a PZT-based ceramic with a piezoelectric strain coefficient of d_{31}=(-300) pC/N, design a shim-less bimorph with a total length of 30 mm (where 5 mm is used for cantilever clamping) which can produce a tip displacement of 40 μm under an applied voltage of 20 V. Calculate the response speed of this bimorph. The mass density of the ceramic is ρ=7.9 g/cm³ and its elastic compliance is $s_{11}^E=16\times10^{-12}$ m²/N.
3. A unimorph bending actuator can be fabricated by bonding a piezoceramic plate to a metallic shim. The tip deflection, δ, of the unimorph supported in a cantilever configuration is given by:

$$\delta = \frac{d_{31} E l^2 Y_c t_c}{(Y_m [t_o^2 - (t_o - t_m)^2] + Y_c [(t_o + t_c)^2 - t_o^2])}$$

Here E is the electric field applied to the piezoelectric ceramic, d_{31}, the piezoelectric strain coefficient, l, the length of the unimorph, Y, Young's modulus for the ceramic or the metal, and t is the thickness of each material. The subscripts c and m denote the ceramic and the metal, respectively. The quantity, t_o, is the distance between the strain-free neutral plane and the bonding surface, and is defined according to the following:

$$t_o = \frac{t_c t_m^2 (3t_c + 4t_m) Y_m + t_c^4 Y_c}{6 t_c t_m (t_c + t_m) Y_m}$$

Assuming $Y_c=Y_m$, calculate the optimum (t_m/t_c) ratio that will maximize the deflection, δ under the following conditions:

(a) a fixed ceramic thickness, t_c, and

(b) a fixed total thickness, $t_c + t_m$.

4. Calculate the electromechanical coupling factor k_{ij} of a piezoelectric ceramic vibrator for the following vibration modes (see Fig. 29):

(a) Longitudinal length extension mode (∥E): k_{33}

(b) Planar extension mode of the circular plate: k_p

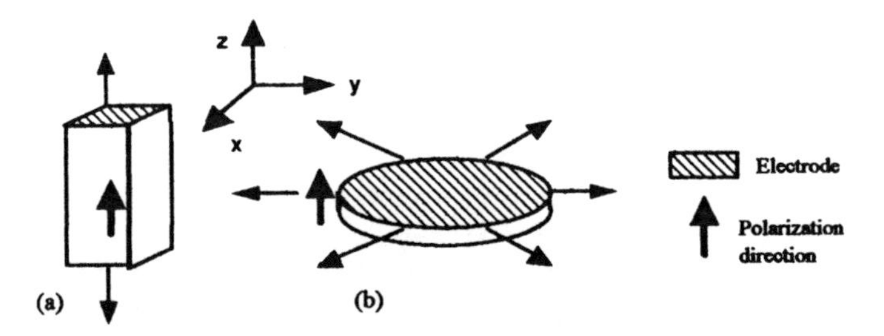

Fig.29 Two vibration modes of a piezoelectric device: (a) the longitudinal length extension and (b) the planar extensional modes.

5. Using a rectangular piezoelectric plate and the transverse d_{31} mode, consider the design of a flight actuator.

(a) Assuming a rectangular negative pulse ($-E_o$) is applied to the plate, which is installed normally and rigidly fixed at one end, verify that the velocity of sound, v, at the other end is given by $[2|d_{31}|E_o v]$, which is independent of the length.

(b) Suppose this velocity is acquired by a small steel ball of mass, M, with no loss in energy. Calculate the maximum height the steel ball will attain, when the impulse is applied entirely in the upward direction.

6. The transfer function for a piezoelectric actuator is given by:

$$U(s) = G(s)\, E^{\sim}(s)$$

$$G(s) = A\, c\, d \,/\, (M s^2 + \zeta s + Ac/l)$$

Calculate the displacement response to the step voltage: ($E^{\sim}(s) = 1/s$).

REFERENCES

[1]K.Uchino, *Ceramic Actuators and Ultrasonic Motors*, Kluwer Academic Pub., Boston (1996).

[2]K.Uchino, *Ferroelectric Devices*, Marcel Dekker, New York (2000).

[3]J. Kuwata, K. Uchino and S. Nomura, "Phase Transitions on the $Pb(Zn_{1/3}Nb_{2/3})O_3$-$PbTiO_3$ System," Ferroelectrics, **37**, 579 (1981).

[4]K. Yanagiwawa, H. Kanai and Y. Yamashita, Jpn.J.Appl. Phys., **34**, 536 (1995).

[5]S.E.Park and T.R.Shrout, "Relaxor Based Ferroelectric Single Crystals for Electromechanical Actuators," Mat. Res. Innovt., **1**, 20 (1997).

[6]X.H. Du, J. Zheng, U. Belegundu and K. Uchino, "Crystal Orientation Dependence of Piezoelectric Properties of Lead Zirconate Titanate: Theoretical Expectation for Thin Films," Jpn. J. Appl. Phys., **36**, 5580 (1997).

[7]K. Uchino and T. Takasu, "Evaluation Method of Piezoelectric Ceramics from a Viewpoint of Grain Size," Inspec., **10**, p.29, (1986).

[8]K. Abe, K. Uchino and S. Nomura, "The Electrostric-tive Unimorph for Displacement Control," Jpn. J. Appl. Phys., **21**, p.L408, (1982).

[9]A. Hagimura and K. Uchino, "Impurity Doping Effect on Electrostrictive Properties of $(Pb,Ba)(Zr,Ti)O_3$," Ferroelectrics, **93**, p.373, (1989).

[10]S. Takahashi and S. Hirose, Jpn. J. Appl. Phys., **32**, Pt.1, p.2422, (1993).

[11]K. Uchino, J. Zheng, A. Joshi, Y.H. Chen, S. Yoshikawa, S. Hirose, S. Takahashi and J.W.D. de Vries, "High Power Characterization of Piezoelectric Materials," J. Electroceramics, **2**, 33 (1998).

[12]S. Takahashi, Y. Sakaki, S. Hirose and K. Uchino, "Electro-Mechanical Properties of $PbZrO_3$-$PbTiO_3$-$Pb(Mn_{1/3}Sb_{2/3})O_3$ Ceramics under Vibration-Level Change," Mater. Res. Soc. Symp. Proc. **360**, 305, Mater. Res. Soc., Pittsburgh (1995).

[13]K. Uchino, J. Kuwata, S. Nomura, L.E. Cross and R.E. Newnham, "Interrelation of Electrostriction with Phase Transition Diffuseness," Jpn. J. Appl. Phys., **20**, Suppl.20-4, 171, (1981).

[14]Q.M. Zhang, J. Zhao, K. Uchino and J. Zheng, "Change of the Weak Field Properties of $Pb(Zr,Ti)O_3$ Piezoceramics with Compressive Uniaxial Stresses," J. Mater. Res., **12**, 226, (1997).

[15]Y. Sugawara, K. Onitsuka, S. Yoshikawa, Q.C. Xu, R.E. Newnham and K. Uchino, "Metal-Ceramic Composite Actuators," J. Amer. Ceram. Soc., **75**(4), 996 (1992).

[16]A. Dogan, K. Uchino and R.E. Newnham, "Composite Piezoelectric Transducer with Truncated Conical Endcaps "Cymbal"," IEEE Trans. UFFC, **44**, 597 (1997).

[17]H. Goto, K. Imanaka and K. Uchino, "Piezoelectric Actuators for Light Beam Scanners," Ultrasonic Techno. 5, 48 (1992).

[18]T. Sashida, Mech. Automation of Jpn., 15 (2), 31 (1983).

[19]M. Kasuga, T. Satoh, N. Tsukada, T. Yamazaki, F. Ogawa, M. Suzuki, I. Horikoshi and T. Itoh, J. Soc. Precision Eng., 57, 63 (1991).

[20]J. Cummings and D. Stutts, Amer. Ceram. Soc. Trans. "Design for Manufacturability of Ceramic Components", p.147 (1994).

[21]B. Koc, A. Dogan, Y. Xu, R. E. Newnham and K. Uchino, "An Ultrasonic Motor Using a Metal-Ceramic Composite Actuator Generating Torsional Displacement," Jpn. J. Appl. Phys. 37, 5659 (1998).

[22]B. Koc, P. Bouchilloux, and K. Uchino, "Piezoelectric Micromotor Using a Metal-Ceramic Composite Structure," IEEE Trans.-UFFC, 47, 836 (2000).

[23]B. Koc, J. F. Tressler and K. Uchino, "A Miniature Piezoelectric Rotay Motor Using Two Orthogonal Bending Modes of a Hollow Cylinder," Proc. 7th Actuator 2000, p.242-245, Axon, Bremen (2000).

[24]K. Uchino, "Materials Issues in Design and Performance of Piezoelectric Actuators: An Overview," Acta Mater. 46[11], 3745 (1998).

[25]H. Aburatani, K. Uchino, A. Furuta and Y. Fuda, "Destruction Mechanism and Destruction Detection Technique for Multilayer Ceramic Actuators," Proc. 9th Int'l Symp. Appl. Ferroelectrics, p.750 (1995).

[26]NEC, "Thickness Mode Piezoelectric Transformer," US Patent No. 5,118,982 (1992).

[27]S. Kawashima, O. Ohnishi, H. Hakamata, S. Tagami, A. Fukuoka, T. Inoue and S. Hirose, "Third order longitudinal mode piezoelectric ceramic transformer and its application to high-voltage power inverter," IEEE Int'l Ultrasonic Symp. Proc., Nov. (1994).

[28]J. Zheng, S. Takahashi, S. Yoshikawa, K. Uchino and J.W.C. de Vries, "Heat Generation in Multilayer Piezoelectric Actuators," J. Amer. Ceram. Soc. 79, 3193 (1996).

[29]K. Nagata, "Lifetime of Multilayer Actuators," Proc. 49th Solid State Actuator Study Committee, JTTAS, Tokyo, Jan. (1995).

[30]K. Uchino and H. Aburatani, "Destruction Detection Techniques for Safety Piezoelectric Actuator Systems," Proc. 2nd Int'l Conf. Intelligent Mater., p.1248, (1994).

[31]K. Uchino, "Reliability of Ceramic Actuators," Proc. Int'l Symp. Appl. Ferroelectrics, Vol. 2, 763 (1997).

[32]K. Uchino, B. Koc, P. Laoratanakul and A. Vazquez Carazo, "Piezoelectric Transformers –New Perspective–," Proc. 3rd Asian Mtg. Ferroelectrics, Dec. 12-15, Hong Kong (2000) [in press].

[33]N. O. Sokal, "RF Power Amplifier, Classes A through S," Proc. Electron. Ind. Forum of New England, p. 179-252 (1997).

[34]Motorola, "Actuator Drive and Energy Recovery System," US Patent No. 5,691,592 (1997).

[35]S. Sugiyama and K. Uchino, "Pulse Driving Method of Piezoelectric Actuators," Proc. 6th IEEE Int'l Symp. Appl. Ferroelectrics, p.637 (1986).

[36]K. Uchino and S. Hirose, "Loss Mechanisms in Piezoelectrics," IEEE UFFC Trans. 48, 307-321 (2001).

Designing a Bionic Cat

L.L. Hench

CERAMIC ENGINEERING DESIGN: DESIGNING A BIONIC CAT

Larry L. Hench
Professor of Ceramic Materials
Imperial College of Science, Technology and Medicine
Prince Consort Road, London, England, SW7 2BP

ABSTRACT

Students and engineers can best learn design principles and practise by example. This article provides an example of Professor George designing and building Boing-Boing, a bionic cat, for his neighbour - a boy named Daniel – who is allergic to real cats. The eight design features incorporated into Boing-Boing illustrate the compromises needed to make an innovative product. The design process used by Professor George is related to the four primary stages of design: identification, information, creative, and optimisation stages.

INTRODUCTION TO DESIGN CONCEPTS

Design, defined as, "The creation of a material, form or object to achieve a specific function" can be by either innovation or by evolution. In the series of children's stories written by the author and published by the American Ceramic Society, Professor George creates Boing-Boing the Bionic Cat [1] using both innovative and evolutionary design concepts. In the first book the bionic cat is built with eight design features (Figures 1 and 2). Boing-Boing has bionic eyes, legs, whiskers and tail and fibre optic fur that transmit sunlight through the fibres to photoelectric cells that charge 9-volt batteries. There is a small microprocessor within the bionic cat's abdomen that controls the bionic components and the voice box simulator. The first story illustrates each of the four phases involved in a systematic approach to design: 1) identification stage, 2) information stage, 3) creative stage, and 4) optimisation stage. These phases are described in more detail in a later section.

The first story also illustrates the impact of an error in the creative feedback process when the bionic cat's voice simulator is improperly programmed. The cat says "Boing-Boing" instead of "Meow-Meow" when the ceramic sensor in its nose is rubbed. The consequence of this error is a major part of the story and serves to demonstrate that the unpredicted, unplanned flaw in the programming has produced an unexpected effect. The error has a positive effect on the behaviour of the child, Danny, who is given the cat by Professor George because Danny is allergic to real cats. Danny loves it when the bionic cat says "Boing-Boing". Children who read the book think this is the best design feature of the cat. Thus, in this case there was no flaw in the design but a flaw in the execution of the design. The consequence was positive but could also have been negative. In fact, to Professor George the designer, the error is considered to be negative and continues to be so throughout the series of books. The message to the design engineer is the following: It is impossible to predict accurately all consequences of a design because there is always the potential for human error in the execution of your design. Thus, nearly all designs are required to be a compromise.

Also, most designs require several iterations of improvements. You must allow time for the modifications. In the subsequent series of eight stories that follow Boing-Boing the Bionic Cat, Professor George adds a new design feature or sometimes even more than one new feature to the bionic cat. At times he does so as a surprise to Danny. Other times his new design features are installed to correct a problem inherent in the original bionic cat design. I will return to some of these features in a later section. At present, I want to emphasise that Professor George's design modifications are sometimes evolutionary and other times innovative, depending upon the need. The design engineer must make this distinction when he/she starts on a project or confusion will make completion very difficult, especially within a deadline, which is almost always the case. I will use as an example later the difficulties experienced in making a prototype Boing-Boing that meets the design specifications and contrast my experience to that of three classes of high school physics students from Westerville, Ohio, who attempted to do the same thing, i.e. build a Boing-Boing within a six weeks time limit and a limited budget, also a condition commonly imposed upon a designer. Before completing the Boing-Boing design analysis let me first describe what I feel is a conceptual foundation for approaching a design problem and then I will use the concepts to discuss the design steps used by Professor George.

DESIGN BY EVOLUTION OR INNOVATION

Design is a critical step in development of a new technology. Design involves the integration or synthesis of many factors into a prototype of a new or improved product or process. The approach taken towards a design can be either innovative or evolutionary.

- Evolutionary design is change through small variations over successive generations.
- Innovative design is change by abrupt, radical or complete substitution of concepts or approaches.

Historically, most improvements in technology occurred in very small steps, i.e., were evolutionary. This resulted in a long time constant for change (see Figure 1 in refs. 3, 4,5). In the past, many factors promoted slow change. At the present time information exchange is nearly instantaneous. Consequently, change occurs very much more rapidly than in the past. The time for both innovative and evolutionary change has become very short. Some factors associated with innovative change are

- Innovation requires people with a special ability or training to see a problem and design and produce a workable solution.
- Innovation involves a high risk of failure, financial and personal.
- Innovation involves uncertainty, threats to security and to the jobs of those who resist the change.
- Innovation requires faith in one's own ideas.

- Innovation requires the establishment of some sort of patent system to create incentives and protection for the inventor.
- Innovation requires people who can put together or recognise the relationship between unusual or unexpected parts.
- Innovation usually requires risk capital.
- Innovation needs time to make trials, which may fail.

The most important differences between evolutionary and innovative design improvements are

1) The relative degree of change involved,
2) The time span over which the change occurs,
3) The relative investment in manpower and capital required to effect the change,
4) The relative risk,
5) The incorporation of a new or different scientific principle in the change.

Innovative change is, by definition, not obvious to a person skilled in the field. Innovation usually involves an unexpected discovery or invention. In contrast, evolutionary changes are often considered to be obvious extensions of previous discoveries and therefore not inventions. Often there is no clear distinction between the two. The Patent Office and the Courts provide the final judgement as to the validity of an invention. This does not necessarily mean that all inventions are innovations. Many inventions are approved because of technical loopholes in the claims describing the innovative invention. Thus, we can say that a patentable invention is a necessary but not sufficient criterion to determine whether a change is an innovative one. Both evolutionary design and innovative design are vital to technological progress. The two approaches often compete for the same market. Figure 1c in ref. 3 shows that new technology starts from an innovation. There is a period of slow growth as the new products compete for market share with already existing products. The existing products experience additional evolutionary change to improve their competitiveness. If the innovative change is sufficiently good, the new products will overtake the old due to economic advantages, which are 1) lower cost, 2) higher performance or 3) both. This results in a period of very rapid growth where large profits occur.

After a period of time, termed the technological half-life, the growth begins to slow and eventually the new technology becomes a stable, mature technology, which may continuously incorporate evolutionary changes. Growth in this mature state is nearly equivalent to growth in the gross national product (GNP) and the technology can then be considered to be a commodity. Profits are relatively small (8-15%) and the emphasis is on maintaining sales volume. Eventually a period of stagnation may occur where sales volume declines. This is generally the case if: 1) a superior technology appears with better margins of profit, or 2) consumers no longer desire the old products, 3) government regulations restrict production, or 4) raw materials, labour, or energy required to maintain the product become diminished. We now live in a worldwide economy where innovative or evolutionary changes in technology are quickly assimilated in the market place regardless of country of origin. Consequently, technology transfer times and design times must be very rapid (Figure 2 in ref.3). Slow evolutionary changes are likely to be obsolete before they can be assimilated due to rapid competitive development of innovative changes. This leads to an important principle: "Design times and technology transfer times must be less than technological half-lifes in order to be valuable".

A SYSTEMATIC APPROACH TO DESIGN

There are four distinct phases associated with designing, described by Hubel and Lussow [(2)]:

1) Identification Stage (What is the problem?)
   - Recognise the problem
   - Determine the objectives
   - Establish the boundary problems (requirements, restrictions, and resources)
   - Prioritise the boundary conditions (relative importance)
   - Establish time limits for each project stage
   - Write a preliminary design plan including all of the above
2) The Information Stage (What can be done?)
   - Determine strengths and weaknesses of existing designs
   - Locate calculational models suitable for design objectives
   - Obtain data required for the calculational models
   - Literature and patent review of alternative solutions
   - Prepare a final concise problem statement
   - Prepare a plan of attack with specific objective
   - Complete within budgeted time
3) The Creative Stage (What are alternative solutions?)
   - Conceive evolutionary approaches
   - Conceive innovative approaches
   - Calculate alternative design concepts
   - Synthesise design concepts into preliminary alternative solutions
   - Present alternative solutions
4) The Optimisation Stage (What is the best solution?)
   - Use calculational models to evaluate alternative solutions
   - Compare relative performance vs. design criteria
   - Select most promising alternative
   - Revise and refine design against design objectives
   - Finalise design
   - Present final design

There are several points to consider throughout a design project.

1) All resources, especially time, must be budgeted and carefully monitored throughout the project. A rule of thumb for time allocation is: Stage 1: 15%; Stage 2: 25%; Stage 3: 40%; Stage 4: 20%

2) Each stage is interdependent and feedback between Stages 1-2-3 can be very important. Therefore, preliminary design concepts (Stage 3) should be evaluated as soon as they are generated by creative insight. Do not be bound by rigid rules.

3) Hubel and Lussow point out, "*The designer should never be held to a rigid sequence of distinct stages. That would reduce the benefit of uninhibited jumps and leaps of the imagination (lateral thinking) On the other hand, it is necessary to sequence and break work and information into small pieces to make optimum use of logical thinking. The human brain can deal with intricate detail only in small units. There is little doubt that good designers operate using both modes - leaps of insight and sequentially - with neither inhibiting the other.*"

"*Insights can always be abandoned if they do not fit the facts. A logical sequence also can be abandoned if it appears to be leading in an unfruitful direction. The ideal situation is, of course, when both insight and sequential methods point in the same direction. Any process or method should be used or retained only so long as it accomplishes the given task. If it fails, it should be replaced with a more functional procedure. A god description of this whole process might be that it is a network which a designer both explores and creates as he or she searches for a satisfactory solution to the design problem.*"

POTENTIAL DIFFICULTIES

<u>Stage One: Identification</u>

The most important aspect of Stage One is "getting started." In many cases it is not clear initially what the design objective is. If a clear definition does not exist you must create one. Confirm from your client or supervisor that your definition is acceptable. If it is not, seek input until a definition is written that it is mutually agreed and understood.

Do likewise with establishing the boundary conditions. Make sure you and your supervisor have agreement as to the performance goals: identify any restrictions, the resources available, manufacturing or capital limitations. Prioritise the performance criteria and establish what trade-offs can be tolerated.

- Obtain agreement on the overall time for the project and budget.
- Manage your time very carefully using the rule of thumb as a guideline (15%- 25% - 40% -20%).

- Take the time to write out a preliminary design plan. It will help clarify your approach to the problem. Discuss your plan and review comments and criticism with an open mind. It is easy to modify your approach at this stage.

Stage Two: Information Stage

The biggest problem faced here is usually too much information from too many sources. Hubel and Lussow call this "Info-Pressure." Their advice for handling "info-pressure" is as follows:

1) Determine your information priorities
2) Be selective, scan and be decisive
3) Eliminate trivia. Do not load yourself with unimportant facts.
4) Avoid procrastination. Act on things or throw them out and forget them. Do not load your brain with things undone.
5) Eliminate distractions - first personal and then environmental. Relax.
6) Be efficient; choose "byte-rich" activities
7) Make notations. Write down things that matter
8) Use other sources to store and retrieve information - a computer, an assistant or a friend.
9) Evaluate your personal information needs and methods of processing. Try to increase your information capacity.

Because of an unlimited supply of information, it is essential to put a priority on what you are seeking and impose a severe time limit. Otherwise there will be too little time for Stages Three and Four. All too often, people try to avoid starting the creative stage by insisting on just one more literature review. On the other hand, a rush into Stage Three without a reasonable level of information is usually a waste of time. A good criterion of information gathering is the following: Highest Priority - Obtain data for calculational methods

Stage Three: Creative Stage

It is important to approach the design problem from both an evolutionary and an innovative design perspective. For an evolutionary design you need to understand current approaches thoroughly and then examine which design element has the most influence on performance or cost. Then explore conceptually how that design element can be altered to achieve the design objective. Usually performance depends synergistically on all of the design elements; i.e., if you change one you change them all. Consequently an evolutionary approach usually involves modification of multiple design elements. Computer based calculations are especially helpful in predicting the effects of simultaneous changes of several design elements. An innovative approach to design usually involves completely fresh thinking. It is useful to emphasise the ultimate function of the material or device and try to conceive a completely new means of achieving that function. That often means use of a different scientific principle from the previous one. Velcro vs. a zip vs. buttons are examples of innovative approaches to temporary fasteners. Professor George designed the bionic cat by himself in the story. However, as Director of a Robotics Institute at the University nearby he often incorporated many multidisciplinary innovations in his designs. The importance of cross-discipline interactions, which lead to innovative concepts, is an important feature of the later books in the Boing-Boing series. It is important to prepare a professional synopsis of the design alternatives at the end or Stage Three. These alternatives (usually no more than Two or Three) should be discussed with your client or supervisor to obtain instinctive reactions or identify design flaws. Prior to this preparation it is essential to compare projected performance and costs with the design objectives.

Stage Four: Optimisation Stage

In many ways this is the most difficult stage emotionally. It means giving up an idea. Not all of the preliminary concepts from Stage Three will perform comparably; if they do, something is usually wrong. It is critical to eliminate all but the best option as soon as possible so that the design can be "fine-tuned." Time also must be retained for a professional quality final report, preparation of a prototype etc. This cannot be done at the last minute or the work will look haphazard. The temptation to do "one more calculation" or "one more drawing," etc is enormous, but must be overcome early enough to prepare the final report and presentation.

EVALUATION CRITERIA

Does your design lead to improved performance? If so it is likely to be satisfactory. However, it is important to recognise there are a number of aspects of technology, materials or products that can be improved. They include:

1) Function
2) Efficiency
3) Reliability
4) Safety
5) Simplicity
6) Compactness
7) Cost of manpower
8) Cost of capital equipment
9) Cost of energy
10) Cost of raw materials
11) Yield (% of rejects)
12) Environmental impact

Function means that a thing works. It does its specific job. It performs as expected. Efficiency means that required performance rates or output are always retained with pre-specified effort or input. Dependability means that a product or system works and performs with expected efficiency any time it is required to do so. Safety is a must for human

operation. Unsafe products are incomplete in their design and therefore unstable. Simplicity is a sign of good design. It is very easy to be complicated. It takes a lot more thinking and effort to simplify things and still retain function and efficiency.

Quite often things are too complicated and overburdened with unnecessary features, parts, and superfluous function. These can add to higher initial costs, unnecessary breakdowns, and costly repairs. Simplicity aids repair and maintenance. Compactness means small, light in mass, or thin. A combination of these qualities results in simplicity, reduction in mass and ease of handling.

DESIGN ELEMENTS FOR CERAMIC PRODUCTS

There are five categories of design elements for ceramic products that determine the form-function relationship. These are listed in the following table. The function of a ceramic material depends synergistically on its physical properties and its form. Form in turn is determined by a combination of factors: composition, microstructure, phase state, surface characteristics, and shape.

Table I: Design Elements

Form	Shape
Composition Microstructure - Number of phases - Percentage of phase - Distribution of phases - Size of phases - Coupled properties - Connectivity of phases	 Function - Mechanical - Thermal - Electrical - Magnetic - Optical
Phase State - Crystal Structure - Defect structure - Amorphous structure - Pore structure	Surface - Flatness - Finish - Composition gradients - Second phases - Porosity

DESIGN ANALYSIS OF CREATING A BIONIC CAT

Stage 1: Identification stage

In the first story [(1)] Professor George learns that his neighbour boy, Danny is unhappy. The reason is that Danny is allergic to cats, so his mother will not let him have one. Professor George decides to build Danny a bionic cat that will have most of the characteristics of a real cat. He knows that a bionic cat will not produce an allergic reaction because it does not shed proteins from its fur.

The challenge facing Professor George in the identification stage is to determine his objectives and establish the practical boundary conditions for making the bionic cat similar

enough to a real cat to satisfy Danny. In the first stage he must also decide how much time to devote to the project, which will have a major impact on his design objectives. He decides to build it that evening. An important feature of the story is the unexpected consequences of Professor George not giving himself sufficient time to achieve his design objectives. Underestimating time requirements is the number one problem in engineering design. It makes establishing priorities of design objectives especially important so that when time runs out the most important features are not compromised.

Stage 2: Information stage

Professor George has great experience in the design and construction of robots, directing a research centre in robotics at the university nearby. He goes to his computer terminal in his basement workshop, a key feature in all nine stories, and sets up a computer model to calculate the boundary conditions for the bionic cat design. He is aware of the trade-off of weight versus mobility and power. The computer model provides him with the quantitative information that he needs to begin creating the cat.

Stage 3: Creative stage

Professor George begins with assumed design objectives from stage 1 that the cat needs to be similar in size and weight to a real cat. In the second stage of design he made the decision that the bionic cat must be able to walk, run and avoid objects in a manner similar to a real cat. Therefore, his creative computer simulation in stage 3 begins with the cat having four bionic legs terminated in paws (Figure 1). The computer calculations show that each leg can be moved by a small electric motor powered by a nine-volt battery. Operation of each

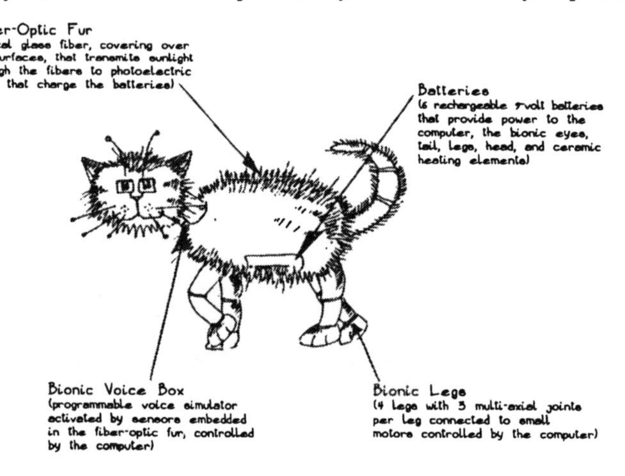

Figure 1. Four of Boing-Boing the Bionic Cat's design features.

leg is controlled by a small microprocessor (Figure 2). The calculations show that the microcomputer and its associated battery (Figure 1) will be the heaviest component in the cat. Thus, they need to be located in the cat's abdomen in order to provide access to the various peripheral ports and also to keep the cat's centre of gravity low and provide a weight balance for the four legs. At this point Professor George does not know how fast the cat can run or how easily it can move around objects. He has an estimate from the computer but decides, for the sake of time to wait until he builds a prototype in stage 4, the design optimisation stage, to determine the actual values of mobility achieved

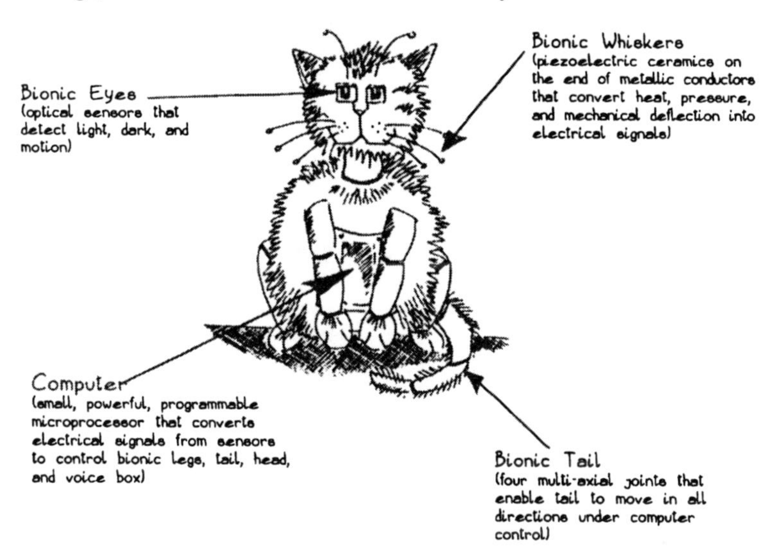

Figure 2. Four additional design features.

by his design. The second creative element in Professor George's design is to incorporate components that enable the bionic cat to avoid objects. He uses piezoelectric ceramic whiskers (Figure 2) to provide input to the computer to send signals to the motors driving the legs telling them to slow down, stop or reverse. The signals come from two whiskers on each side of the cat's head. A third whisker, located between the motion detectors, is included to signal to the computer whether an object is hot enough to be dangerous. Photoelectric cells are used as bionic eyes (Figure 2) to signal to the computer whether objects correspond to a particular colour or type of motion. The photoelectric sensors in the cat's bionic eyes provide input to the computer that controls motion of the legs. This design feature is included in the first version of the bionic cat described in the first book but the function does not appear in a story until the third book. This is an example of an innovative design concept that is in anticipation of evolutionary applications. It is always more efficient and cost effective to incorporate within an original design a capacity for flexibility and improved performance without requiring major modifications.

Professor George decides to incorporate a fifth motor to move a bionic tail (Figure 2). His decision is based on the fact that a cat's mobility in unusual circumstances, such as climbing trees, requires extra balance that is provided by a tail. In a subsequent story, Boing-Boing the Bionic Cat and the Jewel Thief, Professor George adds a piezoelectric sensor to the tail that sends electrical impulses to the computer proportional to the amount of pressure applied to the tail. This is an evolutionary change that is critical to the story and the adventure that takes place. The electrical impulses are converted by the computer to sounds with the magnitude of the sound proportional to the amount of pressure applied. A key creative design feature in the first story is Professor George's decision to incorporate a bionic voice box inside the bionic cat (Figure 1). This feature is central to every story. He recognises that one of the charms of a cat as a pet is its purring when petted. Therefore, he puts a row of small sensors along the back of the cat with connections to the microprocessor and the voice box. The voice box can be programmed using the computer to make a wide range of sounds. This design feature is the key interest point in the first story for children and continues to be featured through all the stories. Communication between Danny and his bionic cat takes the adventures beyond the realm of most children's stories and demonstrates in a very real way the advantages of technology, the purpose of the series of stories. Danny and his bionic cat interact verbally, first at a simple level but with added complexity in subsequent stories. In this manner the bionic cat evolves alongside the growing intellectual ability of the children reading the books. Professor George's design concept included this capability of evolutionary development of the bionic cat from the outset.

The last design feature in the first story is also a critical component in all the stories. It is the bionic cat's fibre optic fur (Figure 1). Professor George recognises that a fundamental limitation of most children's technological toys is their dependence on batteries. He applies his calculations from stages 1 and 2 and establishes that the bionic cat needs an independent energy source. His solution is highly creative. He is aware that all living creatures obtain their energy for living from the same source-the sun. He provides a coat of fibre optic fur on the bionic cat. The fibre optic fur transmits sunlight into the interior of the cat where the photons are converted to electrons by solar cells. He puts the solar cells in parallel to produce nine volts. The cells are connected inside the cat to a battery charger that keeps the batteries at full power for several hours. Professor George knows that living cats prefer to sit in windows in the sun. His bionic cat uses this same characteristic. During the day when Danny is at school the bionic cat can sit in the window of his bedroom and charge its batteries. In the evening they can play together.

Stage 4: Optimisation stage

Because of severe time restraints Professor George proceeds to build a prototype of the bionic cat based on his first set of design criteria. He assumes that modifications can be made later if necessary. The first story describes Professor George constructing the bionic cat and then proceeding through a systematic check of each design feature in order of their priority. It is quite late at night, in fact nearly morning, when he remembers to programme the computer to operate the voice box simulator. He is very tired and sleepy. He makes a mistake. There is no problem with the cat when he rubs its back. The sensors pick up the motion and send the signals for the voice box to say purrrrrrr-purrrrrrrr, just as the design feature requires. However, when the professor rubs the cat's nose he is startled. The programme was supposed to have the bionic cat say meow-meow, mimicking a real cat. The cat instead says "Boing-Boing"! This mistake is shocking to Professor George. He is not used to making mistakes and is quite irritated. He is also very sleepy and knows he does not have time to correct the mistake. So, he grumbles a little as he heads off to get a shower and ready for a morning lecture. He plans to fix the mistake before he gives the bionic cat to Danny. Late the next afternoon Professor George is returning home from the university when he sees Danny sitting on his front porch step in tears because his mother has the flu and is quite ill. Professor George takes Danny down to his workshop and cheers him up by showing him the bionic cat. Danny is very enthused and asks many questions as the professor explains all the design features and lets Danny put the bionic cat through its paces. The professor is quite pleased because it performs just as his calculations have predicted. The cat can move around objects without any problem and avoids the heater in the corner of the room. It can see to move in the dark. All seems well. Professor George asks Danny to pet the cat and it purrs with a gentle sound. Then Danny rubs the bionic cat's nose. It responds with a hearty "Boing-Boing" Danny laughs and thinks the sound is really good. Professor George is shocked and irritated. He had forgotten his mistake in his tiredness. Danny cannot understand why Professor George is angry. Danny explains that it is much better for a bionic cat to say "Boing-Boing" rather than "Meow-Meow" like ordinary cats. He tells Professor George that his bionic cat is the only cat in the world that can say his own name. That makes it even more special. He is happy and his mother is cheered as well. Only the designer, Professor George, finds the mistake hard to live with, as appears throughout the stories as a reminder that technology always carries risk of an error.

The primary message of the first story is that technology can be used for good. However, there is always the potential for mistakes to occur. It is impossible to predict the consequences of errors. It is critical that a designer has sufficient time to test his/her designs and allows sufficient resources to modify and optimise designs before they are committed to production. In my field of research on biomedical materials and tissue engineering there have many examples of devices going to the clinical market without sufficient testing to ensure reliability. The effects have been highly negative and many people have suffered and companies have gone bankrupt. Engineers have to accept responsibility when this happens and should take a firm stand to prevent it from taking place. In a small way the Boing-Boing the Bionic Cat stories are directed towards exposing children to these complex issues at an early age when they can absorb them as stories rather than lectures.

DESIGNING AND CONSTRUCTING A REAL BIONIC CAT

The American Ceramic Society suggested that it would increase children's interest in the Boing-Boing books if it could be shown that a real bionic cat could be built and demonstrated during readings. The time limit was from November 15, 1999 to February 15, 2000 in order to have the prototype available for National Engineer's Week in Washington, D.C. where the book launch was to be held. My first attempt to make a bionic cat from Lego Mindstorms was a disaster so I sought assistance from Mark Campbell, the Lego sales manager at Hamley's, the oldest and largest toy store in the United Kingdom. Mark is an expert at constructing intricate toys from Lego sets of all types. When I showed him the design features of Boing-Boing, Figures 1 and 2, "No problem." He said "I can build you a bionic cat in a weekend." "How much will it cost?" I asked. "No problem." He said "I can use spare parts from kits in the store." To make a very long design story short, after two months later and more than $1,000 I had a prototype Boing-Boing (Figure 3). Mark and I made many design compromises along the way following, in retrospect, numerous examples of the design stages described above. The first design feature we agreed upon was the same adopted by Professor George. The bionic cat had to walk with all four legs. Most toys do not walk. They may have legs but the legs roll on wheels. It is very much easier to move wheels with small motors than to move legs.

Figure 3. Larry Hench and the functional prototype of Boing-Boing, the Bionic Cat.

The second design feature used was to put a small Lego RCX brick (a microcomputer) from a Lego Mindstorms Robotics Invention Kit in the bionic cat to control the motors operating the legs and the tail. Only two motors were used to drive the walking motion of two pairs of legs. This saved weight and power. The bionic eyes were simple LED's and did not have a function other than to shine. That is sufficient for children, I learned. Two piezoelectric whiskers were used, right and left, along with a pair of fibre optic whiskers. The motion active whiskers were sufficient to give signals to the Lego computer to stop and reverse the legs, thereby acting out the scenes in the first story. The fibre optic whiskers are lit up by a light source with a rotating disc of colours that cause them to flash as the cat moves. This demonstrates the principle in the story of the fibre optic fur without involving the expense of designing and constructing the fibre optics, photo cells and battery charger. Our decision was that such a system would be prohibitively expensive both in time and cost to include in the prototype. Six standard 2A batteries are used in the prototype. The voice box simulator was also lost in the design compromises made during construction. The RCX brick can be programmed to make both purring and Boing-Boing noises but the volume is too low to be heard from inside the cat and time ran out before a miniature amplifier could be located. So, when the story is read the bionic cat's voice is heard from an external miniature tape recorder in my hand. The children do not seem to mind (Figure 4). They in fact enjoy learning that the bionic cat in the story has eight design features but the prototype Boing-Boing only has six of those features. That is close enough for them to accept that Professor George could have made Boing-Boing for Danny, as described in the story. It makes the book believable.

Figure 4. Westerville, Ohio, District 3rd graders listening to a reading and show and tell of Boing-Boing, the Bionic Cat by Larry Hench.

A major design constraint imposed by Mark and me was to make the prototype out of Lego parts, exclusive of the fur which was from an unstuffed toy cat and re-dyed to match the illustrations of Ruth Denise Lear (Figure 5). The Lego parts attract the attention of the children immediately because almost all have various levels of Lego sets at home and at school. A challenge was accepted by three classes of high school honours physics students in Westerville, Ohio to construct a Boing-Boing prototype out of Lego following Professor George's design concepts. Two bionic cats were made and both walked, albeit at a very slow pace (Figure 6). The physics students (Figure 7) approached the design compromises differently from Mark and me and had less success. The major difference was that Mark compromised towards using a simple and lightweight design knowing the limitations of the Lego motors. The physics students ended up with much heavier cats and insufficient power to move the legs. These differences illustrate the importance of innovation in a design project. They also illustrate that often it is not possible to achieve an optimal solution in the time and budget available.

Each of the following stories of Danny and Boing-Boing's adventures involves another design feature added to the bionic cat. The new feature provides the opportunity to teach another scientific concept and how that concept is used in a functional manner. The stories become more complex as the bionic cat becomes more complex. This parallels what happens in our technological age. Hopefully children that read these stories will grow to appreciate the wonder and excitement of ceramics and technology and appreciate the good that they provide in all our lives. We hope that the children will also appreciate the importance of people and creativity and come to respect scientists and engineers as happy, friendly and constructive members of society who make life better for most people. We cannot all make a bionic cat in a basement workshop in one night like Professor George. However, we all can take the time to interact with children and share with them the excitement of the world of ceramics.

Figure 5. Professor George and Danny testing the design features of Boing-Boing, the Bionic Cat. Illustrations by Ruth Denise Lear.

Figure 6. Three prototypes of Boing-Boing the Bionic Cat. The left and right versions were built by Westerville High School honors physics students. The bionic cat in the middle was designed and built by Mark Cambell with design suggestions by L.L. Hench.

Figure 7. Professor Hench, kneeling on the right, discussing the design differences in three Boing-Boing prototypes with Westerville, Ohio, honors physics students.

REFERENCES

[1]L.L. Hench, "Boing-Boing the Bionic Cat", *American Ceramic Society,* (2000).
[2]V. Hubel and D. B. Lussow, "Focus on Designing", *McGraw-Hill Ryerson Ltd.*, Toronto, NY, (1984).
[3]L.L. Hench, "Ceramics and the Challenge of Change", *Advanced Ceramic Materials*, 3 (3): 203-206 (1988).
[4]L.L. Hench, "From Concept to Commerce: The Challenge of Technology Transfer", *MRS Bulletin XV*, 8:49-53 (1990).
[5]L.L. Hench, "Sol-Gel Silica: Properties, Processing and Technology Transfer", *Noyes Publications*, Westwood, N.J., pp: 133-150 (1998).

PROBLEMS

1. Select two of the following fields of technology and describe

   a) The sequence of technological advances over the last 200 years
   b) Identify in each advance whether the improvements were evolutionary
   c) Discuss the role of ceramic products and technology in the advances
   d) Forecast what might be the next step of technological advance

The fields to consider are
- i) Lighting
- ii) Communication
- iii) Auditory entertainment
- iv) Pictorial entertainment
- v) Land transportation
- vi) Sea transportation
- vii) Air transportation
- viii) Medical repair of the body

2. Consider the Centennial Review article by L.L. Hench in the Journal of the American Ceramic Society, "Bioceramics", 1998, Vol. 81 (7), 1705-28.

   a) Identify which of the bioceramics discussed in the article can be considered a) evolutionary, or b) innovative from a design concept.
   b) Select two bioceramics and summarize the critical design features that make the material successful.

c) Discuss the use of ceramic phase equilibrium diagrams in designing a bioceramic. Select one bioceramic from the article and analyze its performance from the standpoint of the relevant phase equilibrium diagram.
d) Suggest design improvements of a bioceramic for use in repair of spinal injuries.

3. In the fifth Boing-Boing book of the series, Boing-Boing the Bionic Cat Saves the Space Station, Professor George needs to make a design modification so that the bionic cat can fix a new steering rocket to an outer support arm of the Space Station. What would you do if you were Professor George? Remember the rocket must be attached with sufficient bond strength to withstand the forces of acceleration when the rocket is fired. Also be aware that Boing-Boing must do it alone in space because of a severe radiation storm from the sun.

4. In the sixth Boing-Boing, book Professor George must make it possible for the bionic cat to hide in the jungle while he is searching for explorers lost in the Amazon and perhaps held captive by cannibals. What design concepts would you consider to solve the problem if you were Professor George? Describe the technological limits of the design, such as power requirements, stability in the jungle, weight, portability, etc.

18

Integrated Process Design at the University of Florida

E.D. Whitney

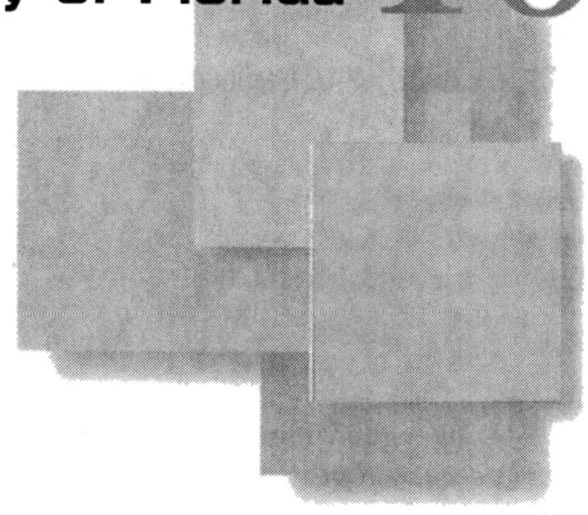

INTEGRATED PROCESS DESIGN AT THE UNIVERSITY OF FLORIDA

E. Dow Whitney
University of Florida
Gainesville, Florida 32611

ABSTRACT

The Integrated Product and Process Design (IPPD) Program is a two-course sequence in which multidisciplinary teams of senior engineering and business students partner with industry sponsors to design and build authentic products and processes. Working closely with industry liaison engineers and a faculty coach, students gain practical experience in teamwork and communication, problem solving and engineering design, and develop leadership, management and people skills.

The IPPD program was launched during the 1995-1996 academic year with five projects, five multidisciplinary design teams and a total of 30 students and seven faculty members. Since 1995, 134 projects have been completed and 766 students have been prepared for professional practice through classroom and laboratory experience.

INTRODUCTION

The Integrated Product and Process (IPPD) Program is an experimental undergraduate engineering education program developed at the University of Florida as part of the National Science Foundation Southeastern University and College Coalition for Engineering Education (SUCCEED) initiative in 1994/1995. The course, first piloted in 1995/1996, is a two-course sequence in which four to six member multidisciplinary teams of senior engineering and business students partner with industry sponsors to design and build authentic products and processes. IPPD is institutionalized at the University of Florida and the course is a substitute for existing capstone design courses and a technical elective. Working closely with industry liaison engineers and a faculty coach, students gain practical experience in teamwork and communication, problem solving and engineering design, and develop leadership, management and people skills. Teams and individuals are evaluated against defined project deliverables and performance in lectures and workshops (1-3).

The IPPD program prepares students for professional practice through classroom and laboratory experiences that show:

- Fundamental engineering science relevance to effective product and process design
- Design involves not just product function but also producibility, cost, scheduling, reliability, quality, customer preference and life-cycle issues
- Major product realization concepts and practices
- A real life design and build project for an industrial customer
- How to complete projects on time and within budget
- Engineering is a multidisciplinary team effort

To be eligible for the IPPD Program, the applicant must:

- have completed the prerequisites for his/her major
- be a senior who is graduating in a spring or summer term
- not have more than 30 semester hours left to be complete for graduation

In Table I are listed the prerequisites/corequisites required of a student entering the IPPD Program according to his/her major department.

Table I. IPPD Program prerequisites/corequisites

Aerospace Engineering, Mechanics & Engineering		
	Aerospace Engineering	
	EGM 3520	Mechanics of Materials
	EAS 4101	Aerodynamics
	Engineering Science	
	EGM 3520	Mechanics of Materials
	EGN 3353C	Fluid Mechanics
Business Administration (MAN 4504 and one of the remaining three courses)		
	MAN 4504	Operations Management
	FIN 3408	Business Finance
	MAR 3023	Principles of Marketing
	ECO 3100	Managerial Economics
Chemical Engineering		
	ECH3203	Chemicals Operations 1
	ECH 4604	Process Design

Computer & Information Science & Engineering

CEN 3031	Introduction Software Engineering
CIS 3020	Introduction to Computer & Information Science
COT 3100	Application of Discrete Structures
CDA 3101	Introduction to Computer Organization
COP 3530	Data Structures & Algorithms
COP 4600	Operating Systems

Electrical & Computer Engineering

EEL 3304	Electronic Circuits 1
EEL 3701C	Digital Logic and Computer Systems

Environmental Engineering (Two of the following four courses)

ENV 4351	Solid & Hazardous Waste Management
ENV 4121	Air Pollution and Control Design
ENV 4514C	Water and Wastewater 2

Industrial & Systems Engineering

EIN 4365	Facilities Design and Material Handling
ESI 4221C	Industrial Quality Control (Corequisite)
EIN 4354	Engineering Economy

Materials Science & Engineering

EMA 3010	Materials
EMA 4717	Materials Selection and Failure Analysis

Mechanical Engineering

EML 3005	Mechanical Engineering Design
EML 4500	Machine Analysis and Design 1 (Corequisite)

PROGRAM MANAGEMENT

A Director, Heinz K. Fridrich and Associate Director, Keith Stanfill, manage the IPPD program. Heinz K. Fridrich is an Industry Professor in the Department of Industrial and Systems Engineering. He joined the University of Florida in 1994 after 43 years with the IBM Corporation. He began his career in Germany in 1950 and held a number of key management positions in Europe and the U.S. including Vice President and General Manager of IBM's largest development and manufacturing site for semiconductors and electronic packages. In 1987 he was elected, IBM Vice President, responsible for worldwide manufacturing and quality until he retired in 1993.

Dr. Keith Stanfill, a Professional Engineer, is also in the department of Industrial and Systems Engineering. Dr. Stanfill has over 10 years industrial experience with United Technologies including 7 years with Pratt and Whitney and 3 years with Carrier Corporation.

In addition to ongoing management of the program, facilities and resources, the Director and Associate Director are responsible for recruiting sponsors, scoping projects, recruiting and training faculty coaches, coaching student teams, securing guest lecturers and keeping the program on track. A part time secretary, a part-time office assistant and student assistant support the administrative work. The support staff handles a large volume of program and individual IPPD teams travel and purchasing transactions.

FACULTY PARTICIPATION

Thirty-four faculty from ten disciplines in the College of Engineering and the College of Business Administration (Decision and Information Sciences) participate in IPPD as coaches and lecturers. Names of the participating faculty and their departmental affiliations are given in Table II. The coaches also assist in recruiting students within their own discipline and assist in evaluating and scoping new projects. Ten faculty members and four industry sponsors provide approximately forty-five lectures. In addition to a weekly faculty meeting, the coaches spend three to ten hours per week with their team. Coaches receive $15,000 for each team they manage. Since IPPD is an undergraduate program, there is no overhead deduction.

TableII. IPPD Faculty/Coaches

DR. JOHN R. AMBROSE	Assoc. Prof. of Mat. Sci. & Engineeering
DR. SHERMAN X. BAI	Assoc. Prof. of Ind. & Sys. Engineering
DR. GIJS BOSMAN	Prof. of Elect. & Computer Engineering.
DR. ANTHONY B. BRENNAN	Prof. of Mat. Sci. & Engineering
DR. THOMAS F. BULLOCK	Prof. of Elect. & Computer Engineering
DR. PAUL A. CHADIK	Assoc. Professor of Environ. Eng. Science
DR. JACOB N. CHUNG	Eminent Scholar of Mechanical Engineering
DR. OSCAR D. CRISALLE	Associate Professor of Chemical Engineering
DR. WILLIAM W. EDMONSON	Assoc. Prof. of Elect. & Computer Engineering
DR. WILLIAM R. EISENSTADT	Assoc. Prof. of Elect. & Computer Engineering
DR. SELCUK S. ERENGUC	Prof. and Chair of Decision & Info. Sciences
DR. NORMAN G. FITZ-COY	Assoc. Prof. of Aero. Eng., Mech. & Eng. Sci.
DR. MICHAEL P. FRANK	Assist. Prof. of Comp. & Info. Sci. & Eng.
MR. HEINZ K. FRIDRICH	Ind. Prof. of Ind. & Sys. Eng. and Dir. of IPPD
DR. KARL GUGEL	Visiting Assist. Prof. of Elect. & Comp. Eng.

DR. ABDELSALAM HELAL	Assoc. Prof. of Comp. & Info. Sci. & Eng.
DR. WILLIAM E. LEAR, JR.	Associate Professor of Mechanical Engineering
DR. SUE M. LEGG	Director of Instructional Resources
DR. DAVID W. MIKOLAITIS	Assoc. Prof. of Aero. Eng., Mech. & Eng. Sci.
DR. RANGA NARAYANAN	Professor of Chemical Engineering
DR. RICHARD NEWMAN	Assoc. Prof. of Comp. & Info. Sci. & Eng.
DR. KHAID. T. NGO	Assoc. Prof. of Elect. & Computer Eng.
DR. TOSHIKAZU NISHIDA	Assoc. Prof. of Elect. & Computer Eng.
DR. JILL PETERSON	Associate Professor of Mechanical Engineering
DR. R. KEITH STANFILL	Associate Director of IPPD
DR. SPYROS SVORONOS	Professor of Chemical Engineering
DR. STEPHEN M. THEBAUT	Assist. Prof. of Comp. & Info. Sci. & Eng.
DR. SULEYMAN TUFEKCI	Associate Professor of Ind. & Sys. Engineering
DR. ASOO J. VAKHARIA	Assistant Professor of Decision and Info. Sys.
DR. LOC VU-QUOC	Prof. of Aerospace Eng., Mech. & Eng. Sci.
DR. E. DOW WHITNEY	Professor of Materials Science & Engineering
DR. GLORIA WIENS	Associate Professor of Mechanical Engineering
DR. JOSEPH N. WILSON	Assist. Prof. of Computer & Info. Sci. & Eng.
DR. JOHN C. ZIEGERT	Professor of Mechanical Engineering

FUNDING

Industry sponsors pay $15,000 to fund an IPPD project plus provide a liaison engineer to work with the students for several hours a week. Sponsors also pay for any prototyping expenses in excess of the $1,000 built into each project's budget. The industry support covers approximately 60% of the program expenses. Matching funds from a variety of sources including SUCCEED, the University of Florida Foundation and the State of Florida, cover program travel, materials and administration. Faculty rewards include course relief and no-strings-attached funding. In Table III is given the IPPD program history beginning with the 1995/1996 academic year.

Table III. IPPD Program History: Industrial Sponsors and Design Projects

	95/96	96/97	97/98	98/99	99/00	00/01
1	Class 1	Class 1	Class 1	Class 1	Class 1	Class 1
2	E-Systems	E-Systems	E-Systems	Raytheon	Raytheon	Raytheon
3	Florida Power Corp.	Florida Power Corp. (2)	Florida Power Corp.			
4	Honeywell Inc.					
5	Paradyne	Paradyne	Paradyne	Paradyne	Paradyne	Paradyne

6		Allied Signal				
7		Cargill Fertilizer Inc.	Cargill Fertilizer Inc. (2)	Cargill Fertilizer Inc. (2)		
8		Energizer	Energizer	Energizer		
9		Eglin AFB	Eglin AFB	Eglin AFB	Eglin AFB	Eglin AFB
10		Florida Power & Light	Florida Power & Light		Florida Power & Light (2)	Florida Power & Light
11		Gainesville Regional Utilities	Gainesville Regional Utilities	Gainesville Regional Utilities	Gainesville Regional Utilities	Gainesville Regional Utilities
12		Intellon	Intellon	Intellon	Intellon	Intellon
13		Lockheed Martin (3)	Lockheed Martin (4)	Lockheed Martin	Lockheed Martin	Lockheed Martin
14		Metal Container Corp.		Metal Container Corp.		
15		Motorola	Motorola (3)	Motorola	Motorola	Motorola
16		Reflectone	Reflectone	Reflectone	Reflectone	
17			Arizona Chemical			
18			Boeing	Boeing	Boeing	Boeing
19			Dow Chemical	Dow Chemical	Dow Chemical	Dow Chemical
20			DuPont	DuPont		
21			Millennium	Millennium	Millennium	
22			PCR			
23			Protel	Protel	Protel	
24			PTI			
25			QMS	QMS		
26			Sensormatic			
27				The Crom Corp.		
28				IBM	IBM	IBM (2)
29				Pratt & Whitney	Pratt & Whitney	Pratt & Whitney
30				Texas Instruments	Texas Instruments	Texas Instruments
31					Cordis	Cordis
32					ERC Innova Tech	

33					Harris Corp (3)	Harris Corp (2)
34					JRS GEO Services	
35					Johnson Controls	
36					PGI	
37						ABB Water Meters
38						Intersil (2)
39						Kimberly Clark
40						Kraft Corp.
41						Mark IV Automotive
42						North American Archery Group
43						RedSea Works Co.
44						Siemens
45						SOCOM (2)
46						Southern Nuclear

Total Industry Sponsors:
5 15 22 22 24 28
Repeat Sponsors:
0 4 13 18 17 16
Design Projects:
5 18 29 23 27 31

( ) Indicates number of projects sponsored if more than one

ASSESSMENT OF PROGRAM

The IPPD course management is evaluated on a regular basis with pre- and post-self assessment of educational objectives. Results are generally very positive (see Table II). The IPPD program has always received positive feedback from industry sponsors.

Since 1995 fewer than 10% of all IPPD projects were less than satisfactory. Self–assessment in these cases showed that major reasons for poor performance were

* weak project management
* inadequate communication between design team and sponsor company
* unrealistic project scope

Most students are motivated and capable to participate in the program. This may be the result of careful student selection criteria but also because of better criteria to assess team and individual performance. Primarily for logistic reasons smaller teams (5-6) seem to work better than larger teams (7+).

Following the Final Design Reviews, at the end of the second semester of the program, a feedback session is conducted with industry sponsors who continue to respond very positively to the IPPD program as an experimental education that benefits students, university and industry.

CONCLUDING REMARKS

This paper outlines the history, goals and structure of the IPPD Program at the University of Florida. What distinguishes the Florida program from similar programs at other institutions is the combination of the following attributes: a lecture and workshop format aimed at improving students' engineering skills, the simulation of a corporate R&D environment, the 10 academic disciplines involved and the final requirements for project completion – namely a complete working prototype. These attributes together with the integrated team approach – interdisciplinary student groups, faculty, and industry liaison engineers – and institutional infrastructure support has not only led to a rewarding educational experience for our students as they embark on their careers but has significantly enhanced their employment potential as well.

This program could not have been initiated without the support of the College of Engineering, administration and industry. Dean Winfred M. Phillips and Associate Dean M. Jack Ohanian were in full support of this new and innovative direction in engineering education. Their help was essential in getting this program started. It is the combined financial support of the National Science Foundation, industry and the State of Florida that have made the University of Florida Integrated Product and Process Design program possible.

REFERENCES

[1] W.R. Eisenstadt, S.S. Erenguc, H.K. Fridrich, R. Narayanan, S. Svoronos, S. Tufekei, E.D. Whitney, J. Ziegert, "Integrated Product and Process Design: A new Paradigm in Engineering Education", The Innovator, The SUCCEED Newsletter, No. 6, Summer 1996.

[2] "Integrated Product and Process Design: How to Set Up a Multidisciplinary Program", SUCCEED Multidisciplinary Design Workshop, Charlotte, NC, March 25, 1998.

[3]"Integrated Product and Process Design: An Undergraduate Program" SUCCEED Multidisciplinary Design Workshop, Charlotte, NC, September 15, 2000.

PROBLEMS

1. Review the structured program for integrated process design as presented in this chapter. Proposed two to three "innovations" to this program that would enhance the design experience for ceramic or materials engineering design students in your program.

2. Prepare a two-page Letter of Intent to the National Science Foundation proposing the initiation of a new design program at your university. Include the major goals, the basic structure and a method for self-assessment of this program. Attach a proposed operating budget for the first five years. Refer to the NSF budget form for the details of how to break down the program costs. Remember to include overhead at a rate comparable to that used by your university.

19

Protecting Property Rights: Patents, Trademarks, and Copyrights

J.A. Calderwood

PROTECTING PROPERTY RIGHTS: PATENTS, TRADEMARKS AND COPYRIGHTS

James A. Calderwood
Zuckert Scoutt Rasenberger
888 17th St. N.W., Suite 600
Washington D.C. 20006-3309

ABSTRACT

In recent years, the legal issues of designing have come to focus as a new design element. Today, not only is product liability a dominant issue as described in the November issue of *Ceramic Bulletin*, but also are copyright, trademark and patent laws. As design becomes the critical step between manufacturer and consumer, designers and manufacturers are protecting their invested interests.

This chapter was first published in 1988 in the Bulletin of the American Ceramic Society.[1]

INTRODUCTION

That area of the law known as "intellectual property law," through its connotation, suggests a certain level of abstractness and is, consequently, one of the least understood and more complicated areas of law. This field of law deals primarily with patents, trademarks, tradenames and copyrights and, therefore, it has many important ramifications for businesses that need to protect the intellectual property. Unfortunately, many entrepreneurs have suffered from their lack of knowledge of the basic principles of intellectual property law.

This area of law is of increasing concern to all those in the ceramics field. Obviously, members of the ceramic industry may encounter certain situations where a basic understanding of this law can help to prevent problems and protect their business interests. Practical questions may arise as to whether a company qualifies for a patent, trademark, tradename or copyright, or whether it might be infringing on another's intellectual property. Often, a company also may wish to protect is corporate name and prevent others from using a similar name.

---

[1]*Ceramic Bulletin*, Vol. 67, No. 12, American Ceramic Society, 1988.

Another question that arises frequently is whether an employer is protected under the law for the intellectual property that is developed by one of its employees. Questions often are asked as to whether the employment contract between the employer and employee should address development of intellectual property and, if so, whether that agreement is enforceable in the courts.

PATENTS

The federal government, under the Constitution of the United States, was empowered with the exclusive authority to grant an individual the exclusive right to make and sell a new invention. Patents are obtained from the federal government through the U.S. Patent and Trademark Office. As of this date, that office, over its history, has issued approximately 4.5 million patents. Even though patents issued by that office are presumed to be valid, issued patents can be found invalid and, therefore, cannot be enforced by the federal courts for various legal reasons.

Obtaining patent protection for new inventions has several requirements. The first requirement is that the device or process must be "novel." Therefore, no patent would exist in the United States for an automobile, although improvements on that automobile (such as a new steering column design, new type of engine or new brake system) would be patentable. An invention to be novel also cannot be obvious. This requirement would prevent an individual from obtaining a patent for the use of a hammer to hammer nails.

The second requirement is that the device or process must serve some useful function or utility. This requirement obviously would eliminate a work of art; something that is purely a work or art by itself would not be patentable for it has only aesthetic value, and does not possess a characteristic of having a utilized purpose.

Finally, the ability to patent a device or process can be lost through the actions of the inventor. In general, where a product, device or process has not been patented, no patent will be granted in the event that the product, process or device has been introduced into the "public domain." An invention is deemed to be in the public domain when it is described in a printed publication or has been in public use for more than a period of one year prior to the patent application. Recognizing the ability to possibly obtain a valid and enforceable patent and acting accordingly within a certain timeframe is important.

Once a patent is obtained for that product, device or process, a valid patent exists for a period of 17 years, which permits the holder of the patent (the patentee) to exclude all others from making, producing, selling or using the product throughout the United States for that 17-year period, without the permission of the patentee. By operation of law, a patent grants a monopoly to the patentee for the period of 17 years and permits that patentee to exclusively derive profits from that invention. Obviously, the exclusive right to sell, license and produce that process or product can create an extremely valuable asset for any company. This right was created by

Congress to provide a sufficient economic incentive for inventors to create and develop new products and processes to benefit society as a whole. To encourage this development, Congress permits the government to grant an exclusive monopoly to the patentee for a period of 17 years as an inducement for invention and to reward that patentee for the benefits of his creation.

While a patent is in effect, any person who uses, makes, sells or licenses any patented invention or process, without permission from the patentee, can be prevented by a court of law through an injunction from continuing to benefit from the patented items and can be help liable for damages. In the event that any business uses a patented device, process or product without permission from the patent holder, that business shall be liable to the patentee for all damages suffered by that patentee. In the event that the infringement is found to be willful or intentional, the court can, as a matter of law, impose damages for up to three times the amount of the profits obtained by the unlawful use.

In the event that a patented process or product possibly is being used in a business, it may be advisable for that business to consult an attorney to determine whether that device or process is the subject of a patent. Obviously, this action will help prevent the business from being named as the defendant in a patent infringement action and, in the event that no patent is found to exist, will permit the business to obtain a patent to protect itself from the misappropriation by others of its novel product or process.

In general, patents are no renewable after the 17-year period has expired. This length of time originally was granted under the Common Law of England so that the patentee would have sufficient time to obtain the benefits of his novel design, process or product. However, extension of the time for exclusive monopolization of that patent fo r more than 17 years generally is denied, because the law deems that length of time sufficient to provide an incentive to invent, without reasonably stifling competition.

It is important to realize that, to effectively protect a valid patent, the product often must be changed significantly, or improved during the 17-year period to protect that patent, as others may attempt to obtain patents with significant changes or improvements from the patented article or process.

Patents are issued most routinely for mechanical inventions and for novel production processes. Although these types of inventions are the most routine for patent attorneys, patents also can be obtained for new, original and ornamental designs for a manufactured article. These patents are known as design patents. These types of patents should be of significantly interest to members of the traditional ceramic industries since products often arise from aesthetic designs and artistic works that exist on a functional article or product, rather than focusing directly on that article or product.

The law of design patents differs somewhat from other areas of patent law. The most obvious difference in obtaining a design patent is that the article or tangible object to which the design is affixed need not be patentable. A ceramic coffee mug with a particular design can be patentable for that design, whereas the coffee mug would not be patentable. However, it is important to realize that a glass figurine or glass artwork, while novel in design, would not be patentable as the requirement of utility would not be met under general patent law. While the design for the coffee mug would be sufficient to obtain a design patent, that same design in the form of artwork or sculpture in and of itself would not be patentable. However, that design or sculpture may qualify for copyright protection under the copyright laws, which will be discussed later.

Often there are disputes concerning a determination of who is entitled to obtain the patent. Many times, individuals will develop novel products or processes while in the employment of a company. The employee may attempt to obtain a patent or hide the discovery from the employer until sufficient time has elapsed after he/she has terminated his/her employment with that company to patent on his/her own without violating an employment contract. The general rule of law is that employers are entitled to obtain the benefits of any patentable invention or processes that are devised by its employees in the scope of their employment and general course of that business.

However, certain exceptions do exist under the law and employers should be cautious about obtaining the benefits of any patent to which that company is entitled. Courts have consistently upheld employment contracts between an employer and its employees, whereby the employee agrees that all intellectual property developed during his/her employment is the property of the employer.

The employer also is entitled to the benefits of any modification or alteration by the employee to a product or process that already has been patented by that employer. Any alteration, modification or improvement is deemed also to be the property of the employer, the patentee of the original invention. In these cases, the employer generally is held to be entitled to all of the benefits and protection of an approved patent.

Clearly, the law involving patents is much more complicated than the issues mentioned above. In the event that an issue involving the patent law may arise or exist within the day-to-day affairs of a business, it is important for that business to consult an attorney to protect the legal interest and financial viability of the business, as a patent clearly can possess extraordinary value.

TRADEMARKS AND TRADENAMES

Another area of intellectual property law involves the use of trademarks and tradenames. A "trademark" is a work, name, symbol or device that is adopted and used by manufacturers or merchants to identify their goods or services and to distinguish them from the goods and services of others. A trademark obviously can take many forms, such as a particular arrangement of letters or a more elaborate design, such as an artistic drawing or other object. Tradenames can consist of words that have no particular meaning, such as Exxon, or an ordinary word which would have not connection with the product or service, such as Levi's bluejeans, which serves only to distinguish the product from other bluejean manufacturers.

It is important to realize that description terms and non-distinctive names that have now acquired a unique and distinctive application are disqualified from trademark protection. Therefore, brand names that are used to distinguish a product or service from similar products or services qualify for trademark protection, but generic descriptions do not.

Trademark laws were enacted by the federal government under the Lanham Act to prevent the public from being confused, mistaken or deceived in the purchase of goods and services, and to protect the trademark owner's product identity. Clearly, any business that has developed and obtained a large amount of consumer goodwill does not want any other business to use a similar mark or name to take advantage of that status, particularly where that competitor would not perform services or produce goods within the standards of the trademark holder. A widely recognized logo or trademark, when used on all invoices, packages, advertisements, transportation equipment, etc., can create and maintain a high demand for the product, thereby constituting a good advertising practice. The trademark, besides being an advertising practice to distinguish one company form a competitor, also endows the business with certain legal rights. The government recognizes the exclusive right of the trademark owner to use that mark in commerce to distinguish their goods or services from other competitors, and prohibits others from using that trademark, or a similar mark, if the use of that trademark would constitute a "likelihood of confusion" in the marketplace for consumers.

To obtain a trademark, the business must adopt a distinguishing mark for the particular product or service by actually affixing that mark to the item, so that the public can associate that mark with the product. Use of a trademark that is first in time creates a right to the exclusive use of that mark. Exclusive use of that trademark, however, can be limited to a particular territory or to a certain kind or class of goods. If a company wants to obtain a trademark for a wide range of products and services, the trademark should be attached to each kind or class of goods or services provided.

Registration with the Patent and Trademark Office will establish the trademark on the principle register. To qualify for trademark protection on that register, the trademark must be used to identify goods or serviced in commerce. The owner of the trademark must file appropriate forms, drawings and the specific

specimens with the Patent and Trademark Office with the registration application form. For instance, in the event that a trademark or tradename is affixed on a tee-shirt and the owner of the trademark wishes to obtain registration of that trademark or tradename, the tee-shirt should be submitted to the Patent and Trademark Office. Registration of that trademark will be refused if the mark is considered to be confusingly or deceptively similar to a mark or name used by another person in the United States or if the mark consists of immoral, deceptive or scandalous matter. For instance, the use of elves to sell cookies could not be registered because those objects or characters already are taken. Something such as "Ronald Reagan Beer" also would not constitute an acceptable tradename.

The Patent and Trademark Office also will register a descriptive term, geographical name or a personal name that becomes distinctive through wide-spread exclusive use. The term takes on a "secondary meaning" referring to works, names, phrases or symbols that have become associated in the minds of the consumer with the source or origin of goods or services. Thus a name or common term may develop a secondary meaning over time. For example, "Piedmont" has become familiar to the public as descriptive of a particular airline and has acquired a secondary meaning, separate and distinct from that geographic area of the United States. All would recognize that tradename as referring to an airline. Therefore, it would be considered deceptive for another person to call a different transportation service "Piedmont Messenger Service." The Patent and Trademark Office most likely would hold such a new trademark to be inappropriate, as consumers would face a likelihood of confusion over the trademark or tradename and naturally would assume that Piedmont Airlines did, in fact, operate the other service.

Although registration, in and of itself, does not create new legal rights to the trademark, it does possess certain advantages over non-registration. Clearly, registration and acceptance of a trademark by the Patent and Trademark Office recognizes the right of that company to use that mark or tradename and appropriate it for its own use. Registration also indicates strong evidence of the belief by the Patent and Trademark Office that the registrant has exclusive control and right to use over the marks on the goods and services specified in the registration. Consequently, the Trademark Act permits injunctions to be used to prevent competitors from violating any right associated with the mark registered and provides that violators may be liable not only for all of their profits derived from the use of that mark, but up to three times the amount of actual damages sustained.

However, this protection is not entirely absolute, because a business that has adopted the trademark prior to registration without knowledge of another's use may continue to use that trademark, but only where continuous use can be established. The registration, therefore, does not grant superior rights in that trademark. Once a trademark is registered, the mark should be included on every article for sale and be used on all of the business' advertisements. The mark also must display a circled capital letter "R" or the words "Registered int eh U.S. Patent and Trademark Office" on the goods, packages and advertisements. Failure to give this notice and use of the trademark on the article will prevent or reduce damages that may be recovered for a violation of the Trademark Act, unless the trademark holder can establish that the violator had actual notice of the registration.

Upon acceptance of the registration by the Patent and Trademark Office, a certificate of registration is issued. The certificate remains effective and in force for 20 years, although the Patent and Trademark Office will request and affidavit within the fifth and sixth years following registration stating that the mark still is being used actively in commerce. Registration can be renewed at the expiration of the 20 years upon filing of a subsequent filing fee and an application for renewal. Unlike patents or copyrights, a trademark can be perpetual, unless it is abandoned for non-use or the company fails to renew its registration. This is commonly known by trademark attorney as the "use it or lose it" rule.

Obvious care must be taken to enforce the exclusive use of a trademark. If a holder of a tradename or trademark permits competitors to appropriate or use that tradename or trademark, the significance of the mark or name will be lost. A tradename or trademark also can become generic in nature as it is used by the public. A perfect example of this particular situation is the use of the term "cola," whereby the term "cola" became generic and Coca-Cola lost its trademark protection over that particular term. Consequently, the term "Coke" has been protected vigorously by Coca-Cola so that the term does not become generic in nature.

Other examples of tradenames that have been lost by becoming generic through public acceptance are aspirin, escalator, thermos and cellophane. As a result, tradename or trademark holders engaged in commerce today specifically protect themselves by inserting the work "brand" in their advertisements. One example of this type of protection is the advertisement concerning Sanka brand decaffeinated coffee or Kleenex brand tissue. The word "brand" has been injected into these ads to remind consumers that, although their products are the best known brands of a common generic product, those names are private trade names and should not be treated as generic in nature.

UNFAIR TRADE PRACTICE

The law of tradenames and trademarks parallels and overlaps the law of unfair competition, which developed under the common law in the United States as an effort to protect the domestic marketplace economy. The general purpose of unfair competition was to prevent one company from passing off its goods or services as those of another. This practice would include using the trademark or tradename of another individual to represent it as being manufactured or processed by that business when it is not. However, the general laws of unfair competition do not necessarily involve the exclusive right to use a name, symbol or device as a trademark or tradename, because lawsuits for unfair competition predominantly seek damages for business loss resulting from the defendant's passing off of goods or services as those of another manufacturer. These laws also will vary, since actions for unfair competition are governed by state law, unlike the federal statutes regarding trademarks and tradenames, which is applicable to all states.

These laws were based on the right to protect the ability of manufacturers or tradespersons to have a cause of action against merchants who use imitative devices or unfair means to copy products, or mislead prospective customers into buying wares or services when they are under the impression that they are buying the wares or services of another. However, under most state laws, a cause of action based on unfair competition is more difficult to win than a trademark or tradename infringement action, because fraudulent or wrongful intent on the part of the person passing these deceptive goods off as those of another manufacturer or competitor is the essential element of the wrong. However, this cause of action often can parallel a trademark/tradename infringement action.

COPYRIGHT PROTECTION

Copyrights normally are obtained to protect either literary or musical works. Copyright protection also is obtainable for graphic, pictorial or sculpture works. Any original decorative design such as decals, pictures, figurines or engravings are protectable under the copyright law. When a work is attached to a particular tangible object, the work is protectable only as to that design or work and not to the mechanical or utilitarian aspect of the object to which it is attached. Once again, where an employee develops a design while working for an employer, that design is considered a work make for hire and the employer initially owns the copyright. A copyright, unlike a patent, has no requirement of utility, only originality.

Copyright protection also differs from the patent area of the law in that a copyright exists automatically once the work is created. Publication or registration is not necessary to assert the protection of a copyright. However, obviously, certain advantages exist with registration of a copyright with the U.S. Copyright Office. Registration first will establish a date upon which the copyright is filed, thereby establishing to some extent an indication of when the work was created. Also, registration usually is necessary to maintain an infringement suit and to recover attorney's fees under the copyright law. Unless registration for that copyright is filed, the plaintiff in an infringement suite only will be allowed to recover actual damages and profits, which may be difficult to establish.

Once a work has been distributed or published, a copyright notice should be placed on all publicly distributed copies. This is the responsibility of the copyright owner. This notice can constitute either the capital letter "C" or the word "copyright," the year of the first publication of the work, or the name of the owner of the copyright. Copyright protection generally endures from the moment of creation until 50 years after the author's death.

LEGAL PROTECTION BENEFITS

Patent, trademark, unfair trade practice and copyright laws provide a benefit to society. These laws provide an incentive for individuals to create and invent, thereby helping to increase societal wealth. Obviously, the disclosure of valid patents on file adds to the technical knowledge of the country and allows others to begin to develop their own competing and non-infringing concepts and ideas based upon those patents. Protection also affords the inventor an economic incentive to reap as many profits as possible from that patent within a 17-year period.

Likewise, the trademark gives its holder a promotional incentive, while protecting the consumer by enabling one to make an informed choice concerning selection of goods or services because of easily identifiable marking. Copyrights give the writer and artist protection for their developmental works and encourage new ideas and productions.

When a new product design or process is invented by a company, or when the company intends to identify a product or service with a particular name or design, one would consider protecting that valuable asset fo the business with the legal rights and remedies available through the federal government's patent, trademark and copyright laws. Moreover, the business would be advised to investigate current applicable registrations regarding intellectual property to determine whether another business already has registered for the exclusive right for the product or trademark that one is developing. In this way, the business may benefit and its marketing may be much easier for purchases by the consumer.

SUGGESTED READING

Mr. Calderwood - Could you provide some texts or other references that would be helpful for this topic?

PROBLEMS/QUESTIONS

If you can think of any thought-provoking situational problems on this topic, it would be most helpful.

Appendix A

Author Biosketches

David E. Clark
Virginia Polytechnic Institute and State University
Department of Materials Science and Engineering
Blacksburg, Virginia
dclark@vt.edu

Dr. David E. Clark is Professor and Head of the Department of Materials Science and Engineering at Virginia Tech since January 2001. He comes from the University of Florida where he taught ceramic and materials engineering in MSE since 1976. His research interests include glass processing and corrosion, nucleation and crystallization of glass-ceramics, nuclear and hazardous waste remediation, and drying and firing of ceramic powders. Most recently, his research has focused on microwave processing of materials, including fundamentals of microwave-material interactions, sintering of ceramics and composites, waste remediation and metal recovery/recycling, self-propagating high-temperature synthesis (SHS) and combustion synthesis of powders, and nano-particulates. He is a past-president and trustee of the National Institute of Ceramic Engineers (NICE), a Professional Engineer, and currently serves as the chair of the Professional Engineering Committee that writes and grades the ceramic engineering licensing examination. He is a past-chair of the Engineering Ceramics Division, a Fellow of The American Ceramic Society (ACerS), has served on the ACerS Board of Directors from 1996 to 2001, and has served as an ABET evaluator for NICE since 1990.

Mary (Missy) Cummings
University of Virginia
Department of Systems Engineering
Charlottesville, Virginia
mlc3u@virginia.edu

Mary (Missy) Cummings received her B.S. in mathematics from the United States Naval Academy in 1988 and her M.S. in Astronautical Engineering from the Naval Postgraduate School in 1994. She spent ten years in the Navy and was one of the Navy's first female fighter pilots. While in the Navy, she also worked as an assistant program manager for the F-14 program in one of the Navy's industrial engineering plants. After teaching at Pennsylvania State University for the

Navy, she joined the Engineering Fundamentals Division of the College of Engineering at Virginia Polytechnic Institute and State University as an assistant professor in August 1999. Her research interests include engineering ethics, learning communities, and engineering education. In 2001, she joined the faculty of the Department of Systems Engineering at the University of Virginia.

Delbert E. Day
University of Missouri–Rolla
Department of Materials Research
Rolla, Missouri
day@umr.edu

Dr. Delbert E. Day is Curators' Professor Emeritus of Ceramic Engineering and Senior Investigator (formerly Director) of the Graduate Center for Materials Research at the University of Missouri–Rolla. After graduating (valedictorian) with a B.S. in Ceramic Engineering from the Missouri School of Mines and Metallurgy, he studied at the Pennsylvania State University where he received an M.S. and Ph.D. in Ceramic Technology. He has been on the faculty at the University of Missouri–Rolla where his teaching and research have dealt with the structure and properties of vitreous solids (glass). He conducted the first experiments on NASA's space shuttle where the containerless melting of glass in microgravity was investigated and is scheduled to conduct additional experiments on the international space station. He is the co-inventor of special purpose glass microspheres, TheraSphere™, which are being used commercially in the United States to treat patients with liver cancer. Currently, he is developing biodegradable glass microspheres that are being evaluated in the United States for treating rheumatoid arthritis. His research on solid waste recycling led to the development of Glasphalt, which has been used to pave roads, parking lots, and airport runways throughout the world. He has published 270 publications and is a registered Professional Engineer. He holds 41 U.S. and foreign patents dealing with glass microspheres, sealing glasses, ceramic dental materials, oxynitride glasses, refractories, and optically transparent composites. He is a Fellow and past-president of The American Ceramic Society.

Dennis R. Dinger
Dinger Ceramic Consulting Services
Clemson, South Carolina
consulting@DingerCeramics.com

Dennis Dinger is a consultant specializing in the processing of ceramics. In May 2001, he retired as Professor of Ceramic and Materials Engineering after teaching ceramic engineering for 13 years at Clemson

University and five years prior to that at the NYS College of Ceramics at Alfred University. His research interests focus on the detailed phenomena that control ceramic production processes. Specifically, he is interested in the influences of particle physics on particle packing, rheological properties, and forming processes of fine particulate ceramic systems. Since the early 1990s, he has been teaching, studying, and implementing the PPC (Predictive Process Control) approach to ceramic processing. He is particularly interested in the application of computer software packages to modeling the properties of ceramic suspensions and bodies, to performing routine ceramic engineering calculations, and to automating the control of ceramic process systems. He is a Fellow of The American Ceramic Society, a Registered Professional Engineer, and a member of NICE.

Kimberly Y. Donaldson
Virginia Polytechnic Institute and State University
Department of Materials Science and Engineering
Blacksburg, Virginia
kdonalds@vt.edu

Kimberly Y. Donaldson is the Associate Director and Technical Manager of the Thermophysical Research Laboratory in the Department of Materials Science and Engineering at Virginia Polytechnic Institute and State University, Blacksburg, Virginia, She received a B.S. in Civil Engineering in 1979 and an M.S. in Engineering Science and Mechanics in 1983, both from Virginia Tech. Her research interests include the theoretical and experimental investigation of the mechanical, thermomechanical, and thermophysical properties of various materials. Research in which she is currently involved focuses on the determination of thermophysical properties of single-phase and multi-phase materials, concentrating on the effect thermal barriers at the interfaces between different materials in close contact have on the overall thermophysical properties of the material or component. These investigations range from examining the thermal barriers encountered between the constituent components of composite materials to those encountered between the different materials used in integrated electronic packaging.

Bonnie J. Dunbar
National Aeronautics and Space Administration
Lyndon B. Johnson Space Center
Houston, Texas
Bonnie.j.dunbar1@jsc.nasa.gov

Dr. Bonnie J. Dunbar earned a B.S. and M.S. in Ceramic Engineering from the University of Washington and a Ph.D. in Mechanical/

Biomedical Engineering from the University of Houston. She worked for both the Boeing Company and Rockwell International Space Division. At Rockwell International, she was a Senior Research Engineer working on the space shuttle thermal protection system. After entering NASA, she became a flight controller for Skylab and the space shuttle and later was selected as a Mission Specialist Astronaut. She has flown five times in space on the space shuttle, performing hundreds of experiments for researchers around the world, operating a robotic arm to capture the research satellite, LDEF, and docking twice to the Russian space station, MIR. Dunbar currently is Assistant Director at the NASA Johnson Space Center, with a focus on integrating university research into NASA's human exploration missions. She is a life member of The American Ceramic Society and was selected to both the Royal Society of Edinburgh and the National Academy of Engineers.

David A. Earl
Alfred University
Department of Ceramic Engineering and Materials Science
Alfred, New York
earlda@alfred.edu

Dr. David Earl is an Assistant Professor of Ceramic Engineering and Materials Science at Alfred University, where he teaches classes in ceramic processing principles and statistical methods and serves as Quality Assurance Manager for the Center for Environmental and Energy Research. He received a Ph.D. in Materials Science and Engineering from the University of Florida (1998) and a B.S. in Ceramic Engineering from Alfred University in 1984. Prior to joining Alfred in 1999, he worked for 14 years in industry for manufacturers of ceramic tile and high-voltage porcelain. His industrial positions included Director of Research and Development, Plant Manager, Corporate Statistics Consultant, and Quality Control Engineer. He was recipient of the 1999 John Marquis Memorial Award presented by The American Ceramic Society's (ACerS) Whitewares and Materials division for the most valuable published paper of the year. He is a Distinguished Mentor for ACerS, a member of NICE, and a member of the Keramos, Omicron Delta Kappa, and Phi Kappa Phi honor societies.

Hassan El-Shall
University of Florida
Department of Materials Science and Engineering
Gainesville, Florida
helsh@mse.ufl.edu

Dr. Hassan El-Shall is Associate Professor at the Department of

Materials Science and Engineering at the University of Florida. He also serves as Associate Director for Research at the National Science Foundations–Engineering Research Center for Particle Science and Technology. He has been a faculty member at the University of Florida since November 1994 where he has been teaching ceramics and materials engineering courses. Before joining the University of Florida, he was the Associate Director for Research at the Florida Institute of Phosphate Research located east of Tampa. His research interests include applied surface chemistry in solid/solid and solid/liquid separation, nucleation and crystallization of ceramic powders, waste remediation and treatment, and dispersion/aggregation characteristics of kidney stones. Currently, he is involved in synthesis and surface modification of particulates for environmental and medical applications.

Diane C. Folz
Virginia Polytechnic Institute and State University
Department of Materials Science and Engineering
Blacksburg, Virginia
dfolz@vt.edu

Diane Folz is a Senior Research Associate in materials science and engineering at Virginia Tech where she works primarily in the area of microwave processing of materials. She graduated from the University of Florida in 1987 with a B.S. in materials science and engineering. She stayed at Florida as a project and laboratory manager until January 2001 when she left to join the MSE department at Virginia Tech. Since 1994, Folz has served as the Executive Director of the National Institute of Ceramic Engineers (NICE). She is Chair of the Public Relations Subcommittee of The American Ceramic Society (ACerS) and is an evaluator for the Accreditation Board for Engineering and Technology (ABET). She has served on the Board of Governors of the American Association of Engineering Societies (AAES) from 1994 through 2001, and is Chair of Link 180 of the Order of the Engineer. She holds memberships in the Ceramic Education Council, ACerS, Keramos–Ceramic Engineering Honor Fraternity, and the Engineering Leadership Circle. She is a Fellow of ACerS and NICE. Folz has served as editor on two books, has more than 25 technical publications and two patents. She served as Technical Vice Chair of the Third World Congress on Microwave and Radio Frequency Processing, held September 2002 in Sydney, Australia.

D.P.H. Hasselman
Virginia Polytechnic Institute and State University
Department of Materials Science and Engineering
and Engineering Science and Mechanics
Blacksburg, Virginia
hasselmn@vt.edu

Dr. D.P.H. Hasselman, Professor Emeritus of Materials Science and Engineering and Engineering Science and Mechanics at Virginia Polytechnic Institute and State University, received a B.Sc. in Engineering Physics from Queen's University, Kingston, Canada, a M.A.Sc. in Metallurgical Engineering from the University of British Columbia, Vancouver, B.C., Canada, and a Ph.D. in Engineering Science from the University of California, Berkeley, California. Hasselman's research interests include the mechanical and thermal properties of structural materials for high-temperature applications. For his work on thermal shock resistance, he was awarded the Jeppson Award and the Gold Medal by The American Ceramic Society as well as the Humboldt Prize by the Federal Republic of Germany. Hasselman is an elected member of the International Academy of Ceramics. More recently, the International Thermal Conductivity Conferences awarded Hasselman the International Thermal Conductivity Award for his experimental and theoretical work on the thermal conductivity of single phase and composite materials. Hasselman has more than 300 technical publications. His theory on thermal shock damage resistance published in 1969 is the 11th most-sited paper in the history of The American Ceramic Society publications.

Larry L. Hench
University of London
Department of Materials
Imperial College of Science, Technology, and Medicine
United Kingdom
l.hench@ic.ac.uk

Dr. Larry Hench is currently Professor of Ceramic Materials in the Department of Materials at Imperial College of Science, Technology, and Medicine, University of London. He is also Director of the Imperial College Center for Tissue Regeneration and Repair and Deputy Directory of the Imperial College Tissue Engineering Center. He assumed the Chair of Ceramic Materials at Imperial College in December 1995 following 32 years at the University of Florida where he was Graduate Research Professor of Materials Science and

Engineering, Director of Bioglass Research Center, and Co-Director of the Advanced Materials Research Center. He completed B.S. and Ph.D. degrees at the Ohio State University in 1964. In 1969 Professor Hench discovered Bioglass®, the first man-made material to bond to living tissues. This unique range of soda-calcia-phospho-silica glasses is used clinically throughout the world for repair of bones, joints, and teeth. This development, together with the accompanying studies of the mechanisms of glass surface reactions and chemical processing of materials, has led to many international awards, publication of nearly 550 research papers, 24 books, and 25 patents. Hench's studies of sol-gel processing of silica has led to the development of a new generation of gel-silica optical components (Gelsil®), including net shape, net surface micro-optics, diffractive optics, and porous optical matrices for environmental sensors tissue engineering, and solid state dye lasers. These products, now commercially manufactured, have led to numerous advanced technology awards in the optics industry. He is also author of a series of children's books, published by The American Ceramic Society, featuring Boing-Boing the Bionic Cat. Current research includes investigation of the osteogenic behavior of bioactive glasses and gel-glasses (US Biomaterials Corp.), development of moldable bioactive paste for bone generation (EPSRC, US Biomaterials Corp.), the application of porous gel-silica optical matrices in environmental sensors (EPSRC), and tissue engineering of a lung lobe (United Therapeutics Ltd.).

W. Jack Lackey
Georgia Institute of Technology
George W. Woodruff School
of Mechanical Engineering
Atlanta, Georgia
jack.lackey@me.gatech.edu

Dr. W. Jack Lackey is a professor at Georgia Tech. He received B.S. degrees in Ceramic Engineering and Metallurgical Engineering from North Carolina State University in 1961. He received an M.S. and a Ph.D. in Ceramic Engineering from North Carolina State University in 1963 and 1970, respectively. He conducted basic and applied research on nuclear fuel fabrication, nuclear waste disposal, and processing of ceramic coatings and composites at Battelle Northwest Laboratory and the Oak Ridge National Laboratory. From 1986 to 1997, while employed at the Georgia Tech Research Institute, he performed research on ceramic coatings and composites. In 1997 he joined the faculty of the George W. Woodruff School of Mechanical Engineering, Georgia Institute of Technology. His current research is in the areas of laser CVD rapid prototyping, processing of fiber-reinforced composites

possessing a laminated matrix for enhancing fracture toughness, and development of an improved process for carbon-coating of mechanical heart valves. He has published 90 refereed papers and has 15 patents. Since 1997, Lackey has developed and taught an undergraduate course on materials selection and failure analysis and two graduate courses emphasizing processing of advanced ceramic coatings and composites and the interrelationships between processing, microstructure, and material properties. He has also taught an undergraduate course in materials science and engineering and a graduate course on nuclear materials.

Thomas D. McGee
Iowa State University
Materials Science and Engineering Department
Ames, Iowa
tmcgee@iastate.edu

Dr. Thomas D. McGee is a professor in the Materials Science and Engineering Department at Iowa State University. He has been active in the Biomedical Engineering graduate program and has a courtesy appointment in the Department of Veterinary Clinical Sciences. His orthopedic research interests include bioceramic materials, composite joint designs, bone replacement, guided bone regeneration, smart biomaterial antibiotic release, and ceramic bone cements. He also has interests in fracture analysis, refractories and their applications, and structural ceramics. He is past-president of the National Institute of Ceramic Engineers (NICE), a Fellow of The American Ceramic Society and of NICE, a past-governor of the American Association of Engineering Societies, a past-president and general secretary of Keramos Fraternity, a past member of the Engineering Commission for Professional Development, and he served on the executive committee of the Engineering Accreditation Commission.

John J. Mecholsky Jr.
University of Florida
Department of Materials Science and Engineering
Gainesville, Florida
jmech@mse.ufl.edu

Dr. John J. Mecholsky Jr. is Professor of Materials Science and Engineering at the University of Florida since January 1990. Prior to that he taught at the Pennsylvania State University and was affiliated with the Applied Research Laboratories at Penn State. He was on the technical research staff at Sandia National Laboratories (1979–1984) and at the Naval Research Laboratory (1972–1979). His research interests include the fracture and quantitative fractography of materials that fail in a brittle manner, fractal geometry, wear and cyclic fatigue, biomimetic structures, and biomaterial composites. Most recently his research has focused on understanding the failure process in biological, hierarchical, laminated structures. He is past chairman of the Subcommittee on Fracture Testing of Brittle Materials [E24.07] for the American Society of Testing and Materials. He is past-chair of the Central Pennsylvania Chapter, and past-vice chair of the New Mexico Chapter, and is current vice chair of the Florida Section of the American Ceramic Society (ACerS). He is a Fellow of ACerS and he received the Carl Schwartzwalder-PACE award from ACerS and the National Institute of Ceramic Engineers.

John H. Selverian
OSRAM SYLVANIA Products, Inc.
Materials Science and Analysis Department
Beverly, Massachusetts
john.selverian@sylvania.com

Dr. John Selverian is a Staff Scientist at OSRAM SYLVANIA Products, Inc., since 1992. His main work is concentrated in modeling and design of materials and structures for lighting applications. Prior to this he worked at GTE Laboratories on ceramic-to-metal joining and the mechanical properties of ceramic thin films. He received the Robert L. Peaslee Brazing Award from the American Welding Society for the best paper on brazing. He is particularly interested in applying design and system optimization techniques early in the design process and in developing tools for engineers to use in their product development work.

Carlos T.A. Suchicital
Virginia Polytechnic Institute and State University
Department of Materials Science and Engineering
Blacksburg, Virginia
ctas@vt.edu

Dr. Carlos T.A. Suchicital is an associate research professor with the Department of Materials Science and Engineering at Virginia Tech where he has been since August 1994. He comes from the University of Illinois at Urbana-Champaign where he earned his doctorate degree. Suchicital conducted research and taught materials engineering in the Department of Chemical, Bio, and Materials Engineering at Arizona State University. Prior to Illinois, Suchicital also did research and

taught materials processing and characterization at the Federal University of Sao Carlos in Sao Carlos, Sao Paulo, Brazil. His research interests include processing and characterization of electronic ceramics and III–V semiconductors, engineering of materials for biomedical applications, and rapid prototyping/free forming technologies.

Kenji Uchino
The Pennsylvania State University
Department of Electrical Engineering
University Park, Pennsylvania
kenjiuchino@psu.edu

Dr. Kenji Uchino, one of the pioneers in piezoelectric actuators, is the Director of the International Center for Actuators and Transducers and Professor of Electrical Engineering at The Pennsylvania State University. After being awarded his Ph.D. from Tokyo Institute of Technology, Japan, Uchino became Research Associate in the Physical Electronics Department at this university. He joined Sophia University, Japan, as an Associate Professor in 1985 and moved to Penn State in 1991. He was involved with the Space Shuttle Utilizing Committee in NASDA, Japan (1986–88), and was vice president of NF Electronic Instruments, USA (1992–94). He is the chairman of the Smart Actuator/Sensor Study Committee sponsored by Japanese MITI, executive associate editor for the Journal of Advanced Performance Materials, associate editor of the Journal of Intelligent Materials, Systems, and Structures, and of the Japanese Journal of Applied Physics. His research interests are in solid-state physics, especially dielectrics, ferroelectrics, and piezoelectrics, including basic research on materials, device designing and fabrication processes, and development of solid-state actuators for precision positioners and ultrasonic motors. He has authored 340 papers, 45 books, and 19 patents in the ceramic actuator area. He is a recipient of the Outstanding Research Award from Penn State Engineering Society (1996), an academic scholarship from Nissan Motors Scientific Foundation (1990), a Best Movie Memorial Award at Japan Scientific Movie Festival (1989), and the Best Paper Award from the Japanese Society of Oil/Air Pressure Control (1987). He is a Fellow of The American Ceramic Society.

Eric D. Wachsman
University of Florida
Department of Materials Science and Engineering
Gainesville, Florida
ewach@mse.ufl.edu

Dr. Wachsman is an associate professor in the Department of Materials Science and Engineering at the University of Florida since January 1997. Prior to that he was a Senior Scientist at SRI International. His research interests are in ionic and electronic conducting ceramics, from fundamental investigations of their transport properties and heterogeneous electrocatalytic activity, to the development of moderate temperature solid oxide fuel cells (SOFC), gas separation membranes, and solid state sensors. He also has worked on the catalytic properties of lanthanide perovskites for NO_x reduction, methane activation, and combustion. Wachsman received his Ph.D. in Materials Science from Stanford University in 1990. He is an associate editor of the Journal of the American Ceramic Society, and editor of Ionics, secretary/treasurer of the High Temperature Materials Division of The Electrochemical Society, and chair of the Florida section of The American Ceramic Society.

E. Dow Whitney
University of Florida
Department of Materials Science and Engineering
Gainesville, Florida
ewhit@mse.ufl.edu

Dr. E. Dow Whitney, a professor in the Department of Materials Science and Engineering at the University of Florida, Gainesville, is also FEEDS (videotaped continuing education graduate engineering instruction program) coordinator and Graduate Student Advisor for the department. He was a NASA/ASEE Summer Faculty Fellow at the Kennedy Space Center, an adjunct professor of ceramic engineering at Clemson University, and adjunct professor of physics and chemistry at Memphis State University. Whitney has numerous fields of interest that include: mining and minerals beneficiation (phosphate ore); thermodynamics and kinetics of high-pressure/high-temperature phase transformations in solids, including diamond synthesis technology; physical chemistry of hard materials; machining and manufacturing with advanced ceramic cutting tools and abrasives; chemical tribology in both metal cutting and hybrid ceramic bearings; chemistry of boron hydride and fluorine-based high-energy rocket propellants and explosives; and fractography and failure analysis of glass and ceramics as related to the technical aspects of product liability litigation. His current research deals with the theoretical calculation of ionization potentials and phase stability of high-temperature inorganic oxides as related to the development of advanced thorium dioxide/uranium dioxide nuclear fuel elements. He is an SME (Society of Manufacturing Engineers) certified manufacturing engineer (Life) and a professional engineer (State of Florida), a Fellow of The American Ceramic Society and Vice President of Materials Consultants Inc., Gainesville, Florida, a professional engineering association dealing primarily with product liability, failure analysis, and accident reconstruction. He is the editor of a book titled *Ceramic Cutting Tools*, has more than 100 technical publications, and holds 13 patents.

George G. Wicks
Westinghouse Savannah River Company
Aiken, South Carolina
george.wicks@srs.gov

Dr. George G. Wicks is a Senior Advisory Scientist at the Westinghouse Savannah River Co. in Aiken, South Carolina, and is considered an "internationally recognized authority on the vitrification of nuclear wastes." His research activities have included defining materials of construction for the joule-heated ceramic melter used to vitrify SRS high-level radioactive waste and developing the first slurry feeding system for the vitrification process. He is co-author of the SRL Kinetic Leachability Model for nuclear glasses and has been actively involved in four different international waste management field-testing programs. Wicks is also co-inventor of sol-gel indicators, a new class of composite materials that can be fabricated into specialized sensors; co-inventor of sol-gel metal hydrides, composites used to store large amounts of hydrogen; and co-inventor of a new hybrid microwave process with the ability to remediate hazardous components effectively. He also published a considerable amount of work in these various areas, including more than 150 publications, and has 10 patents issued. He has served as past-chair of the Nuclear Division of The American Ceramic Society, a member of the U.S. Materials Review Board, and past-president of the National Institute of Ceramic Engineers, as well as a Fellow of The American Ceramic Society. He has also served as adjunct professor to Clemson University and the University of Florida and is currently chairman of the SRS Creativity Committee and SRTC Mentoring Committee. He has been awarded three George Westinghouse Signature Awards for technical excellence, including the Gold Corporate Award and received the Westinghouse Innovators Award.

Warren W. Wolf
Owens Corning
Granville, Ohio
wwolf@insight.rr.com

Dr. Warren W. Wolf was named Vice President, Owens Corning Science and Technology, in 1998. In 2000, the title of Chief Scientist was added. Wolf joined Owens Corning in 1968 as a senior scientist at the company's Science and Technology Center in Granville, Ohio. Since he joined the company, Wolf was named supervisor of the Glass Research and Development Laboratory and later promoted to manager. In 1983, he was named a lab director. Wolf has been involved in directing a variety of activities, including new product and process research with special emphasis on insulation and composite materials. Other activities for Owens Corning included directing glass, polymer, and metals research, as well as the analytical laboratories at Granville. He holds 15 patents in the areas of glass fiber process and composition and has three publications. Wolf was co-editor of the 11th International Congress on Glass and has been a vice president of The American Ceramic Society serving in 1993–94 and in 1997–98. He is a member of the National Institute of Ceramic Engineers and served as its president, 1994–1996. He is also a member of the International Commission on Glass and serves as an American delegate to that commission. He has served on the external advisory board for the Department of Materials Science and Engineering at The Ohio State University for a number of years. From 1995 to 1998 he was co-chair of the technical advisory group for the Governor's Ohio Science and Technology Council. He has also been honored as a distinguished alumnus of the College of Engineering at Ohio State and as a Fellow of the College of Earth and Mineral Sciences at Penn State.

Appendix B

National Academy of Engineering's Top 20 Engineering Achievements List

Ceramic Materials Contribute Heavily in the National Academy of Engineering's Top 20 Engineering Achievements List

The twenty engineering achievements that have had the greatest impact on the quality of life in the 20th century were announced on February 22, 1999 at the National Press Club in Washington, D.C. as a kickoff to National Engineers Week. Astronaut/engineer Neil Armstrong made the announcement on behalf of the U.S. National Academy of Engineering (www.nae.edu). Below are the top twenty achievements with examples of how ceramic materials make these technologies possible.

Top Achievements	How Ceramics Contribute*
1. Electrification	Electrical insulators for power lines, insulators for industrial/household applications
2. Automobile	Engine sensors, catalytic converter substrates, spark plug insulators, windows, engine components, electrical devices
3. Airplane	Anti-fogging/freezing glass windows, jet engine components
4. Safe water supply and treatment	Filters
5. Electronics	Substrates and IC packages, capacitors, piezoelectrics, insulators, magnets, superconductors
6. Radio and television	Glass tubes (CRTs), glass faceplate, phosphor coatings, electrical components
7. Agricultural mechanization	Refractories make melting and forming of ferrous and non-ferrous metals possible.
8. Computers	Electrical components, magnetic storage, glass for computer monitors
9. Telephone	Electrical components, glass optical fibers
10. Air conditioning and refrigeration	Glass fiber insulation, ceramic magnets
11. Interstate highways	Cement for roads and bridges, glass microspheres used to produce reflective paints for signs and road lines.

12. Space exploration	Space shuttle tile, high-temperature resistant components, ceramic ablation materials, electromagnetic and transparent windows, electrical components, telescope lenses
13. Internet	Electrical components, magnetic storage, glass for computer monitor
14. Imaging: X-rays to film	Piezoceramic transducers for ultrasound diagnostics, sonar detection, ocean floor mapping and more, ceramic scintillator for X-ray computed tomography (CT scans), telescope lenses, glass monitors, phosphor coatings for radar and sonar screens
15. Household appliances	Porcelain enamel coatings for major appliances, glass fiber insulation for stoves and refrigerators, electrical ceramics, glass-ceramic stove tops, spiral resistance heaters for toasters, ovens, and ranges
16. Health technologies	Replacement joints, heart valves, bone substitutes, hearing aids, pacemakers, dental ceramics, transducers for ultrasound diagnostics, ceramic scintillator for X-ray computed tomography (CT scans), and many other applications
17. Petroleum and natural gas technologies	Ceramic catalysts, refractories and packing media for petroleum and gas refinement, cement for well drilling, drill bits for well drilling
18. Laser and fiber optics	Glass optical fibers, fiber amplifiers, laser materials
19. Nuclear technologies	Fuel pellets, control rods, high-reliability seats and valves, containerization components, spent nuclear waste containment
20. High-performance materials	Ceramic materials were cited for their advanced properties such as wear, corrosion and high temperature resistance, high stiffness, lightweight, high melting point, high compressive strength, hardness, and wide range of electrical, magnetic, and optical properties.

* *Examples of How Ceramics Contribute were contributed by The American Ceramic Society.*

Appendix C

Code of Ethics of Engineers

Code of Ethics of Engineers

Issued by the National Institute of Ceramic Engineers

The Fundamental Principles

Engineers uphold and advance the integrity, honor, and dignity of the engineering profession by:

I. Using their knowledge and skill for the enhancement of human welfare;

II. Being honest and impartial, and servicing with fidelity the public, their employers, and clients;

III. Striving to increase the competence and prestige of the engineering profession; and

IV. Supporting the professional and technical societies of their disciplines.

The Fundamental Canons

1. Engineers shall hold paramount the safety, health, and welfare of the public in the performance of their professional duties.

2. Engineers shall perform services only in the areas of their competence.

3. Engineers shall issue public statements only in an objective and truthful manner.

4. Engineers shall act in professional matters for each employer or client as faithful agents or trustees, and shall avoid conflicts of interest.

5. Engineers shall build their professional reputation on the merit of their services and shall not compete unfairly with others.

6. Engineers shall act in such a manner as to uphold and enhance the honor, integrity, and dignity of the profession.

7. Engineers shall continue their professional development throughout their careers and shall provide opportunities for the professional development of those engineers under their supervision.

Adopted by the National Institute of Ceramic Engineers, October 25, 1980.

Appendix D

Model Guide for Professional Conduct

Model Guide for Professional Conduct American Association of Engineering Societies

Preamble

Engineers recognize that the practice of engineering has a direct and vital influence on the quality of life for all people. Therefore, engineers should exhibit high standards of competency, honesty, and impartiality; be fair and equitable; and accept a personal responsibility for adherence to applicable laws, the protection for the public health, and maintenance of safety in their professional actions and behavior. These principles govern professional conduct in serving the interests of the public, clients, employers, colleagues, and the profession.

The Fundamental Principle

The engineer, as a professional, is dedicated to improving competence, service, fairness and the exercise of well-founded judgment in the practice of engineering for the public, employers, and clients with fundamental concern for the public health and safety in the pursuit of this practice.

Canons of Professional Conduct

- Engineers offer services in the areas of their competence and experience, affording full disclosure of their qualifications.
- Engineers consider the consequences of their work and societal issues pertinent to it and seek to extend public understanding of those relationships.
- Engineers are honest, truthful, and fair in presenting information and in making public statements reflecting on professional matters and their professional role.
- Engineers engage in professional relationships without bias because of race, religion, sex, age, national origin, or handicap.
- Engineers act in professional matters for each employer or client as faithful agents or trustees, disclosing nothing of a proprietary nature concerning the business affairs or technical processes of any present or former client or employer without specific consent.

- Engineers disclose to affected parties known or potential conflicts of interest or other circumstances which might influence – or appear to influence – judgment or impair the fairness or quality of their performance.

- Engineers are responsible for enhancing their professional competence throughout their careers and for encouraging similar actions by their colleagues.

- Engineers accept responsibility for their actions; seek and acknowledge criticism of their work; offer honest criticism of the work of others; properly credit the contributions of others; and do not accept credit for work not theirs.

- Engineers perceiving a consequence of their professional duties to adversely affect the present or future public health and safety shall formally advise their employers or clients and, if warranted, consider further disclosure.

- Engineers act in accordance with all applicable laws and rules of conduct, and lend support to others who strive to do likewise.

Model Guide for Professional Conduct. Washington DC: American Association of Engineering Societies, 1984. (Revision forthcoming).

Appendix E

Obligation of an Engineer

Obligation of an Engineer
Issued by the Order of the Engineer

I am an Engineer, in my profession I take deep pride. To it I owe solemn obligations.

Since the Stone Age, human progress has been spurred by the engineering genius. Engineers have made usable Nature's vast resources of material and energy for Humanity's [Mankind's] benefit. Engineers have vitalized and turned to practical use the principles of science and the means of technology. Were it not for this heritage of accumulated experience, my efforts would be feeble.

As an Engineer, I pledge to practice integrity and fair dealing, tolerance and respect, and to uphold devotion to the standards and the dignity of my profession, conscious always that my skill carries with it the obligation to serve humanity by making the best use of Earth's precious wealth.

As an Engineer [in humility and with the need for Divine guidance] I shall participate in none but honest enterprises. When needed, my skill and knowledge shall be given without reservation for the public good. In the performance of duty and in fidelity to my profession, I shall give the utmost.

Note: Brackets indicate the original wording of the Obligation. Either wording is acceptable, but new certificates have the newer wording.

Printed with permission of the Order of the Engineer, Inc.

Index

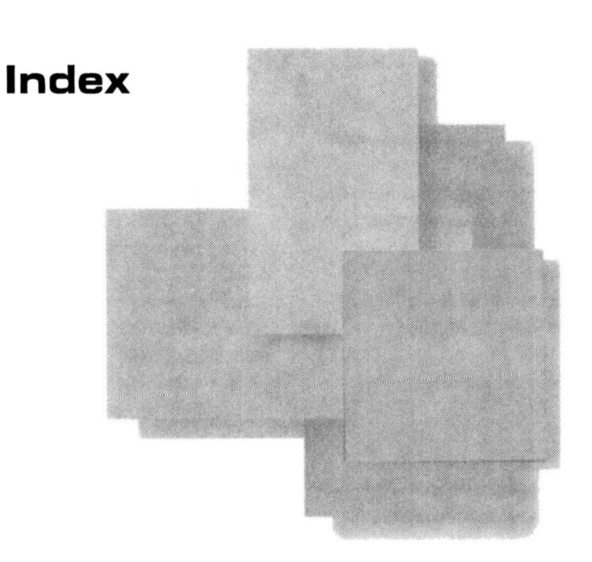

Printed in the United States
127356LV00003B/25/P

9 781574 981315